The Politics of River Basin Organisations

For four especially important members of a new generation,
Laia and Max, Carlijn and Jasper

The Politics of River Basin Organisations

Coalitions, Institutional Design Choices and Consequences

Edited by

Dave Huitema

Professor of Environmental Policy, Institute for Environmental Studies (IVM), VU University Amsterdam and the Faculty of Management, Science and Technology, Open University of the Netherlands

Sander Meijerink

Associate Professor of Water Governance, Institute for Management Research (IMR), Radboud University Nijmegen, the Netherlands

Edward Elgar

Cheltenham, UK • Northampton, MA, USA

Published by
Edward Elgar Publishing Limited
The Lypiatts
15 Lansdown Road
Cheltenham
Glos GL50 2JA
UK

Edward Elgar Publishing, Inc.
William Pratt House
9 Dewey Court
Northampton
Massachusetts 01060
USA

A catalogue record for this book
is available from the British Library

Library of Congress Control Number: 2014932595

This book is available electronically in the ElgarOnline.com
Economics Subject Collection, E-ISBN 978 1 78254 922 2

MIX
Paper from responsible sources
FSC® C013056

ISBN 978 1 78254 921 5

Typeset by Servis Filmsetting Ltd, Stockport, Cheshire
Printed and bound in Great Britain by T.J. International Ltd, Padstow

Contents

Contributors

David Benson is a lecturer in politics at the University of Exeter, UK. Dr Benson's current research, based at the university's Environment and Sustainability Institute (ESI) in Penryn, encompasses a range of issue areas at the interface between political and environmental sciences, most notably European Union (EU) environmental and energy policy, comparative environmental politics and governance, federalism and public participation in environmental decision making.

Christoph Bernhardt is a senior researcher and head of the Department for Historical Research at the Leibniz-Institute for Regional Development and Structural Planning, Germany. He teaches at the Technical Universities of Berlin and Darmstadt (Germany). His main field of research is the European urban and environmental history of the nineteenth and twentieth centuries.

António Guerreiro de Brito is an environmental engineer and Associate Professor with Habilitation at the University of Lisbon, School of Agronomy, Portugal, and former director of a Portuguese river basin organisation. His research interests include water efficiency and resources recovery from wastewater, eutrophication and lake restoration, and water resources planning and governance.

Dan Calvert is a PhD candidate at Oregon State University and lives in Portland, Oregon, USA. His research interests include learning in collaborative watershed partnerships and coupled social–ecological systems thinking.

Daniel Connell teaches courses dealing with transboundary rivers in the Crawford School at the Australian National University. His research focuses on issues related to the institutional design and governance arrangements applying to rivers in federal or multilayered political systems such as Australia, South Africa, the United States, Mexico, the European Union (Spain), India, China and Brazil.

Hadrian Cook is a lecturer and environmental consultant. He is presently at Kingston University in London, UK, but he has also worked in conservation (for the Harnham Water Meadows Trust in Salisbury,

UK) and was a consultant to the UK Research Councils' Rural Economy and Land Use Programme. Originally a hydrologist and soil scientist, Hadrian's academic interests include environmental history and policy and floodplain management. He lives in Salisbury and is a Fellow of the Chartered Institution of Water and Environmental Management.

Ines Dombrowsky heads the Department of Environmental Policy and Natural Resources Management at the German Development Institute, Deutsches Institut für Entwicklungspolitik (DIE), in Bonn, Germany. Her research focuses on various aspects of water governance including transboundary water management, the institutionalisation of integrated water resources management (IWRM) and the governance of the water–energy–food nexus.

Nikki Funke is a senior researcher in the Water Governance Group at the Council for Scientific and Industrial Research (CSIR) in South Africa. Her research interests include understanding the roles of scientists in environmental and water decision-making contexts, as well as the dynamics that hamper or contribute to effective water governance.

Santita Ganjanapan teaches at the Department of Geography and the Regional Centre for Social Sciences and Sustainable Development, Chiang Mai University, Thailand. Her research interests include natural resource management, aquaculture and climate adaptation, gender and space, and peri-urban agriculture.

Joyeeta Gupta is Professor of Environment and Development in the Global South in the programme on Governance and Inclusive Development (GID) of the Amsterdam Institute for Social Science Research, University of Amsterdam and Professor at UNESCO-IHE Institute for Water Education, Delft, the Netherlands.

Nina Hagemann is a scientist at the Department of Economics at the Helmholtz Centre for Environmental Research in Leipzig, Germany. Her research interests include water (infrastructure) governance, transition countries and institutional economics.

Lena Horlemann is a senior researcher at the inter 3 Institute for Resource Management, Berlin, Germany. Her research interests include water governance, organisational development and institutional design in water management.

Annabelle Houdret is a senior researcher at the Department of Environmental Policy and Natural Resources Management at the German

Development Institute, Deutsches Institut für Entwicklungspolitik (DIE) in Bonn, Germany. Her research focuses on water governance, water conflicts and integrated water resources management.

Frank Hüesker is a postdoctoral researcher. Research for contributing to this book was conducted at the Leibniz Institute for Regional Development and Structural Planning in Erkner, Germany. His main field of research is the sustainable governance of water resources, water infrastructures and renewable energies in Europe.

Dave Huitema is a Professor of Environmental Policy who works at the Institute for Environmental Studies (IVM) of the VU University Amsterdam, the Netherlands and the Faculty of Management, Science and Technology at the Open University of the Netherlands. He is interested in the notion of adaptive governance (involving policy learning, change, agency and innovation) and focuses mainly on water and climate change issues.

Frank Jaspers is Associate Professor at UNESCO-IHE, the United Nations Educational, Scientific and Cultural Organization's International Institute for Infrastructural, Hydraulic and Environmental Engineering, in Delft, the Netherlands. His main research interest is design and performance of river basin organisations. He has more than 20 years of experience in the functioning of river basin organisations in Asia, Africa and Latin America.

Alex Inman is a freelance researcher and consultant specialising in socio-economic research within the context of natural resource management problems; particularly relating to land and water conservation. He has worked within the environmental non-governmental organisation (NGO) sector for many years with research interests including the role of civil society organisations within integrated catchment planning.

Denise Lach, PhD, is a Professor of Sociology and Director of the School of Public Policy at Oregon State University, USA. Her research interests have long included the intersection of water institutions and policy as well as the use of scientific information in decision making.

Louis Lebel is Director of the Unit for Social and Environmental Research at Chiang Mai University, Thailand. His research interests include environmental governance, production–consumption systems and adaptation to global environmental changes.

Marco Leidel is a scientist at the Department of Hydrosciences at the University of Technology Dresden, Germany. His research interests include integrated water resources management, capacity development, transboundary water management and water governance.

Rob de Loë is Professor and University Research Chair in Water Policy and Governance at the University of Waterloo, Canada, and director of the multi-university Water Policy and Governance Group (www.wpgg.ca). His research interests relate to water management and governance for source water protection, water allocation, transboundary water resources and climate change adaptation.

Sander Meijerink is an Associate Professor of Water Governance at the Institute for Management Research (IMR) of the Radboud University Nijmegen, the Netherlands. His research interests include institutional analysis, policy continuity and change, leadership and cross-border cooperation. Most of his research focuses on water and spatial governance and climate adaptation.

Richard Meissner is a senior researcher at the Council for Scientific and Industrial Research (CSIR) in Pretoria, South Africa. His research interests include water governance, the linkage between theory and policy, transboundary river basin management and politics, international relations in general, and the role and involvement of interest groups and individuals in water politics and governance.

Michelle Morris is a PhD candidate in the Department of Environment and Resource Studies at the University of Waterloo, Canada. Her research interests include transboundary water governance and collaborative environmental governance.

Andrew Ross is a Visiting Fellow at the Australian National University, and is currently contracted to the Groundwater Section of UNESCO's Division of Water Sciences in Paris. Andre's major current work is on transboundary aquifer management, groundwater governance and conjunctive water use. He is also interested in policy integration and implementation, and the research–policy interface.

Andreas Thiel is Temporary Professor for Environmental Governance at Humboldt-Universität zu Berlin, Germany. His research interests include the role of institutions in social–ecological systems and their transformation, water and land use governance, and climate adaptation.

Vincent Thomas has been working as a research officer for the Afghanistan Research and Evaluation Unit in Kabul, Afghanistan since 2011. His areas of interest are river basin management, water conflicts and transboundary water management. From 2005 to 2010 he worked with the Aga Khan Foundation, leading a European Union (EU)-funded social water management project in Northeast Afghanistan.

Jeroen Warner teaches, trains and publishes on domestic and transboundary water conflict, participatory resource management, and governance issues. Jeroen is Associate Professor of Disaster Studies at Wageningen University, the Netherlands. His main research interests in the disaster studies domain are social resilience and participation, the politics of (flood) disaster risk reduction, and the role of disasters in international relations. He was Visiting Professor at the University of São Paulo, Brazil in Autumn 2013.

Acknowledgements

We would like to thank all of the participating authors for their contributions to this volume, and the many interesting discussions which we had on various occasions. We are grateful to the Royal Netherlands Academy of Arts and Sciences (KNAW) for the conference subsidy which enabled us to organise a meeting in Amsterdam where we discussed draft chapters, and jointly searched for themes running through all contributions. We are much indebted to Simon Verduijn who drafted a report of this meeting, to Ron Wundering who prepared a nice series of maps for this volume, and to Kathrin Ludwig who assisted us with editing the various chapters of this book. Finally, we would like to thank Alex Pettifer, Caroline Cornish, and Cathrin Vaughan at Edward Elgar Publishing, for the pleasant contacts during the preparation of this book.

Dave Huitema
Sander Meijerink

1. The politics of river basin organisations: institutional design choices, coalitions and consequences

Dave Huitema and Sander Meijerink

1.1 SHIFTS IN WATER GOVERNANCE

Water and human development are inextricably linked. Human settlement tends to concentrate along rivers and coasts. This is because water offers fertile soils, opportunities for irrigation, and possibilities for transport and trade. To use the possibilities of the water as much as possible and to reduce the risks associated with human settlement close to water, social organisation and systems of governance are required. Arguably because water is such a crucial element in societal development, many ancient societies had to make decisions about their water management organisations early. The degree to which organisations founded for water management influence later traditions of governing is under debate. Some have claimed that the organisation of water management, which can be centralised and focused on large-scale infrastructure or, alternatively, decentralised and focused on local management, determined the governance system of entire empires (Wittfogel 1957). But others suggest that it is rather the other way around, in the sense that societies with accomplished hierarchical governance structures were better able to develop centralised infrastructures for managing water and thus to control their water environment.

Whatever the protracted history of water management and its importance for broader historical patterns of governance that have emerged since ancient times, the advent of the nation state (depending on the country in question, this took place in most cases in the eighteenth, nineteenth or twentieth centuries) was a significant development, and in most cases a serious break from the traditions of the past. The nation state, built upon concerted efforts to change social cultural preferences ('nation building') through the crafting of new bonds of affiliation and a retelling of history, implied a 'nationalisation' of responsibilities that were previously at the local level, and has, since this took place, shaped the division of responsibilities at what we now appropriately call the 'international level'. Water management was fully implied in the process of nation building. In many countries, new water works served to make fragmented countries more integrated by physically connecting previously separate parts. Various such projects became focal points for nation building by showing highly developed engineering skills. New organisations for water management were founded that operated at the national scale. In the process of nationalisation, previously established communal and private elements of the governance system were overhauled. In many cases this effectively meant an expropriation of rights from private owners to state bureaucracies, which often started using their decision power to further certain economic interests, be this newly emerging industries which were allowed to use the pollution assimilation capacity of water for the release of their waste materials, or agriculture, which today consumes enormous amounts of freshwater for irrigation, often without being fully charged for the costs.

The process of nation state building had two elements that are worth mentioning here. The first is that when it took place government institutions were often designed on the basis of a 'classical modernist' design (Hajer 2003), embedded in respective constitutions. This implied that government tasks and responsibilities would be divided over respective layers of government, usually three or four. Most of these government layers were assigned a wide set of responsibilities (general-purpose governments) and because the boundaries were relatively clear cut, a certain level of jurisdictional integrity was created. Nation states differ in many ways; they can be federal or unitary states, the level of centralisation can differ, the formal leadership can be presidential or monarchical, the role of the judiciary can be expansive or limited, and so on. In many cases, however, water management tasks were allocated to general-purpose government layers, and thus became part of the 'normal' government apparatus. Special-purpose organisations such as river basin organisations (RBOs) did not fit easily in this scheme, although in some cases, such as the Netherlands, the water boards that had existed long before the modern nation state were given a constitutional status equal to that of municipalities. A second element worth mentioning here is that in almost any new nation state, issues of control loomed large. Democratic control, especially in the form of representative democracy, has almost universal appeal and has thus become the norm. There are, however, multiple models of democracy (Held 2006), and some of these models actually convey very little direct citizen control over government institutions. Regardless of the model of democracy that is applied by the elite in the new nation states, the quest for democratic control has almost always been a struggle, as has achieving the adoption of important democratic principles such as transparency, accountability, and the replacement of leaders by the polity.

In the period from the Second World War until roughly the start of the 1980s, the role of governments in many societies across the world expanded, often for reasons associated with development of a welfare state, but usually also because this was an era of confidence in the power of centralised, planned and rational problem solving. And the concept of the nation state, to the chagrin of some who see the nation state's sovereignty as a hindrance for addressing many of the world's problems, is still very much a key aspect in developing responses to water problems. However, the power of most nation states is eroding and power is diffusing to other actors. This change, often described as a shift from government to governance, started taking place around the 1980s and is depicted in Figure 1.1.

The reasons behind these shifts in governance are often debated. There is a strong ideological element, in the sense that neoliberal ideas, which emphasise market initiative and government failure, gained currency in the 1980s. But in the same decade another influential wave of thought emerged: on community governance and self-governance. Both waves of thought have academic roots in economics and the political sciences, and affected each of the shifts signified here in one way or another. Authors

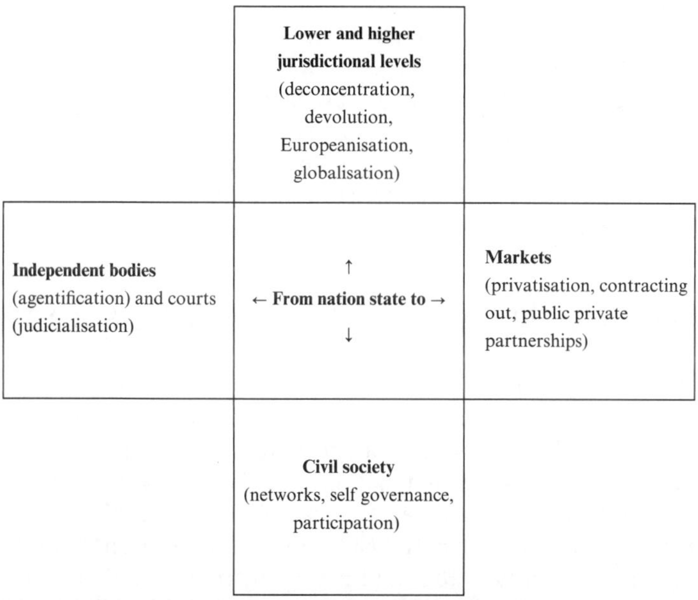

Figure 1.1 Shifts in governance since the 1980s

who write about governance (e.g. Pierre and Peters 2000) also suggest a less prosaic explanation for these shifts: mainly that the nation state had grown too large to sustain itself much longer. Especially in some European countries, public spending amounted to more than half of gross domestic product, and the functioning of the ensuing 'big government', often founded on central control and planning, was considered suspect at best in many areas. 'Red tape', regulations and excessive bureaucracy became rallying cries in the battle to 'hollow out the state' (Rhodes 1996).

The increasingly global discourse on water governance (see e.g. Gupta 2009), bears several traces of these shifts in governance. The central guiding concept of integrated water resources management (IWRM), for instance, suggests greater private sector involvement in water management, and the establishment of pricing mechanisms; it assigns a large role to independent experts by suggesting the greater use of decision-making tools such as cost–benefit analysis, suggests more international collaboration whilst at the same time indicating the possibility for local control, and shows sensitivity to the need for greater public participation and stakeholder involvement. IWRM is often also seen as a corrective to the errors of the 'hydrological mission' (see also Jaspers and Gupta, Chapter 2 in this volume), the spawn of large-scale engineering projects that went hand in hand with nation building and the subsequent growth of nation states and big government. The 'integrated' aspect is often interpreted to denote more attention to the ecological and societal impacts of water management activities, including the typical large-scale projects. It also denotes a geographical emphasis, in the sense that water managers should start paying more attention to the interlinkages between upstream and downstream interventions by working on the river basin scale and founding RBOs.

1.2 THE GUIDING QUESTIONS FOR THIS BOOK

IWRM is clearly a multifaceted concept and implementing each of the aspects may require an extensive job of 'translation' (Mukhtarov 2009) and weighting. Obviously not all aspects are equally important to water managers. It has been suggested that the 'river basin approach' (also called the 'water systems approach', see, for example, Teclaff 1967; Lundqvist et al. 1985; Mitchell 1990; Mostert 2000) is the key innovation that the water governance community is after and preferably in a form that takes away power from 'normal' governments. Schlager and Blomquist (2008: 1) observe:

> For the last 25 years, prescriptions of the water policy literature have centered upon two themes. The first is that 'the watershed' is the appropriate scale for organizing water resource management . . . The second is that since watersheds are regions to which political jurisdictions almost never correspond, and watershed-scale decision making structures do not usually exist, they should be created.

And the plea for river basin organisations has not been without effect. In a worldwide survey, Dinar et al. (2005) found hundreds of transboundary basin organisations.

If one were to analyse the introduction of RBOs from the perspective of the shifts in governance noted above, it is not immediately clear how they would fit in. To some degree, most plans to found RBOs probably constitute a change in the level of governance; for instance, from the national to the international when transboundary organisations are founded. However, there are probably also instances where a river basin organisation is actually a civil society type of organisation, taking the shape of a water user association or watershed organisation, which would imply decentralisation. And it is likely, in certain cases, that RBOs will be dominated by engineers or scientists and would thus effectively constitute independent bodies, which – given the fatigue about holistic planning – will often be 'special purpose' in nature, meaning that they focus on water issues alone. Such design choices, like all shifts in governance that affect the functioning of the nation state and its embedded democratic mechanisms, have multiple consequences and are thus deeply political (see Molle 2009).

Our agenda in this book is to focus exactly on the political dimension of river basin organisations. We do so through a set of four interrelated questions that we will ask in relation to a set of case studies where river basin organisations have been founded across the globe:

1. What are the institutional design characteristics of river basin organisations and how were their geographical boundaries chosen?
2. Who supported and who resisted the founding or reform of river basin organisations?
3. How well do the river basin organisations function in a democratic sense and in terms of realising coordination and environmental effectiveness, and to what degree can this performance be attributed to the institutional design of these organisations and their geographical delineation?
4. What are the implications of these findings for those who are involved in institutional design or reform processes in the water sector?

In this book we shall attempt to answer these questions for river basin organisations that operate within the boundaries of certain nation states, and not for transboundary basin organisations. Water management is but one of the many policy domains in which independent, special-purpose organisations have been created, but for several reasons this policy domain offers particularly fertile ground for an exploration of the research questions posed above. Firstly, there is a long history of special-purpose organisations in water management (see e.g. Molle 2008, 2009; Molle and Wester 2009; Delli Priscoli n.d.). The long and rich history of river basin organisations makes it possible to study their evolution, to trace the impact which various new discourses have had on their institutional design, and to draw lessons from the experiences which water managers have gained with these special-purpose organisations. Secondly, many water problems are pressing issues. The complex relationships between land use and water management and between upstream and downstream water users have caused serious water conflicts in many river basins, and climate change might exacerbate the tensions. Because of the rapid economic development in large parts of the world, along with expected climate change, both the number and the seriousness of water conflicts are expected to increase in the near future. How to design institutions that are able to deal with these issues and conflicts effectively is one of the key challenges for earth-system governance in this century.

With our research agenda, we follow in the footsteps of other governance scholars who have worked on river basin organisations. They have raised several issues in relation to the institution and operation of river basin organisations (Imperial and Hennessey 1999). Firstly, they suggest that the boundaries of river basins are not necessarily so clear or 'natural' (Schlager and Blomquist 2008; Molle 2009). The idea of 'the' river basin suggests a certain simplicity which in reality does not exist, as river basins are connected – sometimes by human intervention – and nested. This means that defining the boundaries of a basin requires choice, and this implies a role for politics. To quote Schlager and Blomquist (2008: 15–16):

> Boundaries are multiple, overlapping, and often contested because people experience and attempt to deal with a host of problems and opportunities that vary in scale from the local to the regional. Drawing boundaries is the first step in determining who decides and how and with what effects. Different boundaries imply different decision makers and different effects.

Some communities may lose local control, whereas others may gain more control. Especially, those who benefit from the current boundaries may object to reshaping the boundaries, for instance, by pleading for a different underpinning of these boundaries (see Molle 2009).

Secondly, there are issues of authority, that is, issues of tasks and responsibilities for the new organisation (Schlager and Blomquist 2008:

20–23). Governance scholars warn that large unitary river basin authorities are just as susceptible to 'bureaucratic pathologies' as any other bureaucracy (cf. Biswas 2004). Schlager and Blomquist (2008) make the point that institution building tends not to follow a pre-established design but can be better described as a patchwork. In composing the patchwork, environmental concerns, for instance, are far from dominant. Instead, economies of scale, the division of skills across organisations, the costs of coordination, and issues of culture and political identity are said to be more important (Schlager and Blomquist 2008: 20–23). Interestingly enough, governance scholars suggest that a patchwork of institutions at various overlapping levels may not only be more feasible, but also more desirable from an environmental performance perspective than a unitary river basin authority, because of the possibility for reorganising the patchwork according to the necessary task (for more explanation, see, for example, Ostrom and Janssen 2004). Supporting the idea of institution building as a patchwork, governance scholars find that the pattern of institution building reflects the importance of governance considerations (politics, institutions) vis-à-vis environmental goals. For instance, Schlager and Blomquist (2008: 4, quoting others) suggest that most examples of river basin organisations in the USA reflect their current institutional contexts, in the sense that they usually do not have formal decision-making powers and sanctioning authority. Conca et al. (2006) analysed a worldwide set of 62 transboundary river agreements. They found, among other things, that many agreements do not include all states in a basin, and that transboundary agreements are concentrated in basins with a tradition of cooperation. They also found that hegemonic states are more likely to participate in such agreements, and that agreements tend to express both the need for responsible management and state rights. Finally, their data tentatively suggest that the content of such agreements depends on power relations between the signatories, with the agreements stressing principles that are advantageous to hegemonic states (Conca et al. 2006: 271–282). Such findings suggest that in terms of institutional change theory (see e.g. Thelen 2004) river basin organisations are normally layered on top of existing institutions, and that they do not displace them.

Thirdly, governance scholars draw attention to the fact that after founding a river basin organisation, it becomes necessary to formulate decision-making arrangements. Two available alternatives mentioned in this respect are consensus and elite decision making. Consensus decision making draws the risk of gridlock, whereas elite decision making may result, among other things, in the exploitation or oppression of minorities (Schlager and Blomquist 2008: 17–18) or in non-implementation of decisions if influential stakeholders have not been involved (cf. Ridder et al. 2005). In practice, decision-making arrangements are a mixture of these options. This point connects to an emerging discussion in the wider governance literature. Although many suggest that the image of a neatly ordered and organised government of yesteryear is probably as much fiction as reality, most authors do seem to suggest that the shifts in governance we have seen in the past decades have led to a fundamentally different situation. Hajer (2003) has suggested the term 'institutional void', which he uses to describe a situation wherein multiple organisations bear responsibility for (parts of) public problems, and wherein their interrelationships have not been prescribed from a central viewpoint. Skelcher (2005) suggests that the many new organisations that have been founded alongside the traditional multipurpose government layers have diminished the sovereignty of government. In a similar vein to Hajer, he observes how certain norms of democratic decision making (representation, transparency and accountability) are much less relevant for the new players in the governance arena, and how this could pose a threat to the functioning of democracy. This critique is equally as relevant for RBOs that operate on the basis of elite decision making as it is for RBOs that use consensus. Thus issues related to the institutional design of RBOs come to the fore.

In the next sections of this chapter, we will provide the conceptual foundations for the rest of this book. In section 1.3 we build the analytical framework for describing the design of river basin organisations. Section 1.4 provides the framework for the analysis of parties who propagate and block the foundation of river basin organisations and the strategies they use to influence their institutional design. In section 1.5 we develop the normative criteria with which we and other authors will evaluate river basin organisations. In section 1.6 we briefly introduce the various chapters that will follow, which helps us look forward to the concluding chapter where the fourth question we asked will be answered.

1.3 THE DESIGN OF RIVER BASIN ORGANISATIONS

As indicated already, it is possible to associate river basin organisations with several different shifts in governance, and it is not immediately clear what is actually being advocated when river basin organisations are being called for. In this book, we will borrow from Elinor Ostrom's framework for institutional analysis to study the way in which river basin organisations have been shaped. Elinor Ostrom's work is mostly associated with the management of common pool resources, and she is often seen as an advocate of community-based management of (small-scale) natural resources. Our reading of Ostrom's work is that she has indeed called for greater attention for a 'third' solution for environmental problems next to the state and the market. However, the more important message she brought is related to the importance of institutions and institutional design on the one hand, and the importance of nuance in discussing the difference between various institutional designs on the other (Kiser and Ostrom 1982). Ostrom's observation is that institutions are actually nested sets of rules. Some rules are merely operational and concern day-to-day decisions, but others are much more fundamental for the actors who work with these rules and might have a constitutional character. Taken together, sets of rules create an 'action situation' that influences the actions that actors can take.[1] Ostrom has suggested that different types of rules exist, and that an analysis of institutional arrangements should take each of these types into account. Ostrom has introduced a rule typology that exists of seven types of rules. Comments by others (see e.g. Heilman 1992: 18; Sabatier 1992: 249; Hupe 1990: 292) suggest that several of these types are overlapping and that certain types of rules alone can carry the analysis. Taking their experience to heart, we posit that five types of rules are particularly relevant for analysing institutional arrangements. These are:[2]

1. Authority rules. These prescribe which positions can take which actions, how actions are ordered, processed and terminated (that is, competencies). These rules imply the creation of positions (for example, a river basin organisation) and identify the scope of the authority (for example, the organisation deals with water quality only) and thus determine which 'states of the world' can be affected and which ones not (Ostrom 1986). The authority of other actors to have a say in the river basin organisation's decisions is also likely to have an impact on outcomes. Some river basin organisations are bound to be influenced greatly by the input of certain privileged professional bodies or interest groups, while it is also likely to matter which 'degrees of citizen power' (Arnstein 1969) exist in river basin organisations when it comes to 'ordinary' citizens (see also the text on boundary rules below).
2. Aggregation rules. These prescribe formulae for weighing individual choices and calculating collective choices. Aggregation rules relate to tensions between individual interests and ways of determining collective choices when these interests are different. At a low level of abstraction, aggregation rules can stress majority or consensus in certain fora (for example, in a municipal council) as an appropriate condition for taking a decision. At a somewhat higher level of

abstraction, aggregation rules are about the philosophical motivation for the decision. One possible aggregation rule is 'the greatest good for the greatest number of people'. In other words, the good decision is the decision that maximises utility for all (utilitarianism). Even though some may lose a bit because of a certain decision, this may be acceptable if the rest of society greatly benefits. Another aggregation rule could stress justice or equity and thus forms of compensation. In this sense, aggregation rules point to the 'type of rationality' that needs to be achieved in the decision process (Diesing 1962).

3. Boundary rules. These set the geographical boundaries around the jurisdiction of the RBO, and the entry and exit conditions for actors. They determine who is in and who is out, and according to which conditions. Legal rights to participate, for instance, on a water abstraction licence or a water management plan, are sometimes assigned to 'anyone', sometimes to 'all aggrieved parties', or sometimes to the 'licence applicant' only. What this means is that the number of actors participating in the process is likely to differ with variations in boundary rules, and subsequently different interaction types may occur.
4. Information rules. These establish information channels, state the conditions under which they are open or closed, create an official language, and prescribe how evidence is to be processed. The management of water may require various 'types' of information such as scientific knowledge, but also local knowledge. Rules can identify which of these types is considered more relevant and where they are to be obtained.
5. Pay-off rules. These prescribe how benefits and costs are to be distributed to participants, given their actions and those of others.

We expect that use of Ostrom's rule types to characterise river basin organisations offers opportunities for greater conceptual clarity and analytical precision in analysing river basin organisations. We feel this is necessary as the current literature on water governance has produced many typologies of river basin organisations, but shows few signs of convergence.

What are some of the typologies that are out there at the moment? Mostert (1998) distinguishes between three models for the organisation of water management. Whereas in the hydrological model special-purpose organisations for river basin management (river basin authorities) need to be established, in the administrative model water management is realised by multipurpose governments, such as provinces and municipalities. In the third, coordinated model, river basin commissions are established which have the task to coordinate action within a river basin. In the first model we may expect conflicts between the river basin authority and general-purpose governments; in the second model other interests may be considered more relevant than river basin management; and in the third model, because the decision-making rule is unanimity, we may expect basin policies to reflect the lowest common denominator (ibid.). Quite similarly, Alaerts (1999) distinguished between secretariat-type river basin organisations, authority-type river basin organisations, and other types of river basin organisations. The secretariat type of river basin organisations are primarily involved in policy and planning, typically have a small staff component, and can be compared with the coordination model described earlier. The authority type of river basin commission has substantial executive and infrastructural operational tasks and a large staff component, and matches Mostert's hydrological model. The other types of river basin organisations are examples of 'cooperation procedures between existing regular technical departments, or through dedicated departments within one central organization' (Alaerts 1999: 13). Since no new organisations have been established, this category resembles Mostert's administrative model.

Whereas Mostert and Alearts use more or less the same typology of three types of river basin organisations, Hooper (2005) presents a taxonomy of no less than nine types of river basin organisations which he inductively extracted from a review of existing river basin organisations in different parts of the world. He distinguishes between advisory committees, authorities, associations, commissions, councils, corporations, tribunals, trusts and federations. Close reading of his works reveals that there is much overlap between these nine types, which makes the typology less useful for empirical analysis.

A final typology that we will discuss here is the one suggested by Molle et al. (2007). They distinguish between basin authorities, basin commissions or committees, coordinating councils and international river commissions. Basin authorities are defined as 'autonomous executive organizations with extensive mandates for their river basin, undertaking most water-related development and management functions. They are regulator, resource managers, and service provider all in one' (ibid.: 610). According to Molle et al. (2007) basin commissions or committees focus more on policy setting and planning and often have specific competences such as defining taxation or effluent standards. Coordinating councils are deliberative decision-making bodies incorporating public and private stakeholders and integrating river basin policies. Finally, international river commissions are set apart as a specific category since more countries are involved in establishing these commissions. A comparison of this typology with those made by Mostert and Alaerts reveals that Mostert's hydrological model and Alaerts's authority type of river basin organisations cover both Molle et al.'s basin authorities and basin committees, and that Molle et al.'s coordinating councils resemble Mostert's river basin commissions.

Both the existence of different typologies and the different naming of RBOs can be confusing. Alaerts (1999: 13) rightly argues that the terminology for naming basin agencies can be misleading: 'Agencies with wide ranging executive tasks are sometimes called "Commissions" or "Councils", such as the Mexican Water Councils and National Water Commission, or "Boards", such as the Dutch Water Boards.'

Therefore, whilst acknowledging the relevance of the empirically based typologies discussed above, we feel it is time for a new typology that is strongly anchored in Ostrom's Nobel Prize-winning work. Here we will work with a typology of river basin organisations that distinguishes between autonomous river basin organisations, agencies, coordinating basin organisations and partnerships. Autonomous river basin organisations have a constitutionally guaranteed independent position and have their own mechanisms for democratic control. Agencies are river basin organisations which are created by the state to perform a limited number of specialised tasks at arm's length from the government; they are accountable only to (parts of) government. Coordinating river basin organisations are collaborations of the founding government partners and respond to them. Partnerships are bottom-up initiated governance arrangements which are accountable to their participants, which include civil society organisations. Table 1.1 presents this typology of RBOs and the accompanying institutional design features.

The typology we propose contains a set of 'ideal types' in the Weberian sense, which means that these are pure descriptions, that will not often be found in reality. To see the types differently would mean a denial of Ostrom's main point: that institutional designs vary in many possible respects and that most of the differences do matter. However, we do believe that scientific analysis cannot occur without a certain level of simplification, and that our typology might occupy fertile territory and a middle ground between treating all river basin organisations the same, and a full-blown application of Ostrom's insights on institutional complexity which would lead to an unmanageable number of types. We find Ostrom's typology useful to 'colour in' these types, but do not completely follow her contention that any small variation in a rule creates a whole new type of institution. Instead, we will regard those variations as a deviation from the ideal type.

One of the goals of this book is to see whether the typology can indeed be applied to real-world examples of river basin organisations – all contributors were asked to apply it to a case study that they are familiar with. But our aims go further as already suggested by our central research questions.

Table 1.1 *This book's typology of RBOs*

Types of RBOs	Typical institutional design features
Autonomous	1. Authority rules. These stipulate the independent position of the organisation vis-à-vis other government organisations. Specify which aspects of the water cycle are exclusively controlled by the organisation. 2. Aggregation rules. The organisation is in full control over the issues over its jurisdiction. In certain cases, control over the decisions of the organisation may rest with an elected body, in others it will rest with an appointed and independent leadership. Natural science, engineering and economics play dominant roles in the decisions of the organisation (rational decision making); technical quality of its work is of the highest importance to the organisation. The organisation may have veto power over decisions by other government bodies that affect water. 3. Boundary rules. The geographical boundaries of the organisation's jurisdiction will be based on the (perceived) boundaries of the river basin. The staff working at the organisation mainly consists of experts in the natural sciences or in engineering. In some cases the political leadership may be elected freely. The organisation is likely to have established procedures for involving (certain) stakeholders and ordinary citizens in its decisions and sees this as a way to build a constituency. 4. Information rules. Arguments from the natural sciences, engineering and from economics are valued very highly. The assumption is that the body has the most relevant expertise on water, and tries to communicate this to other bodies, whose agenda is less water focused. Accountability for decisions rests with the appointed or elected leadership. 5. Pay-off rules. The organisation has its own source of income (for example, has its own tax base); the rules specify which groups contribute to the income of the organisation; these rules may have several bases, including the 'polluter pays' principle. Spending is broadly related to 'water purposes' and payment of taxes will in some cases be related to having a say in the management of the organisation.
Agency	1. Authority rules. These stipulate a mandate that the organisation derives from another body (for example, a ministry or a set of ministries). The organisation has a specific and limited mandate that orientates its mission – this can be economic development-related, but also more focused on ecological goals. Although the organisation can work relatively independently and its actions will not continuously be watched, if performance is not satisfactory the founding organisation can discontinue the river basin organisation. The organisation is relatively free to seek out ways of achieving its goals. 2. Aggregation rules. The organisation has attained delegated control over one issue, but is likely to have little leverage over organisations that have similar or related tasks, and coordination mechanisms might be lacking. Control over

Table 1.1 (continued)

Types of RBOs	Typical institutional design features
	the decisions of the organisation rests with an appointed leadership, which depends on fulfilment of that mission for its continuation. Natural science, engineering and economics play dominant roles in the decisions of the organisation (rational decision making), but cost-effectiveness is the main guiding principle.
	3. Boundary rules. The geographical boundaries of the organisation's jurisdiction will be derived from the government that establishes it. The staff working at the organisation mainly consists of experts in the natural sciences or in engineering. The organisation has few legal obligations when it comes to involving stakeholders and ordinary citizens in its decision processes, but it may opt to organise stakeholder processes in support of attempts to attain its goals.
	4. Information rules. The organisation has a duty to report to its founding organisation and this line of communication is the most important one. The organisation must view its tasks in a narrow way, and only information that is relevant to its task is collected and considered. In debates about its decisions, cost-effectiveness arguments play a major role, although engineering and natural science arguments are also weighted. The agency sees members of the public as customers rather than citizens and approaches them in this fashion.
	5. Pay-off rules. The organisation has a set budget, provided by the founding organisation to achieve goals which are stipulated in a specific way. There may be rewards for (some) staff members of the organisation if the stipulated targets are being met. Such targets may include 'customer satisfaction'.
Coordinating	1. Authority rules. The organisation has no authorities of its own, but it has been founded as a facilitator of coordination between organisations that do have responsibilities in the field of water management. These bodies need to consent to the existence of the coordinating body and their agreement will delineate the responsibilities that the coordinating basin organisation can take up. In most cases this will include a research and plan function. If the organisation performs well in the eyes of its founders, the tasks may be expanded.
	2. Aggregation rules. The coordinating river basin is founded to better plan policy making, to avoid counterproductive measures by the various founding bodies and to stimulate synergy. As the river basin organisation has multiple founders, diverging interests are present; the founding organisations will have required decision making either by consensus or by large majorities (for example, two-thirds of the required votes).
	3. Boundary rules. The coordinating river basin organisation is likely to include organisations that are orientated towards very different geographical scales, but the organisations involved cannot cross the jurisdictional boundaries of their combined territories. As the organisation is fundamentally about administrative rationality (see under 'Aggregation

	rules'), only government bodies participate in the basin organisation; there might be contacts with certain stakeholders, but the general public is not involved in decisions. 4. Information rules. The coordinating river basin organisation does its own research, which generates information about potentially conflicting measures or creates a larger, cross-jurisdictional perspective on problems and issues. To achieve good levels of coordination, the participating organisations need a safe environment in which they can resolve their conflicts without too many onlookers. Access to information about what is being discussed will therefore be restricted in many cases. 5. Pay-off rules. All participants contribute to the functioning of the coordinating basin organisation, but keep most of their own budget.
Partnership	1. Authority rules. The organisation is a bottom-up initiative and has control over a resource that is commonly owned or controlled by the partners. The partnership has a conservation or a sustainable-use purpose, and controls use. 2. Aggregation rules. Deliberation and consultation between the partners are key; decisions are made on the basis of consensus. 3. Boundary rules. Joining the partnership is on a voluntary basis. Social control within the partnership implies, however, that membership does imply a set of obligations. As such, social control is essential for the functioning of partnerships, most operate on the local or regional level, which is the level where participants can know each other. 4. Information rules. Partnerships put the concerns of the participants first. Local knowledge and local memory guides decisions, but not to the complete exclusion of scientific knowledge which might also be available. The key point is that such information is processed through the prism of the interests of the partners. Partnerships are under no obligation to share the information they have with those outside the partnership. 5. Pay-off rules. All partners involved are likely to have a direct stake in the decision process and they will seek decisions that protect the interests of those involved. In some cases such decisions may entail costs for non-members, but such costs are considered less relevant.

Like Schlager and Blomquist (2008) we feel that more attention should be paid to the political aspect of institutional building that occurs when river basin organisations are founded. Schlager and Blomquist (2008) point to the technocratic tendencies in the debate about water governance in general, and about river basin organisations in particular. River basin organisations are often propagated as an alternative to existing institutions that are considered to be too fragmented, too much driven by other concerns than water management, lacking in expertise, and incapable of necessary action by politics. Often an implicit assumption is made that these existing institutions can simply be 'displaced' (Thelen 2004) with new ones such as RBOs, which then bring order and rationality to water governance. Schlager and Blomquist (2008) make several important points in connection with this. They suggest that replacing existing institutional arrangements with river basin organisations has important implications for the nature of public decision making (in line with our discussion of shifts in governance). Moreover they suggest that open debate about water governance – stimulated by various diverging organisational perspectives – is necessary in these times of rapid change, whereas letting one organisation control water management issues could lead to lock-in effects. Finally they suggest that 'watershed politics' simply cannot be avoided, even in a system designed to evade it, and that politicking is simply the way things get done (ibid.: 195). Our book takes these notions further by asking questions about the effects that politics has on the feasibility and institutional design of river basin organisations; in shedding light on these aspects we contribute to a better-informed debate in situations where river basin organisations are considered as an option for improving the governance regime.

1.4 THE POLITICS OF DESIGN

As told by Molle, the history of river basin organisations suggests that since the 'discovery' of the river basin and the 'invention' of the river basin organisation in Western thinking (the Chinese were much earlier, he claims), the concept has been 'used and appropriated by various constituencies to reorder configurations of power' (Molle 2009: 485). The point here is that proposals to introduce river basin organisations tend to come from parties who are motivated to change existing institutions and reorder powers in decision making so as to better fit with their subjective purposes. This insight is not uncontroversial, as many advocates of river basin organisations see their ideas as politically neutral and based on objective evidence, suggesting that river basin organisations are more effective, efficient, and so on. Our contention is that such neutrality does not exist in reality. Indeed, suggesting that river basin organisations will bring improvement of sorts, in comparison with the existing situation, requires a normative position on what would constitute such improvement. Terms such as 'effectiveness' and 'efficiency' suggest neutrality, but applying them actually necessitates the expression of goals: effective in which respect, and efficient in achieving what, exactly? Indeed, the history of the discourse on river basin organisations shows that these organisations have been put forward to accomplish quite a variegating set of goals. (Molle 2009: 484):

> Associated with utopian ideas of the late nineteenth century, it supported ideas of full control of the hydrological regime and multipurpose dam construction in the 1930–1960 period, then partly faded and was revived to address water-quality problems, before re-emerging in the 1990s as the cornerstone of integrated water resources management (IWRM), enriched and blended with watershed and ecosystem management approaches.

RBOs have thus been proposed to further holistic economic development and poverty alleviation, to serve agriculture and power generation, but also to further ecological goals. In all of those cases, the proposal to establish a river basin organisation is thus not neutral at all, as certain goals would be elevated over others when the proposals are adopted. Moreover, the history of river basin organisations shows that discussions about river basin organisations are not only inspired about what is substantively best, but are indeed also about decision power. In discussions about the reorganisation of the state in post-revolutionary France, for instance, '[the] concept was captured by political interests and served as a weapon against centralization, as inherited from the Revolution, and was supported by the land-owning aristocracy and by monarchists/conservatives who sought to re-establish the pre-eminence of the local' (Molle 2009: 486). Perhaps better reflecting currently ongoing discussions, Molle (2009: 491) also argued that:

> The Ministries of Environment have often invested the river basins as a crucial scale for grounding their legitimacy and assuring their role among pre-existing layers of the administration for which, in turn, RBOs have become sites where decision-making power over financial resources and investments might potentially be relocated at their expense.

The Global Water Partnership and the International Network of Basin Organizations (INBO), which have both established regional branches, tend to favour the establishment of 'powerful' or 'strong' river basin organisations (Molle 2009; Huitema et al. 2009). For instance, the World Resources Institute (2006) states that: 'the levels of authority that governments grant to RBOs are obviously critical to their abilities to manage their respective basins. The most successful RBOs have strong bases of support among basin governments, and high levels of authority through formal instruments like legislation.' In terms of Ostrom's rule typology, the expectation is apparently that especially basin governments are relevant constituents for river basin organisations (boundary rule), and that river basin organisations should be equipped with, especially, the power to regulate (authority rules). But we know from the literature on institutional dynamics that creating strong river basin organisations requires the displacement of existing institutions, and that this usually does not occur without resistance. Indeed, the 'incumbent authorities' – general-purpose organisations, such as regions, provinces and municipalities – often have an interest in limiting the authority of RBOs, as any increase in such authority would come at the expense of their own power. Thelen (2004) suggests that institutional change therefore often takes a different character than proponents of a wholesale makeover envision. Rather than full-blown displacement, she and others suggest that more common patterns of institutional design are 'layering' or 'conversion'. Layering assumes that a new institution, such as a river basin organisation, will simply be added as a new layer to an existing regime without replacing it; whereas conversion is a process where existing institutions reinvent themselves and start taking up tasks that they did not perform previously, thereby perhaps eliminating the need for a new institution. The politics of river basin organisations therefore pertain not only to the actual design of the basin organisation, but also to the redesign of other organisations. As a consequence, grand designs and dreams of institutional engineering can dissolve in practice.

However, we should be careful not to always equate the push for the foundation of river basin organisations only with 'strong' river basin organisations, even if these are often preferred in the global water management discourse. Molle (2009: 491) suggests that more recent collaborative approaches at the watershed level have risen to prominence, for example, in the United States (see also Lach and Calvert, Chapter 4 in this volume):

> The watershed is becoming a central feature of new philosophies or strands of activism like bio-regionalism that emphasizes civic responsibility and ecological stewardship . . . and is concerned with developing patterns of governance that befit natural units defined on ecological and community grounds . . . As a result, the river basin-level, traditionally invested by state bureaucracies, is challenged . . . by the proliferation of watershed oriented groups that emphasize the principle of subsidiarity and signal the fragmentation of the basin approach.

River basin organisations are often the locus of conflicts between environmental and economic interests, such as agriculture and industry. Where the former aim to expand the scope of decision making to ecological and environmental issues, the latter often try to limit the scope of decision making to water availability for economic purposes only. Other conflicts may be about the information and aggregation rules, with some arguing open decision processes, whilst others are more inclined to keep information in closed circles; and some suggesting decisions should be mainly based on expert knowledge, whilst others favour democratic deliberation involving lay knowledge.

Discussions may, however, entail any other design choice and rule. More empirical research is needed to learn more about the main controversies in designing RBOs, and the key actors and coalitions involved. As stated, we proceed on the premise that institutional design choices are political in nature. Such choices define the authority of a river basin organisation, the decision-making rules or the penalties for those who fail to live up to agreements made. Thus, if there is a debate about introducing basin organisations, or altering the existing organisations, we consider it likely that parties involved in water resources management and planning will try to influence the institutional design of RBOs in a way that serves their interests best. In this vein, decision making on the architecture of RBOs can be conceptualised as a process of strategic interaction in which actors with diverging ideas about the preferred architecture try to gain support for their ideas. Such actors have diverging preferences for various institutional designs. They may have an interest in either limiting or expanding the authority of a river basin organisation, in consensus or majority decision making, in the inclusion or exclusion of specific parties, and so on.

How do we expect the political discussion on design to unfold? It will be clear by now that we do not expect such discussions to be resolved by convincing arguments alone (be they scientific or otherwise in nature). Seeing this as a political game means that we need to pay attention to the actions of proponents and opponents of river basin organisations, and the different possible varieties of such organisations. Like Thelen (2004) and Mahoney and Thelen (2010) we assume a high degree of agency, that is, deliberate action by those involved in the discussion. Unlike them, we are not so much interested in labelling proponents of institutional change (they use the terms 'subversives', 'parasitic symbionts' and 'insurrectionaries', depending on the level of change the players want to achieve; see Mahoney and Thelen 2010). Instead, we are more interested in the way in which proponents and opponents create and use opportunities to push their case (cf. Huitema and Meijerink 2009). Disasters, such as floods or periods of severe drought, are important windows of opportunity which parties may exploit to launch proposals for changing the institutional design of RBOs. They may try to enlarge their scope (for example, by the inclusion of water quantity issues in their mandate), to give them more authority, or they may question the established criteria for their performance. As we have learned from previous research on the roles and strategies of policy entrepreneurs, framing and narrative strategies are particularly relevant here (Meijerink and Huitema 2010). In the context of this research the relevant questions, then, are which narrative about the causes of the disaster and possible remedies becomes dominant, and more specifically, whether RBOs are presented as a means to prevent new disasters. Political windows, such as a change of government or regime, may offer opportunities for establishing, abolishing or changing river basin organisations as well, since political parties may have different ideas on the authority and other characteristics of river basin organisations. Finally, parties may try to gain support for their proposals in the multiple venues that are present in most modern societies. If they do not succeed in gaining support for a particular reform proposal in one venue, they may try to gain support elsewhere by shifting attention and action to another forum. Since national, regional and local governments and different ministries may have different ideas about the necessity or ideal institutional design of river basin organisations, they may be more or less receptive to particular reform proposals. It has been suggested that the game between proponents and opponents of RBOS has a different flavour in various countries. Molle (2009: 492), pointing to the differences between the developed and the developing world, argues that:

> In the early years of the twentieth century, river-basin planning and management were associated with centralism and state-dominated large-scale infrastructural development typical of the 'hydraulic mission', this is still largely the case in developing countries, with some exceptions. Elsewhere, river basin authorities have tended to gradually morph into coordinating agencies and had to reconfigure themselves in order to accommodate local scales and processes, and the diversity of stakeholders and interests.

We expect to encounter such differences in this book as well.

1.5 EVALUATING RIVER BASIN ORGANISATIONS

Part of the political game surrounding the foundation and design of river basin organisations relates to the purposes they would serve and the criteria that should be used to evaluate success or failure. The various parties involved in the discussions on the design of river basin organisations, including different ministries, levels of government and a wide range of interest groups, all have different demands and will highlight diverging goals for water resources management, hence they tend to use different evaluation criteria. There is a host of evaluation criteria which these parties may use. These include environmental, agricultural or industrial objectives, but also questions of fit into the overall governance regime. Here, we want to limit our evaluation to a more generic set of criteria that may be used to evaluate shifts in governance. The questions that we are interested in are to what extent the RBO enhances coordination, accountability, legitimacy and environmental effectiveness. In the following, we will discuss these evaluation criteria and their relevance for the types of RBOs which we have distinguished previously.

Coordination is here interpreted as pertaining to public decision and policy making. The discourse about governance and shifts in governance came up in an era when government bureaucracies had inflated greatly. Proponents of smaller government successfully argued that the real issue was no longer market failure, but government failure. Government failure was seen specifically in the impossibility to coordinate the multiple government interventions and policies through instruments such as rational planning. It was relatively easy to point out the existence of contradictory policies, for instance, in the field of environmental policy where a lot of policy activity was aimed at undoing the policies of competing ministries such as agriculture. As the debate on governance went on to become a debate about the existence of policy networks and multilevel governance, it became increasingly clear that coordination also has a 'vertical' dimension, that is, that policies from different jurisdictional levels also need coordination. As the belief in planning has waned considerably since the 1970s, there was a need for other mechanisms to achieve coordination. As for the government tasks that were not privatised or hived-off, the belief in planning was replaced with a belief in self-organising networks that would spontaneously create order out of chaos. Public policies – or their governance equivalent, voluntary agreements – were expected to emerge from a process in which governments were an equal partner to other parties (co-management), and the government task was one of facilitation rather than regulation (for instance, through subsidising water user associations). Obviously, the ensuing process of consultations and dialogue does require effort and time, and one may wonder whether the coordination challenge may have become all that smaller. In fact, in many cases where authors plead for 'strong' river basin organisations, their ambition is to do away with a lot of 'red tape' and coordination, and simplify or streamline decision making by concentrating power in the hand of one organisation. This thinking is probably most prominent amongst those pleading for autonomous or agency-type basin organisations. Regardless which type of basin organisation is founded, however, water managers will always remain (highly) dependent on the cooperation of other policy sectors and other actors. Although from a water management perspective the focus on hydrological units is a logical one, the key to solving water management issues is mostly with organisations which work on very different geographical and jurisdictional scales.

The dilemma thus is that one may develop strong and powerful basin organisations, which enhance spatial fit, but that these organisations remain dependent on others for realising their objectives. The issue has been analysed from the perspective of the term 'institutional interplay': 'Institutional interplay refers to boundary problems of a different kind. The boundaries at stake here relate not to physical boundaries, but to political responsibilities and social spheres of influence' (Moss 2004: 5). The establishment of large basin organisations with many competencies means that just another jurisdiction is added, that institutional complexity is increased, and that it may become more difficult to realise the necessary interplay between all relevant jurisdictions (at various levels of government). So the questions remain: which type of basin organisation, then, is better at avoiding contradictory policies, conflicts about geographical and jurisdictional boundaries, and so on? Which one is better able to enhance synergies between water management and other policy fields? Are these the strong forms of river basin organisations, or rather the weak type such as partnerships and coordinating basin organisations?

Accountability is a concept that is increasingly discussed in the debate on the shifts in governance that have taken place in the past decades (see, for example, Bovens 2007, 2010 for a discussion of the relevance of the concept in the setting of the European Union; and Dryzek and Stevenson 2011, or Biermann and Gupta 2011, for a discussion of the relevance at the global level). In water management, the term 'accountability' is heard less often. Water management is regarded by many as an exclusively technical challenge, which is apolitical in character and amenable to finding the best solutions through engineering. This view has permeated the history of water management in many countries and it has had implications for the way decisions have been set up. If water management is a purely technical challenge, access to decision processes should be limited to those who have technical expertise. If it is a purely apolitical challenge, there is little need for fora of contestation. Indeed, the models that engineers and other natural science disciplines use are then probably seen as objective representations of the truth, and there is little sense in arguing about their underlying assumptions and philosophies. In such a situation, demands for accountability can remain limited. However, we interpret the foundation of river basin organisations here as a shift in governance. And we have already demonstrated how authors such as Skelcher (2000, 2005, 2006) develop critical questions about the various shifts in governance that have taken place and the fulfilment of the criterion of accountability. His sombre reading is that shifts in governance have implied greater fragmentation and subsequently less clear decision structures. As there are very few centripetal forces in the system, collective decision making becomes problematic. It becomes murky to understand who is responsible for what, and when there are problems, all governors involved have an incentive to withdraw from sight and to claim not to be responsible.

But what does accountability mean? The term has Anglo-Norman traditions and the word derives from the bookkeeping (accounting) that property holders had to do for the Norman kings. However, according to Bovens (2007), in the English language the term has transcended the realm of financial administration. It currently refers to reporting by governors to their polity rather than the opposite, and conveys the promise of fair and equitable governance. It can therefore be defined as 'a relationship between an actor and a forum, in which the actor has the obligation to explain and justify his or her conduct, the forum can pass questions and can pass judgement, and the actor may face consequences' (Bovens 2007: 450). Bovens (2007, 2010) has suggested that the concept can be used in a broad and a narrow sense. When used in the broader sense it is an evaluative concept, and when applied in the narrower sense it refers to the mechanisms for account giving. Here we are obviously more interested in the accountability as an evaluative concept. From that perspective it is important to discern between three rationales for accountability: accountability can be concerned with the democratic means to monitor and control government conduct, the prevention of power concentrations, and the enhancement of the learning capacity of public administration (Bovens 2007: 462–463). Here the logic we will follow is that of democratic control. The key questions in such a line of reasoning concern the degree to which institutions have been effectively linked to the democratic chain of delegation, which means that questions need to be asked about the degree to which democratically legitimised bodies (or the polity) can monitor and evaluate organisational behaviour, and enable such bodies to influence organisational behaviour so that it is consistent with their preferences. Bovens (2007: 462) suggests that assessments of the accountability provided by agencies or sectors are often a matter of degree and 'follow the logic of more-or-less'. This is because assessments can result in the conclusion that there is a lack of accountability ('accountability deficit'), but there can also be too much accountability, especially when a 'dysfunctional accumulation of a range of accountability mechanisms' (ibid.: 462) has come into existence.

If we connect this to the debate about river basin organisations, we can conclude that the foundation of river basin organisations (almost irrespective of the type) will affect the accountability of government. However, it is imaginable that the foundation of a river basin organisation is a simplification of water management and that the organisation makes water management rather more accountable compared to a situation where water management tasks were part of the general public administration. It can therefore not be assumed beforehand that river basin organisations threaten accountability. We should also not lose sight of the fact that transparency and accountability can be very unevenly applied to various parts of the general public. In fact, the problem of collusion suggests that water managers are extremely accountable to one segment of the public (for example, farmers), but not to other constituencies that might be equally relevant. This obviously relativises the absoluteness of the imperatives of transparency and accountability. What is more, the issue of accountability is different for different types of RBOs. The agency and coordinating RBOs, which we have defined in Table 1.1, are primarily accountable to their constituent government agencies. If these RBOs are given too much freedom in realising their tasks, accountability may become an issue indeed. On the other hand, a limitation of their discretionary powers may come at the cost of their effectiveness, which suggests that there may be a trade-off between accountability and effectiveness in some cases. For autonomous RBOs the accountability issue is more complicated. On the one hand one may argue that for this type of RBO accountability is less of an issue, since these RBOs do have their own (in) directly elected board. On the other hand, because of the existence of such a board there often is considerable institutional overlap with general-purpose governments which also address specific water issues. Although such institutional overlap may contribute positively to the adaptive capacity of the water sector, it may also lead to a situation in which citizens no longer know which organisation should be held accountable for specific water policies. Finally, for the partnerships, accountability is about their relationship to the participating parties. Whereas some may judge these partnership arrangements negatively because of poor accountability to the general public, others may emphasise their merits because they offer opportunities for developing new practices of deliberative democracy or because they offer new avenues for influencing water management practices for actors who used to be marginalised.

Legitimacy refers to the public perception of water management. Countries differ in the degree to which the general population is aware of water management activities, but the treatment of water management as merely a technical issue could have a logical correlate in lower public awareness. Once initial investments in large-scale water infrastructure are made, it takes quite some determination to keep their maintenance on the agenda and long-term investment is often an easy political target for budget cuts. And, especially with flood infrastructure, long periods without floods might instil a false sense of security and this in turn may invite risky investment and settlement behaviour. Obviously, such periods of low public awareness might be punctuated by 'extreme weather events' such as protracted periods of drought or rain. It can thus be argued that it is important to maintain a certain level of public dialogue on water issues so that water problems stay in the public view and mind. Dialogue implies a bilateral communication process, and much of the writing on public participation in water management suggests that water managers can improve their operations if they use the collective wisdom of the general public (although experts will always remain important). This touches upon various other aspects: using local or lay knowledge might help achieve ecological objectives, improve cost-effectiveness and enhance legitimacy. We will be the last to suggest that such legitimacy only depends on dialogue, but especially one form of legitimacy, known as input legitimacy, will benefit from such dialogue as it refers to the way in which institutions effectively deal with feedback from the wider public. Other forms of legitimacy include output legitimacy and systematic legitimacy; these refer, respectively, to the delivery of goods and services and the broader feeling of support for the institutional set-up. It is difficult to say beforehand which type of RBO contributes most to the legitimacy of water resources management as this strongly depends on the way in which a specific RBO involves the general public in developing and/or implementing basin policies. From the perspective of input legitimacy, the partnerships and autonomous RBOs probably do better than the agencies and coordinating committees, but from the perspective of output legitimacy this may well be the other way round. The large variety and number of parties involved in most watershed partnerships, in combination with a decision-making rule of unanimity, may easily lead to a stalemate or to decisions which reflect the 'lowest common denominator' (Schlager and Blomquist 2008). That is why partnerships can be assumed to often not deliver, and subsequently have a low output legitimacy.

The criterion of environmental effectiveness refers to goal attainment in the sphere of ecological issues. As suggested above, most modern-day proponents of river basin organisations do expect river basin organisations to take ecological issues more seriously than the existing institutions, and hence deliver better performance. But what are these ecological issues and how can we classify them? Certainly since the 1950s and 1960s water pollution has continued to be a staple element in the debate about water management. In many countries, water quality issues that emanated from petrochemical and other previously existing industries (such as textile and paper production) became increasingly controversial when their effects on nature became clear (one can think of Rachel Carson's 1962 book *Silent Spring* here). The impact of agriculture, through water abstraction, through the use of fertilisers and pesticides, should also not be overlooked. But there are more issues. The hydrological mission – often hinging on the erection of massive infrastructural constructions – has had (and is having) enormous impacts on the landscape, on the flora and fauna connected to the water system, and the livelihood of countless people living in affected areas. The environmental movement that started in the 1960s has battled against such developments in many countries, often in combination with other civil society groups. And then there is the issue of climate change; although climate change is often thought of as a global phenomenon, the local effects can differ greatly. Whereas the predictions are that some climate zones will get drier, in others more precipitation can be expected. And even if averages in rainfall and other weather conditions were to remain constant in various regions, the frequency with which extreme weather events (be they droughts or high levels of precipitation) occur is expected to rise. Coupled with a pattern of economic investment and human settlement in vulnerable regions, this has led to increased damage levels in the last decades (for a discussion on the various factors affecting such risks see, for instance, Bouwer et al. 2007; Bouwer 2010). An obvious correlate of the variable impacts of climate change is the fact that such damage can vary greatly between localities and jurisdictions, including those that are upstream and downstream of rivers. River basin organisations are expected to help address such problems by connecting such jurisdictions, but also by focusing more squarely than general-purpose governments would do on the ecological aspects. But whether such expectations really come true is still an open question (for attempts to answer aspects of it, see Dinar et al. 2005; Schlager and Blomquist 2008). However, from our brief summary of ecological problems that water managers face, it is already becoming clear that such issues are now more rather than less challenging. Climate change cuts through and across jurisdictions, sectors (such as water and energy), and does not stop and start at the boundaries of a river basin. Whether autonomous river basin organisations handle such challenges better than other types – as is often assumed – remains to be seen.

1.6 LOOKING AHEAD

In the rest of this book the four central questions that guide our analysis – about the design of river basin organisations; their proponents and opponents and their effects on design; the impacts that these organisations have; and whether we can develop recommendations for those engaged in the discussion – will be addressed. The heart of our book consists of 11 chapters on river basin organisations within a range of countries, each addressing the first three questions. These chapters are preceded by Chapter 2 which provides a general overview of the global discussion on river basin organisations, which helps to further contextualise the work. Chapter 14, the last chapter, provides a synthesis of the observations from all 11 chapters, draws general conclusions, and addresses the fourth question about the implications of the findings from this book for the debate on RBOs.

It is probably good to offer some explanation of why we are focusing here on river basin organisations within countries and why we subsequently selected the organisations and countries that were included in this book. The focus on national discussions on river basin organisations stems from the personal preferences of the editors, but is also related to the fact that basin organisations within country borders are logically a more common and frequent form of organisation than the 'special' case of transboundary RBOs, which have in fact attracted a large – perhaps disproportionate – amount of attention already from water governance scholars (e.g., see Beach et al. 2000; Rieu-Clarke 2005; Conca 2006; Molle et al. 2009; Schmeier 2013).

The selection of countries and cases presented in this book has not been steered as much by conceptual considerations as one would normally like. In the ideal setting one would select such cases on the basis of either the independent or the dependent variable, and the chances they offer to reject explicit hypotheses surrounding these key variables. Here the issues we are grappling with cannot be framed so neatly, however. Our contribution is, firstly, to see whether the new typology of RBOs can be applied to real-world cases (this is not a matter of rejecting hypotheses; except perhaps of course that we expected the typology to work). Secondly, we wish to see how proponents and opponents of RBOs tried to influence RBO design; thirdly, to see whether certain types of RBOs perform better than other types; and finally, to derive recommendations. Only for the performance issue might one say that there is a working hypothesis in the literature – which would be that autonomous RBOs will perform better (see above), but this hypothesis has certainly also been challenged extensively. The implication is that the contributions to this book should be seen largely as inductive empirical studies, and only to a limited degree as focusing on deductive hypothesis testing. Despite that being the case, the book makes a contribution to the overall discussion on water governance, by better conceptualising river basin organisations, and by asking questions about the politics surrounding the foundation of RBOs. The latter topic has been treated in the existing literature of course, but is usually quickly dispatched as a problem, as something irrational in the way of an otherwise brilliant plan, and something that should be done away with rather than analysed more profoundly (exceptions include Schlager and Blomquist 2008). In bringing together 11 country analyses, we hope to show that politics always plays a role in these discussions and simply cannot be treated as the irrational part of the equation. In methodological terms, our collection presents a 'most different systems approach', which implies that when the relation between two variables (say politics and institutional design) is found across a number of very different systems (or countries) this can be seen as corroboration that the variables are indeed causally related; in fact, the more countries tested for and the greater the differences between them, the greater the confidence in the finding of a connection (Hopkin 2002).

The set of three questions that we ask about every country may seem simple at first, but actually requires quite a bit of in-depth knowledge of the case studies. In combination with the fact that we mostly have an inductive and exploratory agenda, this means that in this book we opted for collaboration with colleagues who intimately knew the case studies they describe beforehand. By collaborating with us, and applying our framework, they were forced to present this knowledge in a way that was fitting with our purpose, and they helped us to explore the questions we asked. We found the colleagues with the knowledge and skills to perform the type of analysis we wanted by posting several open calls within relevant networks and websites (see e.g. www.watergovernance.eu), by the snowballing technique, and by repeatedly presenting our agenda at relevant scientific conferences such as the yearly meetings of the Earth System Governance project. Our contention is that in the end we have found an excellent, highly informed and able set of authors to describe the discussion on river basin organisations in various countries (or rather, parts of these countries, as many chapters zoom in on the discussion about RBOs at the regional or even local level). We acknowledge that working with authors who are familiar with certain case studies has potential downsides. It might introduce certain biases (some authors may have become attached to 'their' case studies) and it could have implied an exclusive focus on parts of the world for which many analyses already exist. We have worked to erase any biases emanating from familiarity with the case study by submitting all book chapters to peer review (all authors read each other's work and critically commented on previous versions in a meeting in Amsterdam, the Netherlands, in November 2012). It is obviously up to the reader to decide whether these biases were indeed effectively removed, but we believe that to be the case. And we do feel that a book that contains chapters on water management in Mongolia, Afghanistan and Ukraine can hardly be accused on focusing on the same old case studies; in fact several case studies are here described in the English language for the first time.

Another potential criticism of our case study selection could focus on the fact that the RBOs studied in the various countries do differ quite considerably in geographical scale. This is a consequence of our way of selecting cases (based on the expertise of the authors, not on the comparability of the case studies). And to our mind this variation offers mainly advantages from a methodological perspective (see the comment about the most different systems approach) and from a conceptual perspective, as the variation in geographical size does offer an opportunity to see whether certain types of RBOs (such as partnerships) are indeed confined to the very local scale, as is often assumed in the literature (perhaps this is another rare example of a testable hypothesis).

Before we now turn to the various chapters included in this book, we briefly introduce each of them. Figure 1.2 provides an overview of the countries where study was made for the purpose of this book.

This book contains multiple chapters on coordinating-type RBOs. Chapter 3 by Rob de Loë and Michelle Morris is on the Mackenzie River Basin Board (MRBB) in Canada. Chapter 9, by Nina Hagemann and Marco Leidel, discusses the situation in the Western Bug basin in Ukraine, and specifically the operation of the River Basin Administration and a River Basin Council there. Chapter 10 by Jeroen Warner and Vincent Thomas focuses on the Lower Kunduz and Taloquan River Basin Councils in Afghanistan (and another part on the River Basin Agency, which is of a different type). Chapter 11, by Ines Dombrowsky, Annabelle Houdret and Lena Horlemann, focuses on Mongolia, and an analysis is presented of the fleeting general discussion on river basin organisations there. Chapter 12, by Santita Ganjanapan and Louis Lebel, discusses the foundation of new river basin organisations in the Mae Kuang watershed, the north of Thailand, and specially the Ping River Basin Committee and the Mae Kuang sub-basin group there. Chapter 13, finally, by Andrew Ross and Daniel Connell, focuses on the Murray–Darling Basin Ministerial Council in Australia and the Basin Community Committee there (they also discuss the Murray–Darling Basin Authority, but this is an agency type of organisation).

Several common threads weave the stories on the coordinating type of RBOs together. In terms of institutional design, these organisations all have an indirect mandate, derived from the organisations that have founded them. This means that the pattern of institution building can usually best be described as layering (although a case such as Mongolia shows greater

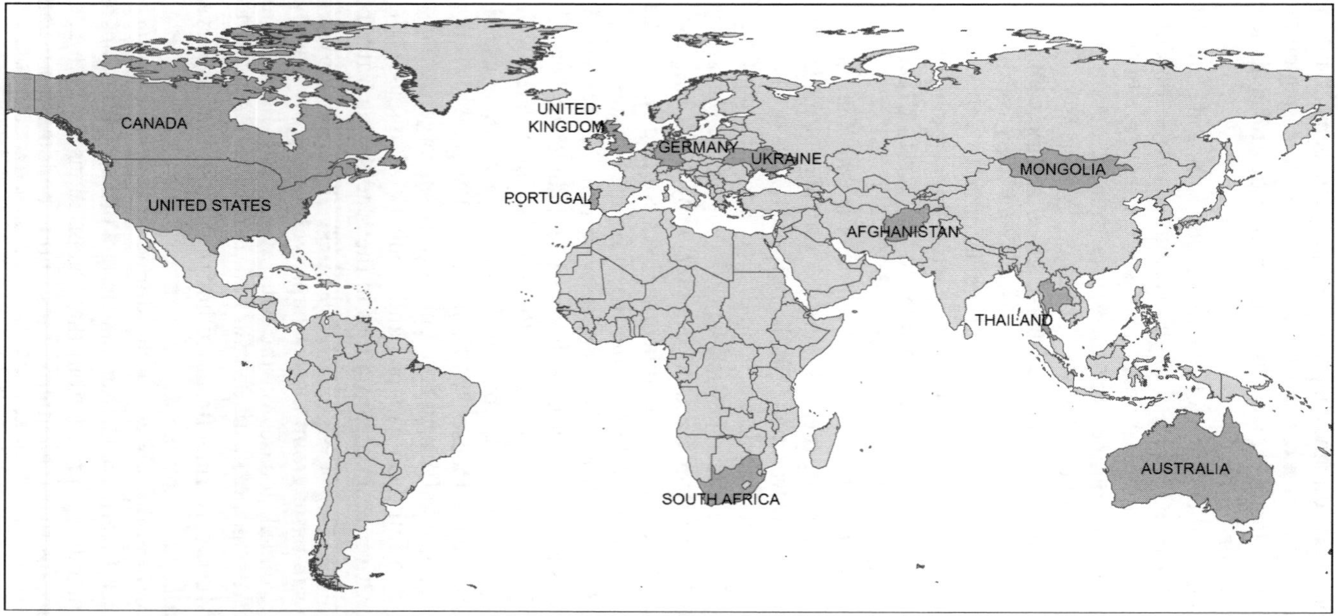

Figure 1.2 Overview of countries as the locations of the cases presented

tendencies towards displacement perhaps), meaning that the new institution is just another element in an already complex setting. Two points become clear about the process of layering: one about bureaucratic politics, and one about the relation to more traditional institutions. As for bureaucratic politics, the founding organisations – be they ministries, provinces or other forms of government – are not in the game wholeheartedly. This means that there are usually organisations that try and limit the amount of resources that the new organisations will obtain (in terms of finances, information flows, decision power) either by institutional design, or by not collaborating with the RBOs once they have been founded. As for the relation to more traditional institutions (visible in the Afghan and Thai cases), it is noteworthy that RBO design often does not fully involve these institutions, or even neglects their existence. Part of the reason is that the idea for the foundation of the RBO comes from outside the circles that were managing water before. Especially in developing and transition country case studies (such as Afghanistan, Mongolia and Ukraine), foreign parties such as consultancies from Western countries carry the international thinking on RBOs to these countries, usually accompanied by national allies that stand to gain with the RBO or support the idea intellectually. Such parties enter through windows of opportunity created by various recurring factors such as the dependency on foreign subsidies, broader societal and political changes, and an active political leadership that supports innovation. The dependency on other parties, but also the fact that the initiative comes from outside traditional water circles, goes a long way to explain the relative weakness of almost all coordinating RBOs described in this book in terms of resources and authority. As a correlate, performance is often judged to be relatively poor by our authors, in some cases with traditional institutions actually needed to help out in terms of coordination in times of crisis (which underpins the case for institutional diversity and polycentricism, perhaps).

There are also multiple analyses of agency-type RBOs in this book. The first we encounter is described by Denise Lach and Dan Calvert in Chapter 4, and it is the Oregon Watershed Enhancement Board from the US (which works through local partnerships). Then follows the Erftverband (Special Law Water Board) case from North Rhine-Westphalia in Germany, described by Frank Hüesker and Christoph Bernhardt, in Chapter 6. As per the authors, this organisation combines an agency-type profile, with certain elements of an autonomous RBO. Chapter 8, by Richard Meissner and Nikki Funke, analyses the functioning of the Breede–Overberg Catchment Management Agency in South Africa, which also has elements of a coordinating and partnership type of RBO. The Ukrainian, Afghan and Australian chapters discuss agency-type organisations too (Chapters 9, 10 and 13, respectively). In each of these countries, agencies co-exist with coordinating-type RBOs in one way or another, often as the 'implementing branch' of the institutional set up that also contains a coordinating organisation.

It is interesting to observe that especially the US and German RBOs discussed here have a relatively narrow mandate (focusing on salmon restoration, and water quality and provision, respectively), which fits with our typology, but that the other agency-type RBOs actually have a much broader mandate. We can perhaps take the US and German cases as relatively extreme examples of the spectrum of agency type of RBOs. Whilst both RBOs are (almost) unique within their state or country, the Oregon, USA RBO essentially works on the basis of a voluntary approach with local restoration groups (but under the shadow of the threat of federal legal action), whereas the German example is on the basis of the power to regulate. These are obviously two very different ways of handling collective action dilemmas, and both sets of authors point to the importance of framing in explaining the difference. Both RBOs emanated from crises with water quality, but in the German case much more emphasis was placed on the victims of bad water quality, whereas in the Oregon case the emphasis was placed on the willingness of farmers and landowners to care for their environment. Of the two, the German example seems to be the most stable, with nobody calling for the abolishment of the RBO; whereas the Oregon RBO appears to be vulnerable to changes in its funding scheme. For all agency-type RBOs described in this book, the political sensitivity of their establishment and the key importance of having sufficient resources is confirmed. The South African chapter details how, over the course of 14 years, only two out of 19 projected Catchment Management Agencies (CMAs) in South Africa were actually founded. This is to do with opponents to the decentralisation drive implied in their foundation, but the authors also claim that a lack of resources (funding, capable personnel) is also to blame. The issue of resources also comes forward very prominently for the Western Bug case in Ukraine – here the River Basin Administration does have the power to collect taxes, but such taxes need to be handed back to the central government that then redistributes them, essentially earmarking them for other purposes. In several countries, notably South Africa and Mongolia, the thinking about RBOs also appears to be hesitant and evolving. In South Africa, CMAs are meant to be very much under the wing of the relevant minister, until they have proven they can function independently (when they become more purely an agency). But given the paucity of actual experience with the CMA model, their performance is actually quite hard to judge. In Mongolia, the emphasis has shifted somewhat from a coordinating type of RBO to an agency type of RBO, in the expectation that it will perform better.

There is only one example of discussion on an autonomous type of RBO in this book, and this can be found in Chapter 7, where Andreas Thiel and António Guerreiro de Brito describe the national discussion on water governance in Portugal. Connecting very much with the point just raised, the emphasis on autonomous RBOs (in the form of River Basin District Authorities) was relatively short-lived – essentially from 2008 to 2011. Before this period, the Portugese system was focused more on coordination-type RBOs, but currently the desire is to create a system of agency-type RBOs. The change in direction was very much connected to the ebb and flow of electoral politics (although the legislation of the European Union is also influential), and the authors suggest that none of the more recent models has been able to prove itself in practice; although their interviewees suggest that the system of River Basin District Authorities seemed to be working well before it was replaced by a new model.

Our book contains also just one example of a partnership-type RBO. In Chapter 5, Hadrian Cook, David Benson and Alex Inman describe the case of the Westcountry Rivers Trust from the Southwest of England. They suggest that the rise of partnership-type basin organisations in the UK should be seen in the light of low trust in the formal system of water governance, which is dominated by agency-type organisations. The partnership nature of this RBO is reflected in its focus on consensual decision making, local expertise, a holistic approach and a desire to find win–win solutions. In this particular case, the initiators were successful in attracting outside resources (such as from the European Union) and the authors consider its performance to be relatively good.

Having briefly introduced the various contributions to this book, we now move to Chapter 2, which was added to better contextualise the various country studies. In this chapter, Frank Jaspers and Joyeeta Gupta further detail the global discussion on RBOs. Their discussion is partly historic, deepening insights presented already in this chapter; and partly ideational and discursive, describing the various ways in which RBOs have been conceptualised, talked about and understood over the years. They contend that the conversation on RBOs is becoming increasingly globalised, and also increasingly 'nexus' oriented, meaning that a more holistic treatment of water issues and broader authorities for RBOs are strived for. They conclude by discussing various contextual factors that are important for the effective functioning of RBOs, especially the fit between RBO design and context, and the need to keep RBOs 'aquacentric', but also relatively simple aspects such as resource availability. They also suggest that climate change will present a major challenge for both existing and new RBOs. Let us now see how RBOs have coped with this and other challenges – an important input to the recommendations that we discuss in the final chapter, in line with the fourth question we seek to answer.

NOTES

1. The action situation is also influenced by natural circumstances and culture, but these are factors that Ostrom has not elaborated upon.
2. Ostrom has indicated that rules work configurationally. This implies that if the 'setting' of one type of rule changes, the total set of rules is affected, and the rules will have a different effect (Ostrom 1986: 14). This insight is profound, but also potentially very disruptive of our possibilities for generalisation. Imagine an institutional arrangement wherein 50 positions are specified, each associated with 50 boundary rules, 50 information rules, and so on. If it is true that a change in one of these rules affects the entire setting and effectively a new institutional arrangement is created, we would end up with a massive amount of possible institutional configurations, with assumedly all very different outcomes. Such a complex web of options goes beyond the human capacity of analysis and understanding. This is probably why Elinor Ostrom is often associated 'simply' with community-based management (which could mean a range of things if one applied her typology). It is probably also why the impressive work Ostrom has done leads to a set of relatively simple design guidelines (for instance, that the enforcement of rules and thus the design of enforcing positions and authorities is important). For us, the conclusion is that the level of sophistication in the analysis is better bounded and 'templates' or 'shorthand' such as the market or the state do have some value.

REFERENCES

Alaerts, G. (1999), *Institutions for River Basin Management. The Role of External Support Agencies (International Donors) in Developing Cooperative Arrangements*, Washington, DC: World Bank.

Arnstein, S.R. (1969), 'A ladder of citizen participation', *Journal of American Institute of Planners*, **35**, 216–224.

Beach, H.L., J. Hamner, J. Hewitt, E. Kaufman, A. Kurki, J. Oppenheimer and A. Wolf (2000), *Transboundary Freshwater Dispute Resolution. Theory, Practice and Annotated References*, Tokyo: United Nations University Press.

Biermann, F. and A. Gupta (2011), 'Accountability and legitimacy: an analytical challenge for earth system governance', *Ecological Economics*, **70**, 1854–1855.

Biswas, A.K. (2004), 'Integrated water resources management: a reassessment', *Water International*, **29**, 398–405.

Bouwer, L.M. (2010), 'Disasters and climate change: analyses and methods for projecting future losses from extreme weather', dissertation, Amsterdam: VU University Amsterdam.

Bouwer, L.M., R.P. Crompton, E. Faust, P. Höppe and R.A. Pielke, Jr (2007), 'Confronting disaster losses', *Science*, **318**, 753.

Bovens, M. (2007), 'Analysing and assessing accountability: a conceptual framework', *European Law Journal*, **13** (4), 447–468.

Bovens, M. (2010), 'Two concepts of accountability: accountability as a virtue and as a mechanism', *West European Politics*, **33** (5), 946–967.
Carson, R. (1962), *Silent Spring*, Boston, MA: Houghton Mifflin.
Conca, K. (2006), *Governing Water: Contentious Transnational Politics and Global Institution Building*, Cambridge, MA: MIT Press.
Conca, K., F. Wu and C. Mei (2006), 'Global regime formation or complex institution building? The principled content of international river agreements', *International Studies Quarterly*, **50**, 263–285.
Delli Priscoli, J.D. (n.d.), 'River basin organizations', unpublished paper, http://transboundarywater.geo.orst.edu/research/case_studies/Documents/RiverBasinOrganizations.pdf.
Diesing, P. (1962), *Reason in Society: Five Types of Decisions and their Social Conditions*, Urbana, IL: University of Illinois Press.
Dinar, A., K. Kemper, W. Blomquist, M. Diez, G. Sine and W. Frau (2005), 'Decentralization of river basin management: a global analysis', Policy Research Working Paper 3637, Washington, DC: World Bank.
Dryzek, J. and H. Stevenson (2011), 'Global democracy and earth system governance', *Ecological Economics*, **70**, 1865–1874.
Gupta, J. (2009), 'Driving forces in global fresh water governance', in D. Huitema and S. Meijerink (eds), *Water Policy Entrepreneurs: A Research Companion to Water Transitions around the Globe*, Cheltenham, UK and Northampton, MA, USA: Edward Elgar, pp. 37–57.
Hajer, M. (2003), 'Policy without polity? Policy analysis and the institutional void', *Policy Sciences*, **36**, 175–195.
Held, D. (2006), *Models of Democracy*, Cambridge: Polity.
Heilman, J.G. (1992), 'Some thoughts on a rule based conceptual framework for institutional analysis in water policy research', in J.Th.A. Bressers and L.J. O'Toole (eds), *International Comparative Policy Research: Preparing a Four Country Study on Water Quality Management*, Enschede: CSTM, pp. 59–83.
Hooper, B. (2005), *Integrated River Basin Governance. Learning from International Experience*, London: IWA Publishing.
Hopkin, J. (2002), 'Comparative methods', in D. Marsh and G. Stoker (eds), *Theory and Methods in Political Science*, 2nd edn, Houndmills: Palgrave Macmillan, pp. 249–270.
Huitema, D. and S. Meijerink (eds) (2009), *Water Policy Entrepreneurs. A Research Companion to Water Transitions around the Globe*, Cheltenham, UK and Northampton, MA, USA: Edward Elgar.
Huitema, D., E. Mostert, W. Egas, S. Moellenkamp, C. Pahl-Wostl and R. Yalcin (2009), 'Adaptive water governance: assessing adaptive management from a governance perspective', *Ecology and Society*, **4** (1), 26.
Hupe, P. (1990), 'Intenties en interacties; over postmoderne en andere instrumenten van overheidssturing', in J.A.M. Hufen and A.B. Ringeling (eds), *Beleidsnetwerken; overheids-, semi-overheids-, en particuliere organisaties in wisselwerking*, Den Haag: VUGA, pp. 223–234.
Imperial, M.T. and T. Hennessey (1999), 'Environmental governance in watersheds: collaboration, public value and accountability', paper presented at the Twenty-First Annual Research Conference of the Association for Public Policy Analysis and Management, Washington, DC, 4–6 November.
Kiser, L.L. and E. Ostrom (1982), 'The three worlds of action: a meta-theoretical

synthesis of institutional approaches', in E. Ostrom (ed.), *Strategies of political inquiry*, Beverly Hills, CA: SAGE, pp. 179–222.

Lundqvist, J., U. Lohm and M. Falkenmark (1985), *Strategies for River Basin Management: Environmental Integration of Land and Water in a River Basin*, Dordrecht: Reidel.

Mahoney, J. and K. Thelen (2010), 'A theory of gradual institutional change', in J. Mahoney and K. Thelen (eds), *Explaining Institutional Change: Ambiguity, Agency, and Power*, Cambridge: Cambridge University Press, pp. 1–37.

Meijerink, S. and D. Huitema (2010), 'Policy entrepreneurs and change strategies: lessons from sixteen case studies of water transitions around the globe', *Ecology and Society*, **15** (2), 21; http://www.ecologyandsociety.org/vol15/iss2/art21/.

Mitchell, B. (1990), *Integrated Water Management; International Experiences and Perspectives*, London: Belhaven Press.

Molle, F. (2008), 'Nirvana concepts, narratives and policy models: insights from the water sector', *Water Alternatives*, **1** (1), 131–156.

Molle, F. (2009), 'River-basin planning and management: the social life of a concept', *Geoforum*, **40** (3), 484–494.

Molle, F., T. Foran and M. Käkönen (eds) (2009), *Contested Waterscapes in the Mekong Region: Hydropower, Livelihoods and Governance*, London: Earthscan.

Molle, F. and P. Wester (eds) (2009), *River Basin Trajectories; Societies, Environments and Development*, Wallingford: CABI.

Molle, F., P. Wester and P. Hirsch (2007), 'River basin development and management', in D. Molden (ed.), *Water for Food, Water for Life: A Comprehensive Assessment of Water Management in Agriculture*, London: Earthscan, pp. 585–624.

Moss, T. (2004), 'The governance of land use in river basins: prospects for overcoming problems of institutional interplay with the EU Water Framework Directive', *Land Use Policy*, **21** (1), 85–94.

Mostert, E. (1998), 'River basin management and planning', paper presented to the 4th National Congress on Water Resources by the Portuguese Water Resources Association, Lisbon, 27 March.

Mostert, E. (ed.) (2000), *River Basin Management*, Proceedings of the International Workshop on River Basin Management, 27–29 October 1999, The Hague, the Netherlands, Paris: UNESCO.

Mukhtarov, F.G. (2009), 'The hegemony of integrated water resources management: a study of policy translation in England, Turkey and Kazakhstan', dissertation, Budapest: Central European University.

Ostrom, E. (1986), 'An agenda for the study of institutions', *Public Choice*, **48**, 3–25.

Ostrom, E. and M.A. Janssen (2004), 'Multilevel governance and resilience of social and ecological systems', in M. Spoor (ed.), *Globalisation, Poverty and Conflict*, Dordrecht: Kluwer, pp. 239–259.

Rhodes, R.A.W. (1996), 'The new governance: governing without government', *Political Studies*, **44**, 652–657.

Ridder, D., E. Mostert and H.A. Wolters (eds) (2005), *Learning Together to Manage Together: Improving Participation in Water Management*, Osnabrück: University of Osnabrück.

Rieu-Clarke, A. (2005), *International Law and Sustainable Development: Lessons from the Law of International Watercourses*, London: IWA Publishing.

Sabatier, P.A. (1992), 'Book review of *Governing the Commons: The Evolution of Institutions for Collective Action* by E. Ostrom', *Public Administration Review*, **86** (1), 248–249.

Schlager, E. and W. Blomquist (2008), *Embracing Watershed Politics*, Boulder, CO: University of Colorado Press.

Schmeier, S. (2013), *Governing International Watercourses: River Basin Organizations and the Sustainable Governance of Internationally Shared Rivers and Lakes*, London: Earthscan.

Skelcher, C. (2000), 'Changing images of the state: overloaded, hollowed-out, congested', *Public Policy and Administration*, **15** (3), 3–19.

Skelcher, C. (2005), 'Jurisdictional integrity, polycentrism, and the design of democratic governance', *Governance: An International Journal of Policy, Administration, and Institutions*, **18** (1), 89–110.

Skelcher, C. (2006), 'Does democracy matter? A transatlantic research design on democratic performance and special purpose governments', *Journal of Public Administration Research and Theory*, **17**, 61–76.

Teclaff, L.A. (1967), *The River Basin in History and Law*, The Hague: Nijhoff.

Thelen, K. (2004), *How Institutions Evolve: The Political Economy of Skills in Germany, Britain, the United States, and Japan*, Cambridge: Cambridge University Press.

Wittfogel, K. (1957), *Oriental Despotism: A Comparative Study of Total Power*, New Haven, CT: Yale University Press.

World Resources Institute (2006), 'Transboundary environmental governance: the ebb and flow of river basin organizations', Washington, DC: World Resources Institute, http://earthtrends.wri.org/text/environmental-governance/feature-46.html.

2. Global water governance and river basin organisations

Frank Jaspers and Joyeeta Gupta

2.1 INTRODUCTION

There is considerable literature on river basin organisations (RBOs) at the national and international level (Blomquist et al. 2005; Conca 2006; Hooper 2005; Zawahri 2008; Mostert 2003; UNECE 2009; OECD 2011), but very little that actually deals with a global perspective on RBOs. We argue in favour of a global analysis because although rivers are primarily fluvial in nature and not global, water problems are cumulatively severe (Vörösmarty et al. 2010), nearing planetary boundaries (Rockström et al. 2009); they affect a common global hydrological system; and are affected also by global drivers and global change (Pahl-Wostl et al. 2008). A key institution for dealing with water is the RBO. Hence, this chapter sets out to address the questions: what factors have influenced RBO design over time, and what challenges face RBOs?

In order to answer these questions, this chapter reflects on the extensive literature on national and transboundary RBOs in order to generate global reflections on the issue. This chapter examines the history of the establishment of RBOs, the current discourses that influence the establishment of RBOs, and the challenges facing RBOs today.

2.2 THE EVOLUTION OF RIVER BASIN ORGANISATIONS GLOBALLY

2.2.1 Introduction

There is nothing new about RBOs. Through history, society has organised itself to manage its water resources whether informally or formally (Dellapenna and Gupta 2009). However, as society has become more complex and globalisation has affected the way we use our water, such organisations are becoming more and more institutionalised. Whenever integrated management of water resources is needed in a river basin to combat floods and droughts, to preserve aquatic ecosystems or to streamline other upstream–downstream relations, the prevailing manner for doing this is by concentrating water-related functions in a specific single-purpose RBO (EEB 2001). RBOs appear all over the world as existing institutions or as upcoming arrangements in reforms in the water sector. The term 'river basin organisation' is a patchwork quilt covering many different types of organisations, with very different tasks and responsibilities. Some typologies of organisation focus on size (for example, watershed, catchment, river basin and transboundary), while others focus on organisational structure (for example, as in this book).

Worldwide there are thousands of rivers, of which 261 major river basins are shared by one or more sovereign states and even more river basins cross local, state or provincial boundaries. Historically, shared rivers were governed through treaties at the international level or inter-agency compacts at local or state level. The earliest formal river commissions include those established on the Rhine. The agreements on the Rhine first aimed at establishing a commission to promote free navigation on the Rhine, and later on to address quantitative and qualitative issues. In the early twentieth century cooperation on international rivers led to a treaty establishing the International Joint Commission between the United States (US) and Canada, and another treaty led to the International Boundary and Water Commission between the US and Mexico. In the mid-twentieth century, a series of agreements led to the establishment of the Intergovernmental Commission for Coordination between Argentina, Bolivia, Brazil, Paraguay and Uruguay established in 1968. At around the same time, a Permanent Joint Commission was created between Egypt and Sudan by a 1958 agreement. Other African commissions include the Niger River Commission and the Zambesi Rivercourse Commission. The Mekong Committee established in 1957 was followed by the 1996 Mekong River Commission established after the Agreement on the Cooperation for the Sustainable Development of the Mekong River Basin. The Indus is regulated by a Joint Commission between India and Pakistan. Recent commissions include the 1992 Interstate Coordination Water Commission established between Kazakhstan, the Kyrgyz Republic, Tajikistan, Turkmenistan, Uzbekistan and Afghanistan.

At the national level, RBOs date back several thousand years, but institutionalisation in a modern form can be traced back around 300 years ago to Europe when, in the aftermath of the advent of the Industrial Revolution, trade and transport through the waterways became important and the use of water increased considerably. This led gradually to a proliferation of river basin bodies with vastly different characteristics (Hooper 2005). In more recent decades, institutionalisation of water reforms in the developing world led to the establishment of modern-day RBOs in the European Union (EU) and many parts of the developing world. Within the EU, the European Commission recommends incorporating all water bodies, natural and artificial, into one comprehensive system of water resources planning. For this purpose all river basins should be organised into river basin districts (Directive 2000/60/EC, Article 3) along hydrological boundaries. In the meantime all member countries have carried out this exercise. The government of Turkey is also presently involved in a process of identifying river basin districts and empowering them with authority in order to produce integrated river basin plans. This is in line with the requirements for admission to the EU posed in the EU Water Framework Directive. The link between water quantity and water quality is a crucial issue in Turkey (Senter International 2001). In the past few years (2007–2010) the Ministry of Nature, Environment and Tourism of Mongolia has embarked, through the National Water Authority, on a process of establishment of River Basin Councils (Mongolian Water Law 2012, Article 9; for more on Mongolia, also see Dombrowsky et al., Chapter 11, this volume).

Examples from the developing world include the legal water sector reform of Indonesia (2004) which hinges on a rather complex subdivision of the country into nine different types of RBOs. The Water Act of South Africa is based upon the concept of water management on hydrological boundaries through the establishment of Catchment Management Agencies (Jaspers 2001). The International Network of Basin Organizations, established in 1994, is a body that tries to share knowledge and experiences and promotes cooperation between these bodies. It currently has 133 member organisations in 50 countries and this does not include all RBOs at the local and national level (International Network of Basin Organizations, www.INBO-news.org). Most of these organisations can be characterised as intergovernmental commissions to optimise the wise use of international waters. The cooperation is mainly in the technical field and in the preparation of decision making. Effective decision making is often confined to ministerial level (Bakker 2013). This section discusses the phases of RBO development, and the factors and actors that have influenced RBO development.

2.2.2 Phases of RBO Development

History demonstrates the intimate connection between the stability of a group of people, its economic and social development, and the availability and reliability of water supply, often through collective productive and protective water infrastructure. This has led many authors to define the first developed social groupings that arose between 8000 and 3000 BCE as hydraulic civilisations (Wittfogel 1957). A hydraulic civilisation refers not just to the collective dependence on water and water works but also to the related (despotic) social and political processes where power emanates from control over access to food and water. All major human migrations and the birth of towns and communities have been closely related to the search for and the settlements around flood plains, naturally irrigated areas and valleys adequately supplied with water. The control over these resources was not just necessary for survival but also necessary for control over society (Dellapenna and Gupta 2009).

Early sedentary communities settled in or around fertile river basins, for example, Egypt (Butzer 1976), China, Mesopotamia (Kornfeld 2009) and the Indus Valley civilisation (Cullet and Gupta 2009). As soon as human groups settled around a water point or in a river valley, the need arose for a certain level of control over the water in order to satisfy water demands and to ensure water distribution between different uses and users. It is from this need that the earliest local and cooperative organisational systems developed, RBOs *avant la lettre*. Their growth, persistence and character varied and were dependent upon many factors, such as local geophysical and climatic conditions, socio-economic and managerial situations, and the religious and philosophical beliefs of the population concerned (Caponera 1992). The extent to which these early management systems focused on the accumulation of power and control over people (Wittfogel 1957), or on community sharing as promoted by early religions, is contested, but what is clear is that these bodies focused on both goals and water administrative systems. For example, in Roman times river basin development was an integrated part of the Roman administration. Through the classification of water as private goods (*res privados*) or public goods (*res communis omnium*) citizens were directly and/or indirectly involved in the management and development of water.

The phases of RBO development through history can be characterised by six main features (see Figure 2.1). First, in terms of scale, RBOs began as community-based organisations, moved on to catchment-based organisations and have become larger and larger in scale, now defined as more centralised RBOs that operate on national and transboundary rivers. What are called catchment- or watershed-based organisations have a smaller geographical scale, less authority, lower resources and capacity and often an informal legal base, and focus on local issues. Many of the larger RBOs have a larger geographic scope, greater authority, a regional focus and legal authority emanating from national law (for example, the Tennessee Valley Authority in the US, the River Basin Committee in the

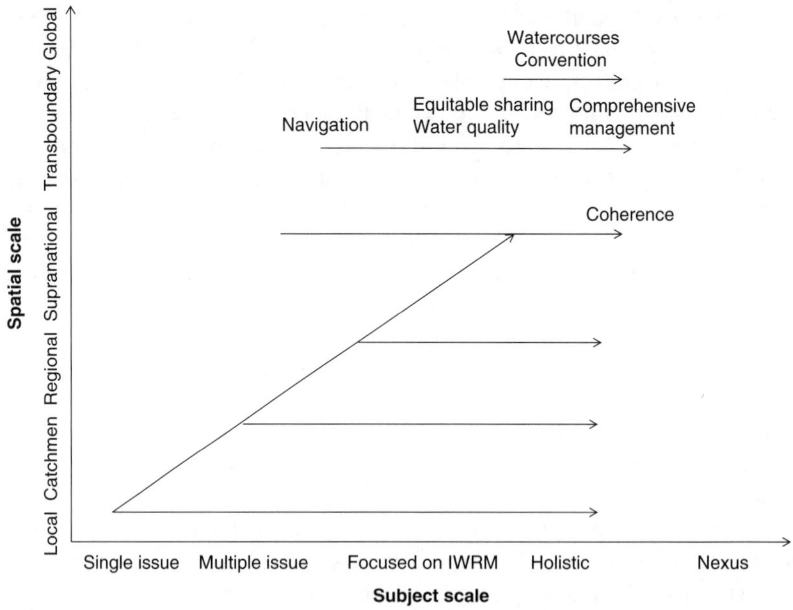

Figure 2.1 The evolution of river basin organisations on the spatial and subject scale over time

San Francisco River Basin) or a transboundary agreement (for example, the Columbia River Treaty Organization) (Delli-Priscoli 2004). Their role has been further institutionalised in supranational agreements such as the EU's Water Framework Directive (European Communities 2000) and at international level through the adoption (though not yet entry into force) of the United Nations (UN) Convention on the Law of the Non-Navigable Uses of International Watercourses of 1997. This Convention states: 'In determining the manner of such cooperation, watercourse States may consider the establishment of joint mechanisms or commissions, as deemed necessary by them, to facilitate cooperation on relevant measures and procedures in the light of experience gained through cooperation in existing joint mechanisms and commissions in various regions' (UN Watercourses Convention 1997: Article 8.2).

Second, in terms of purpose, the goals of RBOs have evolved over time. While in early civilisations such bodies focused on promoting community development, over time RBOs have evolved from single (navigation or quantity management) through dual to multipurpose to integrated and comprehensive water management, and increasingly include ecosystem services. They have evolved from a political model of discussing how to allocate resources, to a more technocratic bureaucratic model which often discusses the same allocation issues but through more complex bureaucratic procedures and optimisation models (Mukhtarov 2007; Hooper 2005; Barrow 1998). The focus is far more on the adoption of the methods and tools of integrated water resource management as opposed to debate and discussion about which paradigms, tools and options are most appropriate especially as resources are quite tight in many parts of the world.

At the international level, RBOs try to promote cooperation and coordination, the exchange of data and harmonisation of policies (Dombrowsky 2007). However, increasingly there are doubts as to whether water is a 'sector' in itself or a cross-cutting issue and whether it should be dealt with in relation to its nexus with other sectors and priorities in society (UNESCO 2012; Gupta et al. 2013). In a nexus approach, overlapping issues and mandates are discussed, negotiated and, if possible, resolved between the relevant sectors on a more or less equal basis. This last issue may create challenges for RBOs and water ministries. Where RBOs attempted to manage water and related issues from a dominantly aquacentric perspective, a nexus approach will have to balance out the competing issues of importance to the different sectors involved; this might have the effect of not necessarily fully protecting and managing the allocation of water resources but, on the other hand, other interests may get more space. This might require RBOs to take on more coordinating tasks. Figure 2.1 shows the evolution of RBOs from single issue to holistic, moving possibly to the nexus approach on the x axis. The y axis shows that there is also an evolution on the spatial scale from community-based, through catchment-based to river basin-based organisations to international river-based organisations all aiming to ensuring greater coherence in water management. Transboundary RBOs have moved from navigational to comprehensive management issues. At the global level, the Watercourses Convention provides guidance.

Third, in terms of water definitions, RBOs initially focused only on surface water river flow, much later on entire watersheds. But increasingly they are focusing on groundwater, grey (waste) water, green (in leaves and plants) water and 'virtual' water too. For example, water organisations in Spain are actively accounting for water use as a first step to see if the concept of virtual water may lead to better policy (Aldaya et al. 2010).

Fourth, in terms of principles of water management, early institutions and the legal systems on which they were based were initially focused on community management principles (see Dellapenna and Gupta 2009). RBOs then began to focus on more top-down management systems, and are now moving in the direction of requiring cooperation, greater public participation in water governance as well as a larger role for the private sector, and more science-based governance. They include water law principles such as sovereignty, equitable sharing of water resources, and the no-harm principle; environmental principles such as environmental impact assessments, sustainable development, the precautionary approach, the polluter-pays principle, subsidiarity and notification; governance principles such as decentralisation, participation, conflict resolution and prior informed consent; and increasingly, human rights principles such as the human right to water and sanitation and the rights of indigenous peoples are being included within the purview of sustainable development principles. Many of these principles emerge from the legal instruments that led to the establishment of these RBOs. Fifth, in terms of resources, some organisations have very limited mandates, capacities and resources; others have comprehensive mandates, capacities and resources.

Sixth, in terms of organisational density, RBOs have become dense. Initially, they were mostly single-level bodies, now they are increasingly developing into multilevel hierarchically integrated organisations. Figure 2.2 shows the increasing density of the organisational framework

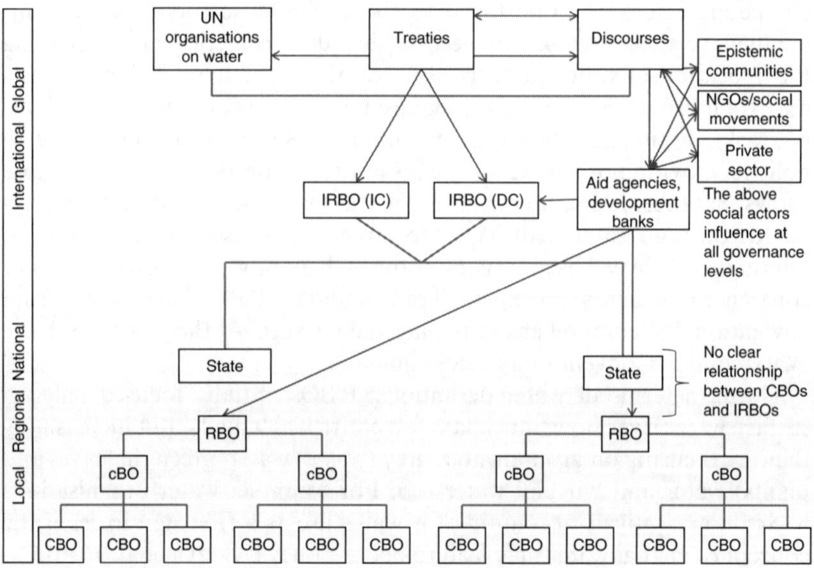

Note: CBO: community/user based organisation; cBO: catchment based organisation; RBO: river based organisation; IRBO: international river based organisation; IC: industrialised country; DC: developing country.

Figure 2.2 River basin organisation organisational framework at multiple levels

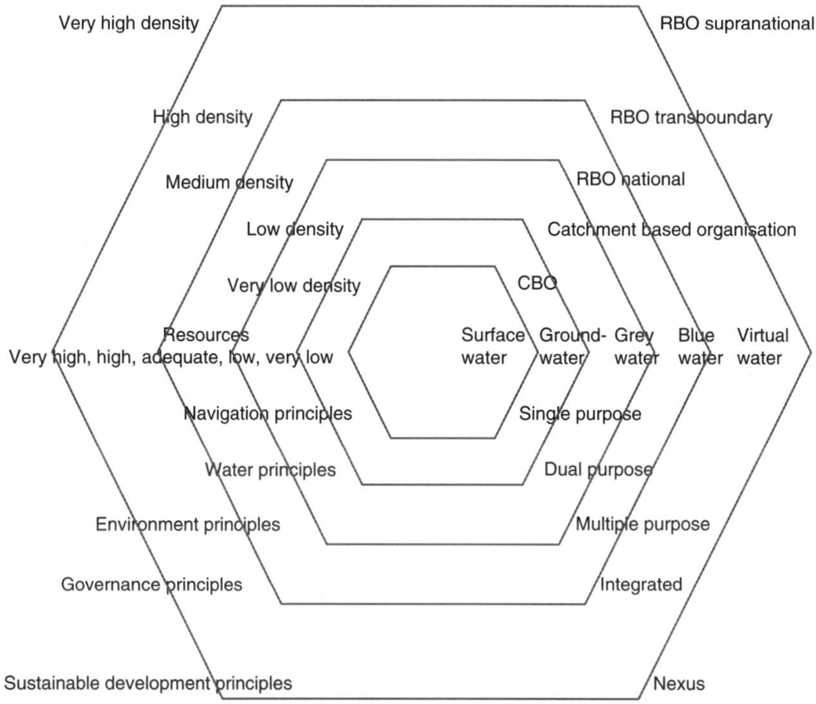

Figure 2.3 The evolving scales of river basin organisations

that operates at multiple levels of governance and has multiple influences from different social actors.

There is nothing linear in the above trends. Each evolution has a space- and time-specific character. Current RBOs are in different stages of the evolutionary process. They could be evaluated in terms of how they have evolved in relation to the six characteristics mentioned above (see Figure 2.3).

Having said that, it is also important to note that the RBOs in developed countries have a longer history of evolution and are at more complex stages in their development, while those in developing countries still have to mature in terms of purpose, resources and organisation.

2.2.3 Factors and Actors Influencing the Current Spread of River Basin Organisations

Currently, we find that the number of RBOs is increasing worldwide. A number of factors have influenced the rise of RBOs.

First, RBOs have arisen through a formalisation of existing customary, informal, autonomous organisations through history. A case in point is the Netherlands where the current water boards are a formalised and consolidated version of customary water boards established as far back as the twelfth century.

Second, RBOs have developed as an international legal instrument in order to deal with international upstream–downstream relations with respect to navigation issues, water quantity sharing and water quality monitoring. These include RBOs on the Rhine, Danube, Plata, Indus and Mekong. Here the reason has often been a specific transboundary problem – sharing water use, flooding, dam construction, navigation, environmental pollution – and the key actors historically have been government officials. Sometimes third parties such as the World Bank have played a key role in promoting cooperation, for example, through the Indus treaty.

Third, RBOs have arisen in response to the rise and evolution of water management discourses and the multiple agents (epistemic communities, political leaders, policy entrepreneurs, development banks and aid agencies, and social movements) that have marketed these ideas through their efforts at evaluation and diffusion of these ideas (see section 2.3). The epistemic communities include water lawyers (for example, the International Law Association), water engineers, water managers and water scholars who, for example, share their collective knowledge during meetings such as the Stockholm Water Week. Political leaders have expressed their views on water discourses at global water conferences from the Stockholm Conference on the Human Environment in 1972 and the Mar Del Plata Conference on water in 1977 right up to the Rio+20 Conference in 2012. Policy entrepreneurs at multiple levels of governance promote discourses on water. In recent decades, organisations such as the World Bank (World Bank 1993), the Global Environment Facility (Uitto and Duda 2002), the United Nations Development Programme, and aid agencies from Sweden, the Netherlands, Germany, the US, Australia and the UK (Schlüter and Herrfahrdt-Pähle 2011; van der Zaag et al. 2007; IWPP 2007) actively promote institutional arrangements for water management where the RBOs play a key role. Global through to local non-governmental organisations (NGOs) also support such organisations as a way to promote water management, address conservation issues and deal with local social challenges (McNally and Tognetti 2002). Many of these organisations support such institutions because of the groundwork by epistemic communities such as the International Law Association, the World Water Council and the Global Water Partnership, and the engineering community.

2.2.4 Inferences: The Relevance of River Basin Organisations

Thus, water resources management based on hydrological boundaries is not a new phenomenon. It has existed since ancient times whenever serious water competition arose and water could be used as a means of control. The necessity for communal flood control delivered a sound breeding ground for institutional 'upstream–downstream' arrangements (such as the development of the originally very local Dutch Water Boards that have subsequently become more regional). In recent times the inability to manage water quality or to preserve environmental integrity and to sustain environmental flows has offered a new dimension. Some argue that it is at present nearly impossible not to organise water resources management in an integrated manner and on hydrological boundaries. Many argue that integrated river basin management and related organisations are the most appropriate tool to deliver integrated water resource management at a basin scale (Hooper 2005), and that these organisations are key to strategic management of river basins (Radosevich and Olson 1999). The need for establishing or restructuring RBOs is today part of almost every contemporary water sector reform and is incorporated in nearly all new water legislation that has been produced over the past ten years (Brazil, South Africa, Nigeria, Indonesia, the EU, the Netherlands, and so on). One may say that RBOs are an accepted reality, a globally applied concept with its advantages and disadvantages. RBOs have institutionalised the politics of water sharing and water management through an arsenal of tools and techniques with a highly technocratic character. At the same time, nation states may have competing national responsibilities – that is, they may need to provide the dry areas within the country which are not necessarily in the watershed with water. Balancing these two issues, among many others, requires debate about how best to organise water management within a country.

2.3 DISCOURSES (AND ACTORS) INFLUENCING RIVER BASIN ORGANISATION DESIGN

2.3.1 Introduction

This section describes and analyses the dominant discourses and actors influencing the key elements of RBO design. Figure 2.4 presents the five discourses influencing water governance today, namely the water discourses, which take place within the context of development discourses, development cooperation discourses, governance discourses and environmental discourses. These are discussed below.

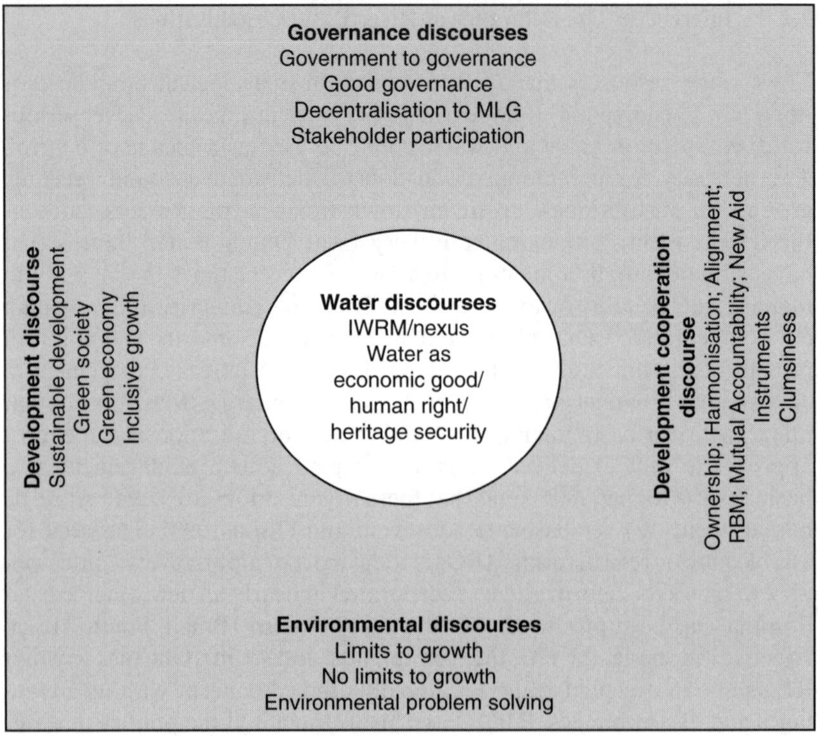

Figure 2.4 Discourses influencing river basin organisations

2.3.2 Water Discourses and Actors

Historically, water treaties were shaped by the need to navigate on shared water bodies, share water and manage water problems. Such treaties go back some 800 years. These treaties formalised some of the discourses on navigation principles (on the right to trade and/or navigate), sovereignty principles (absolute territorial sovereignty, absolute integrity of state territory, limited territorial sovereignty) and equitable sharing of water resources – some of which are now consolidated in the UN Watercourses Convention.

In the twentieth century new water discourses emerged as well. Following the hydraulic paradigm and its focus on water engineering promoted by water engineers (the hydraulic mission; Allan 2006), integrated water resource management has become the key credo. The integrated water resource management concept has become very popular with policymakers and has been included in Agenda 21 (1992), the Dublin Principles (1992), the Global Water Partnership and the 2002 World Summit on Sustainable Development. It calls for integrating all kinds of waters (green, blue, brown; fresh, ocean) and all kinds of interests (sectoral, upstream–downstream). It has a wide arsenal of instruments, namely: water resource assessment, plans, demand management, social change instruments, conflict resolution, regulatory instruments, economic instruments and information management and exchange (UNESCO 2006: 14). However, integrated water resource management is difficult to implement in practice (Agyenim 2011) and the scientific community is also not clear about what this term means precisely (Conca 2006). Nevertheless, reference to the concept has been made in legislation introduced worldwide (such as in Indonesia, Brazil, Nigeria, South Africa and the EU; Jaspers 2001, 2003).

There have also been debates that have promoted water as an economic good that needs to be priced and traded (the Dublin Conference in 1992 and at the World Water Forums of 1997, 2000, 2003 and 2006) rather than as a heritage (European Communities 2000). This has opened up the opportunities for private sector participation in water sector services. Proponents, mostly from the water industry, development banks and neoliberals in general, argue that private sector participation raises resources for water management, increases the efficiency of water services delivery and can provide these services at competitive prices in comparison to the often monopolistic and bureaucratic government (UNESCO 2003: 380; Petrella 2001: 50–51; North 1990; van Dijk 2003). However, in the water sector, the private sector may, in fact, operate as a monopoly and it is usually driven more by profit motives than by public interest (Petrella 2001; Shiva 2002).

A response to private sector participation and water pricing has come through a social movement which has pushed the human rights agenda (Smets 2000; McCaffrey 1992; see also General Comment No. 15 2002). In July 2010, a human right to water and sanitation was adopted by the UN General Assembly (Resolution A/RES/64/292; Gupta and Thompson 2010). This aims to guarantee minimum access to water and sanitation services and imposes on governments and other social actors the responsibility to respect, protect and fulfil their obligations. In recent years, there has also been a focus on defining water as a 'security' issue (Bakker 2012; Grey and Sadoff 2007). Here there are two dimensions. For some countries (for example, Israel) water is seen as a national security issue, as essential to survival. For others water is seen as a human security issue. Defining water as a security issue is seen as one way, albeit controversial because of its national security component, of raising the profile of water on the international agenda.

2.3.3 Governance Discourses and Actors

A number of governance discourses have been influential in shaping water governance. The government-to-governance discourse, ironically supported by both supporters of the lean government discourse as well as those who promote deeper democracy, is a shift that demonstrates a move away from top-down centralised policymaking to a multidirectional, multi-actor approach to decision making (see Huitema and Meijerink, Chapter 1, this volume). Such a shift is seen as more legitimate as it engages social actors in the decision-making process and is more effective as through the engagement of social actors it gains their consent through a process of deliberative democracy and their willingness to implement policies (Kooiman 1997; Rhodes 1997; Hajer and Wagenaar 2003; Hanf and Sharpf 1978; Arts and van Tatenhove 2005; Kickert et al. 1997; Koppenjan and Klijn 2004; Fischer 2003). Imposing governance processes in countries with centralised, top-down policymaking traditions may lead to an incomplete paradigm shift (Li et al. 2013). The government-to-governance discourse supports the idea of RBOs being open to a diversity of stakeholders and the RBO as a forum for deliberation (Amma-Anokye 2013).

A second shift visible in the governance literature is the shift towards decentralisation (Conca 2006). With the development of the nation state and post-independence politics, most countries first engaged in centralised policymaking as a way to control natural resources. However, there has been a shift in the ideology towards decentralisation in order to actively take into account context-specific knowledge and characteristics and to enhance the efficiency and effectiveness of policies (Rondinelli et al. 1984; UNESCO 2006). The shift towards decentralisation can be seen as a shift towards both RBOs within countries as well as international RBOs, where the power of decision making has moved from the state to the local and regional level RBOs (see Huitema and Meijerink, Chapter 1, this volume). Within countries there is often a power struggle between administrative authorities and water authorities (Laster and Livney 2009). One of the first legal reforms incorporating full functional decentralisation in Africa was from Zimbabwe (in 1999; see Jaspers 2001). Similar processes are going on in Tanzania, Kenya and Namibia. The key actors promoting decentralisation have combined international organisations, aid agencies, experts and NGOs which were committed to the idea of deepening democracy through greater local control of local resources. A key characteristic of most RBOs is the devolution of authority. The public administration transfers functions, tasks and competencies to RBOs for a specific period of time. This transfer is aimed at promoting sector-specific, efficient, effective, transparent and accountable policy guaranteed through participation and appeal procedures. Notions of centralisation and decentralisation have now given way to notions of multilevel governance, where level- and actor-specific tasks are organised in a manner that is coherent and complementary. The distribution of tasks can include deconcentration (tasks assigned to lower authorities in the same administration), delegation (reversible transfer of authority to another public or private body) or devolution (irreversible transfer of authority to another administrative body). The degree of autonomy of the body depends on the independence of its decision-making process, its financial independence to collect and allocate resources, ownership of assets, and independence of personnel. Each country has its own typical RBOs. In France the Agences de l'Eau gained in importance and control after 1964 (Alaerts 1995). Delegation can be prompt and definitive (Zimbabwe) or gradual and progressive (South Africa). Progressive delegation is applied over time as the need for delegation arises and on request by the stakeholders. Actual delegation takes place when stakeholders have the necessary capacities and when effective institutional arrangements have been established. In South Africa and Tanzania the delegation of responsibilities to RBOs is progressive (for more on the discussion on RBOs in South Africa, see Meissner and Funke, Chapter 8, this volume). In Zimbabwe delegation has been necessarily prompt (Jaspers 2001). In Turkey, Mongolia and some African countries the process of delegation is still restricted or experimental. Sometimes water management activities are undertaken through co-administration where sub-national agencies cooperate in implementing policy based on instructions and finances received (such as Indonesia: '*tugas perbantuan*'). The decentralisation idea moves water governance from centralised state control into other hands and often is accompanied by a power struggle between parties, expressed through limited mandates and resources accompanying the transfer of authority, especially in developing countries. At the same time, sometimes power has shifted from state authorities to international RBOs and this may have the opposite effect of decentralisation and may lead to greater concessions to the needs and interests of neighbouring riparian countries.

A third evolution in the literature is normative, towards the concept of good governance. This concept promotes the ideas of legitimacy, accountability, transparency, efficiency, effectiveness, responsiveness to new problems, public participation, equity, coherence and the rule of law (see Huitema and Meijerink, Chapter 1, this volume). These general ideas have also been applied in the context of water, and often include the concept of an integrative approach (World Water Development Report 2003: 373). Good scientific ideas of water management are seen as effective only within the context of good governance. This implies that RBOs need to meet good governance standards. However, it is well known that exogenous efforts to create good governance in countries may have counterproductive results (Anders 2005).

A fourth shift is the recognition of the importance of stakeholder participation in policymaking. Stakeholder participation aims to deal with the twin problems of democratic deficit and technocratic decision making. It is increasingly being seen as important not only in the general governance discussions but also in the water-specific discussions (Cosgrove and Rijsberman 2000; Berlin Rules – ILA 2004). Stakeholder participation ranges from providing information to allowing stakeholders to actively engage in the decision-making process (Amma-Anokye 2013). In general, some countries have high stakeholder participation in water (Thailand, Uganda, South Africa, Chile and the US) while some have lower participation (for example, India and Indonesia; UNESCO 2006: 77). Imposing stakeholder participation in countries with generally weak democratic traditions may lead to increased bureaucracy and higher costs, while not really delivering in terms of effectiveness, as people may be afraid to speak and the process may be manipulated. It may also lead to unpredictable and unintended impacts (Keeley and Scoones 2003; Lemon 2001; Stiglitz 2000). Such stakeholder participation can range from being instrumental to empowering in nature (Amma-Anokye 2013).

In general, these discourses bring two kinds of actors together: the first group includes those who see the need to enhance democratic legitimacy; the second group includes those who argue in favour of small government and greater public participation.

2.3.4 Development Discourse

Water is closely linked to economic growth and social development. The development discourse has moved from the focus on income generation through alternative development to the sustainable development discourse. This last discourse has two axes: one focusing on economic, ecological and social dimensions; and the other focusing on time, meeting the needs of present generations without compromising on the ability of future generations to meet their own needs (WCED 1987; Beg et al. 2002; Cohen et al. 1998; Schneider et al. 2000; Banuri et al. 2001; Markandya and Halsnæs 2002; Metz et al. 2002; Munasinghe and Swart 2000; Najam et al. 2003; Smit et al. 2001; Swart et al. 2003; Wilbanks 2003). Hard sustainability does not allow for trading off between the different elements (Barnett 2001; Robinson 2004). Sustainable development has been adopted as a guiding concept by most actors, possibly because it is multi-interpretable.

However, the difficulty in achieving hard sustainability has led to a split in this discourse into three component discourses: green economy, green society and inclusive growth. Of these the green economy discourse has become very popular and has been articulated by the United Nations Environmental Programme (2010) and adopted by the G20 Conference (2010) and the Rio+20 conference (2012). It is a neoliberal argument in favour of promoting the adoption of modern technologies and the internalisation of environmental externalities by industry. The social aspects are not quite so prominent in this discussion. The alternative concept of inclusive growth focuses on enhancing economic development through social policies that aim at including social actors and reducing poverty (Rauniyar and Kanbur 2010: 455); the environmental component is given less importance. The G77 tends to espouse this position. The green society concept focuses on environmental issues and social well-being but questions the perpetual quest for economic growth. The green economy is a concept promoted by ecological modernists, who focus on neoliberal approaches to dealing with environmental problems. Inclusive growth is promoted more by those who see themselves on the left-hand side of the inverted Kuznet's curve. This curve shows that as societies grow they tend to pollute more, but after a point they begin to invest in cleaning up the environment. However, this curve does not hold for global environmental pollution. The green society is promoted by the zero-growth group who see environmental exploitation as resulting from capitalism. An emphasis on social issues has also led to the adoption of the Millennium Development Goals (MDGs) in 2000 and the current discussions of the follow-up to the MDGs via the Sustainable Development Goals.

2.3.5 Development Cooperation Discourse

The global water challenge has a very strong North–South character. This is so because the bulk of water challenges appear to be concentrated in the South (access, sanitation, eutrophication and pollution, overextraction, floods, droughts), while some of the knowledge about, and on how to deal with, such challenges appears to be concentrated in the North. Furthermore, water challenges in the South cumulatively add up to a major global challenge. This has made water cooperation an area of key interest for development cooperation (Scheumann et al. 2008). Aid agencies and the development banks have espoused certain paradigms and marketed these to the developing world through their development cooperation projects. The World Bank has been an active promoter of the concept of integrated water resource management (Hartje 2008). Most aid agencies, including the German Federal Ministry for Economic

Cooperation and Development, have actively pushed the Millennium Development Goals as a key part of their development cooperation strategy (Kipping 2008). Increasingly the development cooperation discourse is also dominated by its own values as espoused in the Paris Declaration on Aid Effectiveness and follow-up meetings which emphasise principles of 'ownership' by recipient countries, donor harmonisation (that all donors should work on similar issues and have coherent roles), results-based management of aid projects, and the use of new and varied instruments for delivering aid. There are others who argue that aid needs to be 'clumsy' and realise that where a genuine stakeholder process precedes project design and implementation, it can lead to unpredictable project design and implementation and not necessarily one that is amenable to principles of efficiency and results-based management (Gupta and Thompson 2010).

2.3.6 Environmental Discourses and Actors

Water is closely connected to many environmental issues. For example, climate change can influence hydrological cycles, lead to a rise in sea levels and the melting of glaciers; environmental pollution can degrade water resources. It is thus inevitable that water discourses will also be influenced by environmental discourses. Thus, initially, the RBO discourse was influenced by conservation paradigms which then evolved into pollution prevention and an issue-by-issue approach to the environment, especially through the adoption of issue-by-issue environmental treaties at global level.

With the rise of water pollution as a key challenge in the post-war years, water quality standards were set up initially in the US and European countries, but very soon thereafter in other parts of the world. A more integrated approach was taken when environmental impact assessment required information on all projects on the water and the environment. With the report of the World Commission on Environment and Development in 1987 attention was focused on the concept of sustainable development. After the adoption of the Rio Principles on Environment and Development in 1992 (Rio Declaration 1992), all major RBOs began to take these issues into account. Increasingly, it will become important for the water discourse to take into account the ecosystem services discourse (MA 2005; UNESCO 2012) and that will shape the nature of the issues that RBOs should also be dealing with.

Today, environmental discourses can be divided into three main groups: those (mostly environmentalists and scientists) who see environmental pollution as showing the limits to growth; those (mostly neoliberals) who argue that there are no limits to growth and that human ingenuity will allow for exploitation of our resources to enhance human well-being; and those who tread the pragmatic middle ground and seek incremental solutions to the environmental challenge. This translates into a water discourse of limiting water use and pollution; competing for and exploiting water resources to advance economic development in the short term; and the application of tools and techniques for incrementally dealing with water.

2.3.7 Nexus Approach: The Widening Discourse

In the meanwhile, the rise of integrated approaches in different issue areas is seen as a contradiction in terms. Integrated water resource management, integrated waste management and integrated pesticide management are examples of ideas developing in a parallel manner in different fields to try and promote integration. But all these fields are interlinked. Water is intimately linked to land, food, energy, transport, industry and environmental management. However, the water sector takes an increasingly 'aquacentric' approach to management. This has led to the rise of the 'nexus' concept (UNESCO 2012; Gupta et al. 2013) which links water issues with other issues but does not prioritise water over other issues. It is increasingly being argued that many of the issues that affect the water sector are actually issues in other social sectors in society – transport, agriculture, industry, technology, demography, and so on – and that a water perspective needs to be incorporated into these other approaches. The drivers of change in the water sector arise often in other sectors with other priorities and operate at multiple levels of governance. The convergence of problems in a variety of sectors has led to water being seen as a cross-cutting issue where it is perhaps easier to examine the issue-by-issue nexus to ensure that problems get solved. This is often referred to as looking outside the 'water' box, calling for a widening framework of action. The question is whether the mandate of existing RBOs and the bodies that fall under them will allow them to look outside the 'water box' and integrate other issues.

2.3.8 Discourses and RBOs

These discourses, through the agency of actors, have helped to shape the modern contours of RBOs. Actors use discourses as a way to organise their vision of how resources should be allocated between different uses. Discourses appeal on the basis of: (1) their scientific merit and through epistemic communities who influence global policies and national policies; (2) the financial assistance and related conditionalities in aid and development projects; (3) the political attractiveness of discourses as matching existing trends and ideological beliefs; (4) their ability to raise resources in the banking and private sector; and (5) their ability to capture the imagination of and engage civil society. Because discourses are often abstract notions they often obfuscate how the power games are being played out by the different actors in the field.

The design of RBOs to include and or reflect dominant discourses is a technocratic route to addressing power struggles. RBOs in developed countries tend to be the source of many of these discourses; these discourses are also internalised into these bodies. However, many of these discourses are imported into developing countries and may, on occasion, even be seen as alien to local cultures. They may even lead to, or aggravate, plural policymaking. Where RBOs are shaped by imported discourses, this bypasses national and local politics to shape a polity and policy that reflect choices made by outsiders under the guise of good science. However, these designs do reflect specific choices that privilege some over others by changing the rules of water access and allocation in specific contexts. Where such RBOs include stakeholder participation, such participation may be limited to issues that are already defined by the design principles put in place when the RBO was initially designed!

Most catchment-based or community-based organisations, including user-based organisations, tend to be decentralised bodies that embody on paper good governance principles and the goal of integrated water resource management and sustainable development. Most see water as an economic good which is marketed as an integral part of the integrated water resource management concept by many aid agencies and the World Bank. The human right to water discourse has yet to penetrate most of the RBOs. The relationship between green economy, green society and inclusive growth and RBOs is also still to develop.

2.4 CHALLENGES FACING RIVER BASIN ORGANISATIONS TODAY

RBO design is thus influenced by a confluence of historical events, past agreements, evolving problems, new ideas and paradigms, and a vast range of actors who have multiple pathways of influencing management.

This section identifies four key challenges with respect to current day RBOs. First, RBOs in the North and the South are different and the changing realities of these countries will also influence these RBOs. There are a number of differences in basin realities. Developed-country RBOs tend to have temperate climates and higher rainfall, while in developing countries, the rainfall could be lower and there may be higher mean temperatures. The population distribution and density is different. Water rights tend to be consolidated in the North, while being plural in the South. The focus is more on blue water and water provision from formal service providers, there are fewer but larger stakeholders, and relatively low costs for monitoring water use and charging water users in the North as compared to the South (Shah et al. 2001). The financial and capacity resources are relatively smaller while the water problems are relatively greater in the South. The South is of course changing, but even while some Southern countries are becoming richer, the influence on the RBOs remains limited. RBOs involve local people and intend to impact on local water policy. Even though they engage stakeholders, if the stakeholders do not represent a range of interests and understand the process effectively, their participation may not yield the needed results (Blackstock et al. 2012). Abers (2007) argues that when RBOs create a collective identity by defining the water problem in a way that local people can identify with the problem, they are better able to organise themselves and achieve results. This depends on a combination of appropriate leadership and network expansion. A singular focus on formal functions is less rewarding than a focus on informal creativity and innovativeness. Importing ideas that have worked in the West, such as stakeholder participation, may be expensive and difficult to implement in the developing countries, although if they were implemented they could perhaps be very empowering (Amma-Anokye 2013). Meaningful participation and communication may be difficult to achieve (Swatuk and Motsholaphekol 2008). The decentralisation process in RBOs is successful under specific circumstances. These include the availability of government financial support, financial responsibility given to users and the existence of dispute resolution options (Dinar et al. 2007; Dinar et al. 2005). However, it is not necessarily true that wealthier RBOs do better than poorer RBOs (Dinar et al. 2007). RBOs that focus on integrated water resource management and holistic approaches have been more attractive in developing-country contexts and marketed there by international bodies and aid agencies because of their systemic approach. However, they have been less strongly marketed in developed countries which were less enamoured by utopian approaches (Barrow 1998). But the application of these approaches in developing countries faces bureaucratic challenges, corruption, conflicting cultural values to the ones being imported, excessive focus on the creation of expensive centralised bureaucracies in these RBOs (as in the Niger Basin Authority); they do not often take existing institutions into account (Goldin and Kibassa 2009), and although attractive have not yielded the results expected (Agyenim 2011).

Second, most RBOs have yet to integrate climate change into their governance processes. With the growing emphasis on climate change as a source of major changes in the global hydrological regime, it is necessary to understand both how RBOs can contribute to mitigating the climate change problem and how they can help with adaptation to climate change. Mitigation can focus on the role of hydropower dams as a source of power, but also on how irrigation and other water use facilities lead to large-scale energy use. Adaptation can focus on coping with the impacts of floods and droughts (Li et al. 2013). Increasingly the latter is very gradually being taken into account in some RBO systems such as on the Mekong (Lebel and Garden 2008).

Third, the growing realisation of the interlocking development, environmental and financial crises, the realisation that issues that were previously seen as self-contained sectors are now cross-cutting issues that affect a wide range of actors, sectors and priorities, has led to the call for adopting a nexus approach. This means looking outside the water sector for developing coalitions and promoting coherence with other policy issues. This may lead to questioning the very basis of the RBO whose fundamental logic is the hydrological entity encapsulated in a river. But if we accept that water is continuously flowing and is not just encapsulated in a river then the logic behind the RBO becomes less clear. If we further accept the logic that the drivers influencing water use and abuse are spread through the multiple levels of society from local to global level, and that water use patterns can have global repercussions, then it becomes even clearer that the logic behind an RBO is less obvious. At the same time, it is impossible to organise society in such a way as to take all interlinked issues into account. This implies that even if the RBO is not the most logical entity, it may be for practical purposes the most useful entity. But clearly the RBO will have to rise above its own limited mandate in order to ensure that it takes into account the multiple nexuses that exist.

Fourth, the story of RBOs is also the story of water-related power structures. In the post-colonial context, an RBO is a special management arrangement. It removes the management of water issues from ministries and government departments dedicated to water management issues, and vests it in a bureaucratic or hybrid body whose purpose is to try and manage the river either in its hydrological totality or in relation to specific catchments. It thus transfers authority from (inter)state governance to an administrative body and is thus part of a global rising trend of administrative law from local to global level. Inevitably, at transboundary level, this implies sacrificing state sovereignty over water resources in favour of a common and optimal governance system over the entire basin. At national level, it means sacrificing 'national' interests in favour of the interests of the catchment area. Where rivers cross over two provinces within a country such basin organisations try to bring the interests of both provinces to the table. The consequence of such organisations is the threat of a creeping loss of sovereignty and/or control over water resources and is possibly an inevitable result of the gradual globalisation of water issues. This has led many to reduce the RBO to a mere technocratic, administrative body with limited power and mandate. Generally, states reserve for themselves the ultimate authority to make decisions regarding navigation, quality and quantity control. They often create RBOs to promote technical cooperation and involve local stakeholders. How the struggle for control over water resources, especially if one recognises that in many parts of the world there are plural claims to water ownership, will resolve itself is difficult to predict but will affect the effectiveness of water governance. RBOs at international level have had some success in controlling conflict, but are seen as often reflecting the power imbalances of their member states (Barrow 1998). Although RBOs are decentralised, they tend to centralise power and act as a monopoly with no external control or accountability (Barrow 1998). Such RBOs often reflect institutional interests and are less interested in promoting cooperation (Keskinen and Varis 2012). The projects promoted may also thus promote the interests of the more powerful actors engaged in the discussion; namely the local elites, the local and foreign contractors, bureaucrats and capitalist farmers (Salau 1986). However, both institutional interests and stakeholder interests can be used to promote cooperation if the policy networks created are interested in such cooperation and can create support outside the network for their perspectives as well as enable those outside to deal with the opposition to their ideas (Lemos and Olivera 2004).

2.5 CONCLUSIONS

This chapter set out to understand the factors that have influenced RBO design over time. It concludes that RBOs are an artefact of water management and have existed in some form or other since ancient times. The evolution of RBOs has taken place along six scales: the scale of geographic size (local to global), the scale of water definition (surface water to all waters), the scale of focus (single to nexus), principle (from few to multiple), resources, and organisational density (from single to multiple layers).

Four factors have influenced their rise. The first is the formalisation of existing informal cultural organisational structures. Second, the need for cooperation to balance complex riparian interests (including upstream and downstream interests) and manage navigation, flooding and water sharing issues. Third, the rise of ideas and technologies has promoted the need for RBOs. Fourth, the rise of modern agents of change has marketed these ideas.

Five bodies of discourses have been influential in shaping RBOs. These are water management discourses (integrated water resource management; water as a heritage, economic good, human right or security issue), environmental discourses, developmental discourses, development cooperation discourses and governance discourses (government to governance, good governance, decentralisation, stakeholder participation).

This chapter argues that RBOs have substantive and procedural aspects and both shape and are shaped by technologies and infrastructure. However, four challenges remain. First, importing Western ideas for RBO design into different cultural and socio-political contexts may lead to unexpected results. Second, RBOs will have to take on the complex task of climate-proofing their work. Third, as we realise that the drivers of water use and abuse are local to global in nature, and that local water uses can have global repercussions, RBOs will increasingly have to adopt the nexus approach if they wish to be effective. Finally, RBOs embody a power struggle – a shift of power from the state to an administrative body at both international and national level – and this power struggle has often led to minimising the mandate of the RBO. This power struggle is reminiscent of the discussion on hydrological civilisations as discussed by Wittfogel. This struggle will ultimately decide if the design of the RBO can be effective enough to address water problems in relation to its context.

REFERENCES

Abers, R.N. (2007), 'Organizing for governance: building collaboration in Brazilian river basins', *World Development*, **35** (8), 1450–1463.

Agyenim, J.B. (2011), 'Investigating institutional arrangements for IWRM in developing countries: the case of White Volta Basin, Ghana', PhD thesis, VU University Amsterdam.

Alaerts, G.J.F.R. (1995), 'Institutional arrangements in the water and sanitation sector, European country studies', IHE Working Paper, Delft.

Aldaya, M.M., P. Martinez-Santos and M.R. Llamas (2010), 'Incorporating the water footprint and virtual water into policy: reflections from the Mancha Occidental Region, Spain', *Water Resource Management*, **24**, 941–958.

Allan, J.A. (2006), 'IWRM: the new sanctioned discourse?', in P.P. Mollinga, A. Dixit and K. Athukorala (eds), *IWRM in South Asia: Global Theory, Emerging Practice and Local Needs, Water in South Asia Series*, New Delhi: Sage, pp. 38–63.

Amma-Anokye, N. (2013), 'Stakeholder participation in water resources

management: the case of Densu Basin in Ghana', PhD thesis, VU University Amsterdam.
Anders, G. (2005), 'Civil servants in Malawi: moonlighting, kinship and corruption in the shadow of good governance', PhD thesis, Law Faculty, Erasmus University Rotterdam.
Arts, B. and J. van Tatenhove (2005), 'Policy and power – a conceptual framework between the "old" and "new" policy idioms', *Policy Sciences*, **37** (3–4), 339–356.
Bakker K. (2012), 'Water security: research challenges and opportunities', *Science*, **337**, 914–915.
Bakker, M. (2013), *Program in Water Conflict Management, Institute for Water and Watersheds*, Corvallis, OR: Oregon State University, available at www.transboundarywaters.orst.edu (accessed 26 March 2013).
Banuri, T., J. Weyant, G. Akumu, A. Najam, L.P. Roas, S. Rayner, W. Sachs, R. Sharma and G. Yohe (2001), 'Setting the stage: climate change and sustainable development', *Proceedings of the Third Assessment Report of the Intergovernmental Panel on Climate Change, IPCC, Mitigation*, London: Cambridge University Press, pp. 73–114.
Barnett, J. (2001), *The Meaning of Environmental Security, Ecological Politics and Policy in the New Security Era*, London: Zed Books.
Barrow, C.J. (1998), 'River basin development planning and management: a critical review', *World Development*, **26** (1), 171–186.
Beg, N., J.C. Morlot, O. Davidson, Y. Afrane-Okesse, L. Tyani, F. Denton, Y. Sokona, J.P. Thomas, E.L. La Rovere, J.K. Parikh, K. Parikh and A.A. Rahman (2002), 'Linkages between climate change and sustainable development', *Climate Policy*, **2**, 129–144.
Blackstock, K.L., K.A. Waylen, J. Dunglinson and K.M. Marshall (2012), 'Linking process to outcomes – internal and external criteria for a stakeholder involvement in river basin management planning', *Ecological Economics*, **77**, 113–122.
Blomquist, W., A. Dinar and K. Kemper (2005), *Comparison of Institutional Arrangements for River Basin Management in Eight Basins*, Washington, DC: World Bank, available at https://openknowledge.worldbank.org/handle/10986/8321 (accessed 1 August 2013).
Butzer, K.W. (1976), *Early Hydraulic Civilization in Egypt: A Study in Cultural Ecology*, Chicago, IL: University of Chicago Press.
Caponera D.A. (1992), *Principles of Water Law and Administration: National and International*, Rotterdam: Balkema.
Cohen, S., D. Demeritt, J. Robinson and D. Rothman (1998), 'Climate change and sustainable development: towards dialogue', *Global Environment Change*, **8** (4), 341–371.
Conca, K. (2006), *Governing Water: Contentious Transnational Politics and Global Institution Building*, Cambridge, MA: MIT Press.
Cosgrove, W. and F. Rijsberman (2000), *World Water Vision: Making Water Everybody's Business. World Water Council*, London: Earthscan Publications.
Cullet, P. and J. Gupta (2009), 'India: evolution of water law and policy', in J. Dellapenna, and J. Gupta (eds), *The Evolution of the Law and Politics of Water*, Dordrecht: Springer Verlag, pp. 157–174.
Dellapenna, J. and J. Gupta (eds) (2009), *The Evolution of the Law and Politics of Water*, Dordrecht: Springer Verlag.
Delli-Priscoli, J. (2004), 'River basin organizations, RBO: US experiences', paper presented at World Bank Training on Building RBOs, Abuja, Nigeria, June.

Dinar, A., K. Kemper, W. Blomquist, M. Diez, G. Sine and W. Fri (2005), 'Decentralization of river basin management: a global analysis', Washington, DC: World Bank Policy Research Working Paper 3637, June.

Dinar, A., K. Kemper, W. Blomquist and P. Kurukulasuriya (2007), 'Whitewater: process and performance of decentralization reform of river basin water resource management', *Journal of Policy Modeling*, **29** (6), 851–867.

Dombrowsky, I. (2007), *Conflict, Cooperation and Institutions in International Water Management: An Economic Analysis*, Cheltenham, UK and Northampton, MA, USA: Edward Elgar.

Dublin Principles (1992), 'Guiding principles, the Dublin statement on water and sustainable development', adopted 31 January by the International Conference on Water and the Environment, Dublin.

European Communities (2000), 'Directive 2000/60/EC of the European Parliament and of the Council of 23 October 2000, establishing a framework for community action in the field of water policy', *Official Journal of the European Communities*, L 327/1, 22 December.

European Environmental Bureau (EEB) (2001), *Making the EU Water Framework Directive Work: Ten Actions for Implementing a Better European Water Policy*, Brussels: EEB.

Fischer, F. (2003), *Reframing Public Policy: Discursive Politics and Deliberative Practices*, Oxford, UK and New York: Oxford University Press.

Goldin, J. and D. Kibassa (2009), 'Past, present and future landscapes of water policy in Tanzania', in D. Huitema and S. Meijerink (eds), *Water Policy Entrepreneurs: A Research Companion to Water Transitions around the Globe*, Cheltenham, UK and Northampton, MA, USA: Edward Elgar, pp. 215–236.

Grey, D. and C. Sadoff (2007), 'Sink or swim? Water security for growth and development', *Water Policy*, **9** (6), 545–571.

Gupta, J., A. Akhmouch, W. Cosgrove, Z. Hurwitz, J. Maestu and O. Ünver (2013), 'Policymakers' reflections on water governance issues', *Ecology and Society*, **18** (1), 35, available at http://www.ecologyandsociety.org/vol18/iss1/art35/ (accessed 26 March 2013).

Gupta, J. and M. Thompson (2010), 'Development and development cooperation theory', in J. Gupta and N. van de Grijp (eds), *Mainstreaming Climate Change in Development Cooperation: Theory, Practice and Implications for the European Union*, Cambridge: Cambridge University Press, pp. 33–66.

Hajer, M.A. and H. Wagenaar (eds) (2003), *Deliberative Policy Analysis: Understanding Governance in the Network Society*, Cambridge: Cambridge University Press.

Hanf, K. and F.W. Scharpf (eds) (1978), *Interorganizational Policy Making*, London: Sage.

Hartje, V. (2008), 'The World Bank's water sector policy reforms', in W. Scheumann, S. Neubert and M. Kipping (eds), *Water Politics and Development Cooperation: Local Power Plays and Global Governance*, Dordrecht: Springer, pp. 301–332.

Hooper, B. (2005), *Integrated River Basin Management: Learning from International Experience*, London: International Water Association Publishing.

Integrated Watershed Planning Process (IWPP) (2007), *Donor Activity in Transboundary Water Cooperation in Africa: Results of a G8-initiated Survey 2004–2007*, Eschborn, Germany: Deutsche Gesellschaft fur Technische Zusammenarbeit, available at http://www.giz.de/Themen/de/dokumente/

en-donor-activity-in-transboundary-water-cooperation-in-africa.pdf (accessed 26 March 2013).

International Law Association (ILA) (2004), 'The Berlin rules on water resources', *Report of the Seventy-First Conference*, London: ILA.

Jaspers, F.G.W. (2001), 'The new water legislation of Zimbabwe and South Africa, comparison of legal and institutional reform', *International Environmental Agreements: Politics, Law and Economics*, **1**, 305–325.

Jaspers F.G.W. (2003), 'Institutional arrangements for integrated river basin management', *Water Policy*, **5** (1), 77–90.

Keeley, J. and I. Scoones (eds) (2003), *Understanding Environmental Policy Processes. Cases from Africa*, London: Earthscan Publications.

Keskinen, M. and O. Varis (2012), 'Institutional cooperation at a basin level: for what, by whom?', *Natural Resources Forum*, **36**, 50–60.

Kickert, W.J.M., E.H. Klijn and J.F.M. Koppenjan (eds) (1997), *Managing Complex Networks: Strategies for the Public Sector*, London: Sage.

Kipping, M. (2008), 'Challenges for German development cooperation in the water sector', in W. Scheumann, S. Neubert and M. Kipping (eds), *Water Politics and Development Cooperation: Local Power Plays and Global Governance*, Dordrecht: Springer, pp. 333–352.

Kooiman, J. (1997), 'Governance and governability: using complexity, dynamics and diversity', in J. Kooiman (ed.), *Modern Governance – New Government – Society Interactions*, London: Sage, pp. 35–50.

Koppenjan, J.F.M. and E.H. Klijn (2004), *Managing Uncertainties in Networks*, London: Routledge.

Kornfeld, I. (2009), 'Mesopotamia: a history of water and law', in J. Dellapenna and J. Gupta (eds), *The Evolution of the Law and Politics of Water*, Dordrecht: Springer Verlag, pp. 21–36.

Laster, R. and D. Livney (2009), 'Israel: the evolution of water law and policy', in J. Dellapenna and J. Gupta (eds), *The Evolution of the Law and Politics of Water*, Dordrecht: Springer Verlag, pp. 121–138.

Lebel, L. and P. Garden (2008), 'Deliberation, negotiation and scale in the governance of water resources in the Mekong region', in C. Pahl-Wostl, P. Kabat and J. Möltgen (eds), *Adaptive and Integrated Water Management: Coping with Complexity and Uncertainty*, Berlin: Springer, pp. 205–225.

Lemon, A. (2001), 'Gender and forestry: integrating local knowledge into environmental planning and management', *International Journal of Sustainable Development*, **4** (3), 265–281.

Lemos, M.C. and J.F.D. Oleveira (2004), 'Lessons learned from Cambodia's Tonle Sap Lake. Can water reform survive politics? Institutional change and river basin management in Ceara, Northeast Brazil', *World Development*, **32** (12), 2121–2137.

Li, H., J. Gupta and M.P. van Dijk (2013), 'China's drought strategies in rural areas along the Lancang river', *Water Policy*, **15**, 1–18.

MA (2005), *Ecosystem Services and Human Well-being: Policy-Responses*, Millennium Ecosystem Assessment, Vol. 3, Washington, DC: Island Press.

Markandya, A. and K. Halsnæs (eds) (2002), *Climate Change and Sustainable Development: Prospects for Developing Countries*, London: Earthscan Publication.

McCaffrey, S.C. (1992), 'A human right to water: domestic and international implications', *Georgetown International Environmental Law Review*, **5**, 1–23.

McNally, R., and S. Tognetti (2002), 'Tackling poverty and promoting sustainable development: key lessons for integrated river basin management', World Wide Fund for Nature Discussion Paper, London.

Metz, B., M. Berk, M. den Elzen, B. de Vries and D. van Vuuren (2002), 'Towards an equitable climate change regime: compatibility with Article 2 of the climate change convention and the link with sustainable development', *Climate Policy*, **2**, 211–230.

Mostert, E. (2003), *Conflict and Cooperation in the Management of International Freshwater Resources: A Global Review*, Paris: UNESCO.

Mukhtarov, F. (2007), 'Intellectual history and current status of integrated water resources management: a global perspective', in C. Pahl-Wostl, P. Kabat and J. Möltgen (eds), *Adaptive and Integrated Water Management: Coping with Complexity and Uncertainty*, Berlin, Heidelberg: Springer, pp. 167–185.

Munasinghe, M. and R. Swart (eds) (2000), *Climate Change and its Linkages with Development, Equity and Sustainability*, Geneva: Intergovernmental Panel on Climate Change.

Najam, A., A.A. Rahman, S. Huq and Y. Sokona (2003), 'Integrating sustainable development into the fourth IPCC assessment', *Climate Policy*, **3** (1), 9–17.

North, D.C. (1990), *Institutions, Institutional Change and Economic Performance*, Cambridge: Cambridge University Press.

Organisation for Economic Co-operation and Development (OECD) (2011), *Water Governance in OECD Countries: A Multi-level Approach*, Paris: OECD, available at http://www.oecd.org/document/13/0,3746,en_2649_37429_48896205_1_1_1_37429,00.html#downloads (accessed 3 July 2013).

Pahl-Wostl, C., J. Gupta and D. Petry (2008),'Governance and the global water system: a theoretical exploration', *Global Governance*, **14** (4), 419–436.

Petrella, R. (2001), *The Water Manifesto: Arguments for a World Water Contract*, London: Zed Books.

Radosevich, G.E. and D. Olson (1999), *Existing and Emerging Basin Arrangements in Asia: Mekong River Commission Case Study*, Third Workshop on River Basin Institution Development, 24 June, Washington, DC: World Bank.

Rauniyar, G. and R. Kanbur (2010), 'Inclusive development: two papers on conceptualization, application, and the ADB perspective', January draft, Manila: Independent Evaluation Department, ADB.

Rhodes, R.A.W. (1997), *Understanding Governance. Policy Networks, Governance, Reflexivity, and Accountability*, London: Open University Press.

Rio Declaration (1992), 'Rio Declaration and Agenda 21', Report on the UN Conference on Environment and Development, Rio de Janeiro, 3–14 June, UN doc. A/CONF.151/26/Rev.1.

Robinson, J. (2004), 'Squaring the circle? Some thoughts on the idea of sustainable development', *Ecological Economics*, **48**, 369–384.

Rockström, J., W. Steffen, K. Noone, Å. Persson, F. Stuart Chapin, et al. (2009), 'A safe operating space for humanity', *Nature*, **461**, 472–475.

Rondinelli, D., J. Nethis and S. Chemma (1984), *Decentralization in Developing Countries: A Review of Current Experience*, Washington, DC: World Bank.

Salau, A.T. (1986), 'River basin planning as a strategy for rural development in Nigeria', *Journal of Rural Studies*, **2** (4), 321–335.

Scheumann, W., S. Neubert and M. Kipping (eds) (2008), *Water Politics and*

Development Cooperation: Local Power Plays and Global Governance, Dordrecht: Springer.

Schlüter, M. and E. Herrfahrdt-Pähle (2011), 'Exploring resilience and transformability of a river basin in the face of socioeconomic and ecological crisis: an example from the Amudarya river basin, Central Asia', *Ecology and Society*, **16** (1), 32.

Schneider, S.H., W.O. Easterling and L.O. Mearns (2000), 'Adaptation: sensitivity to natural variability, agent assumptions and dynamic climate changes', *Climatic Change*, **45** (1), 203–221.

Senter International (2001), 'Implementation of EU Water Framework Directive in Turkey', Terms of Reference, The Hague.

Shah, T., I. Makin and R. Sakthivadivel (2001), 'Limits to leapfrogging: issues in transposing successful river basin management institutions in the developing world', in C. Abernethy (ed.), *Intersectoral Management of River Basins*, Colombo: International Water Management Institute, pp. 89–114.

Shiva, V. (2002), *Water Wars: Privatization, Pollution and Profit*, Cambridge, MA: South End Press.

Smets, H. (2000), 'The right to water as a human right', *Environmental Policy and Law*, **30** (5), 248–250.

Smit, B., O. Pilifosova, I. Burton, B. Challenger, S. Huq, R.J.T. Klein and G. Yohe (2001), 'Adaptation to climate change in the context of sustainable development and equity', *Climate Change 2001: Impacts, Adaptation and Vulnerability. Contribution of Working Group II to the Third Assessment Report of the IPCC*, Cambridge: Cambridge University Press, pp. 877–912.

Stiglitz, J. (2000), 'Vers un nouveau paradigme du développement', *L'Economie Politique*, **5**, 6–39.

Swart, R., J. Robinson and S. Cohen (2003), 'Climate change and sustainable development: expanding the options', *Climate Policy*, **3** (1), 19–40.

Swatuk, L.A. and M. Motsholapheko (2008), 'Communicating integrated water resources management: from global discourse to local practice – chronicling an experience from the Boteti River Sub-basin, Botswana', *Physics and Chemistry of the Earth*, **33**, 881–888.

Uitto, J.I. and A.M. Duda (2002), 'Management of transboundary water resources: lessons from international cooperation for conflict prevention', *Geographical Journal*, **168** (3), 365–378.

United Nations Economic Commission for Europe (UNECE) (2009), *River Basin Commissions and other Institutions for Transboundary Water Cooperation: Capacity for Water Cooperation in Eastern Europe, Caucasus and Central Asia*, New York and Geneva, Switzerland: UN.

United Nations Educational, Scientific and Cultural Organization (UNESCO) (2003), 'Water for people, water for life', *United Nations World Water Development Report*, Oxford: Berghahn Books.

United Nations Educational, Scientific and Cultural Organization (UNESCO) (2006), 'Water: a shared responsibility', *United Nations World Water Development Report 2*, Paris: UNESCO.

United Nations Educational, Scientific and Cultural Organization (UNESCO) (2012), 'Managing water under uncertainty and risk', *United Nations World Water Development Report*, Paris: UNESCO.

United Nations Watercourses Convention (1997), *UN Convention on the Law of the Non-Navigational Uses of International Watercourses*, UN Doc. A/51/869; 37 ILM 719.

van der Zaag, P., F.G.W. Jaspers and C.H. Komakech (eds) (2007), *A Comparative Assessment of Transboundary River Basin Organizations in Southeast Asia, Europe and Africa: Towards Institutional Strengthening of Transboundary River Basin Organizations*, Delft: UNESCO-IHE Institute for Water Education.
van Dijk, M.P. (2003), *Liberalization of Drinking Water in Europe and Developing Countries*, Delft: UNESCO-IHE Institute for Water Education.
Vörösmarty, C.J., P. McIntyre, M.O. Gessner, D. Dudgeon, A. Prusevich, P.A. Green, S. Glidden, S.E. Bunn, C.A. Sullivan, C.R. Liermann and P. Davies (2010), 'Global threats to human water security and river biodiversity', *Nature*, **467**, 555–561.
Wilbanks, T. (2003), 'Integrating climate change and sustainable development in a place-based context', *Climate Policy*, **3** (1), S147–S154.
Wittfogel, K.A. (1957), *Oriental Despotism, a Comparative Study of Total Power*, New Haven, CT, USA and London, UK: Yale University Press.
World Bank (1993), 'Water resources management', Policy paper, Washington, DC.
World Commission on Environment and Development (WCED) (1987), *Our Common Future*, Oxford: Oxford University Press.
World Water Development Report (2003), *Water for People, Water for Life*, Oxford: Berghahn Books.
Zawahri, N.A. (2008), 'Designing river commissions to implement treaties and manage water disputes: the story of the Joint Water Committee and Permanent Indus Commission', *Water International*, **33** (4), 464–474.

3. Cooperative transboundary water governance in Canada's Mackenzie River Basin: status and prospects

Rob de Loë and Michelle Morris

3.1 INTRODUCTION

Canada is a party to one of the world's better-known institutions for transboundary water management, the Boundary Waters Treaty of 1909 and the organisation it created, the International Joint Commission (e.g. Fischhendler and Feitelson 2005). Less well known internationally are institutions for transboundary water management within Canada. Internal transboundary experiences are pertinent in the context of this book because, we argue, the challenges of governing water across jurisdictional boundaries within countries organised as federations can be as profound as those facing sovereign countries. For instance, conflicts such as the 'Tri-State Water Wars' among Georgia, Florida and Alabama (Jordan and Wolf 2006) point to the need for effective transboundary water management in the United States. Australian experiences with transboundary water governance in the Murray–Darling Basin also offer numerous insights relevant for federal states (e.g. Bhat 2009; see also Ross and Connell, Chapter 13, this volume).

Like Australia and the United States, Canada is a federation where responsibility for water is divided among jurisdictions at multiple levels. Under Canada's constitution, the federal government and the ten provincial governments share responsibility for water. The division of responsibility in Canada is complex because water is not mentioned specifically in the Canadian constitution (Bowal 2006). Rather, the division is based on shifting interpretations (by governments and by the courts) of the clauses relating to subjects such as trade, property and other matters. In broad terms, the federal government is responsible for fisheries, navigation, water pollution, water that flows across boundaries, Aboriginal peoples[1] and water on federal lands (including in the territories located in Canada's north). The provinces, in turn, are primarily responsible for the

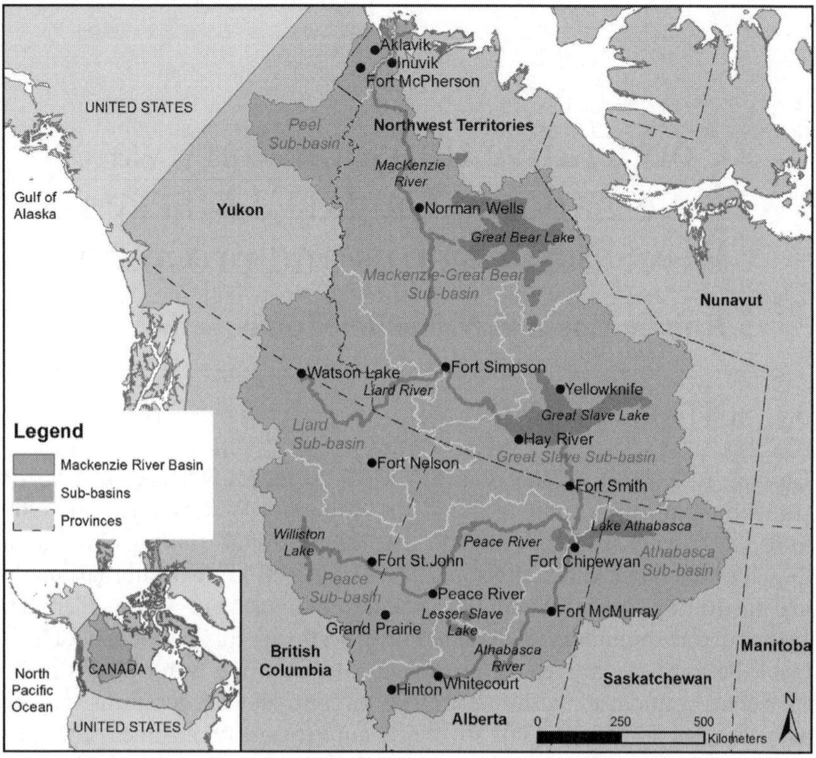

Figure 3.1 The Mackenzie River Basin, Canada

management of water within their boundaries. As a result of shared jurisdiction over water, special mechanisms (such as the one discussed in this chapter) have been developed to deal with issues in internal transboundary basins.

In this chapter, we explore governance for water in an internal transboundary basin in northern Canada that accounts for almost a fifth of the country's land area: the Mackenzie River Basin (MRB) (Figure 3.1). This basin contains globally significant ecosystems, and plays a large role in the country's future economic development plans due to its natural resources (in particular, energy). The MRB is a complex setting for transboundary water governance not only because of the jurisdictional complexity described above, but also because the basin is home to Indigenous peoples who have critical roles in governance akin to those of government actors due to their constitutional and treaty rights, and because of the presence of land claims agreements in the northern territories. The Mackenzie River

Basin Board (MRBB), a coordinating river basin organisation, was established in 1997 under the Mackenzie River Basin Transboundary Waters Master Agreement to provide a basin-wide perspective. The MRB Board's role in transboundary water governance has been limited to date, and pressure is building to deal with transboundary issues more effectively. Thus, how Canadians address the challenge of cooperative governance in this critical shared basin provides rich insights for both scholars and practitioners of transboundary water management.

In section 3.2, we provide an overview of the physical and socio-economic setting for transboundary water governance in the MRB. In section 3.3, we address the first two guiding questions underlying the book through exploring the design of the MRB Board, and the politics and strategies that resulted in this design. Our evaluation of the effectiveness of the MRB Board in section 3.4 addresses the third question underlying the book. Questions of coordination, accountability, legitimacy and environmental effectiveness are linked to the design of the organisation. Insights for our evaluation are based on three sources: (1) analysis of key documents and published studies; (2) interviews conducted specifically for this chapter with nine key informants currently or previously involved in governance for water in the basin; and (3) the personal experiences of the first author through his position as a former member of the Northwest Territories' Strategic Advisory Group of Experts regarding transboundary negotiations. We conclude the chapter with implications from this case study for transboundary water governance in federal countries (section 3.5).

3.2 CASE AND CONTEXT

Even in a country recognised globally for its considerable water resources, the MRB (Figure 3.1) stands out. It contains Canada's longest river system, at 4241 km. With an area of approximately 1.8 million km^2, the basin drains approximately one-fifth of Canada's land area (MRBB 2003). The basin's headwaters are the Peace and Athabasca rivers, which begin in British Columbia and Alberta, respectively. Numerous lakes – including the largest lake entirely in Canada (Great Bear) and the deepest lake in North America (Great Slave), along with internationally significant deltas and vast wetlands – are part of the basin's hydrology. The waters of the basin ultimately discharge from the Mackenzie River into the Beaufort Sea on Canada's northern coast.

The basin's terrestrial and aquatic ecosystems are nationally and internationally significant. In addition to containing some of the last expanses of nearly pristine wilderness on the North American continent, the MRB also has globally significant wetland, tundra and forest ecosystems; these provide habitat for numerous species that reside for part or all of the year in the basin. Deltas provide critical staging and breeding zones for millions of migratory birds; examples include tundra swans and geese that use the Peace–Athabasca Delta on Lake Athabasca (Figure 3.1). The Mackenzie Delta on the Beaufort Sea, Canada's largest active delta at approximately 9000 km^2 (Prowse et al. 2009), is a critical habitat for species that include fish, beluga whales, moose and beaver (MRBB 2003). The fish and wildlife resources of the basin are critical not only for their role in ecosystems, but also as a source of food for a significant portion of the basin's people, especially its Aboriginal peoples (Government of the Northwest Territories 2010).

Canada is a northern country, but the bulk of its population is located in the south, along the Canada–United States boundary. With a population of 41 462 people, the Northwest Territories is the most populous of the three territories, which collectively had 107 265 of Canada's 33.48 million people in 2011 (Statistics Canada 2012a). Hence, it is not surprising that the MRB is sparsely populated. Its population in 2001 (the most recent date for which basin population data have been calculated) was an estimated 397 000 people, the majority of whom resided in the Alberta portion (MRBB 2003). As in northern Canada as a whole (Prowse et al. 2009), population growth and demographic change have been closely linked to resource development. Development of Alberta's oil sands, near the town of Fort McMurray (Figure 3.1) accounts for much of the basin's recent population growth. For example, between the 2006 and 2011 national census years, the population of Fort McMurray grew from 47 705 to 61 374 people (Statistics Canada 2012b). Much of this population growth is due to the labour demands of oil sands mining and construction (Regional Municipality of Wood Buffalo 2012). Aboriginal peoples, who comprised 15 per cent of the 2001 population, have lived in the MRB for millennia. The proportion of the population that is Aboriginal is highest in the less densely populated northern portions of the basin (MRBB 2003), where communities tend to be small (fewer than 500 people) and located along coastlines, lakeshores and rivers (Prowse et al. 2009).

Jurisdiction over the basin is shared and fragmented, with key actors including provinces, territories, the federal government and Aboriginal peoples. Portions of the basin are located in three of Canada's provinces (Alberta, British Columbia and Saskatchewan) and three territories (Yukon, Northwest Territories and Nunavut – although Nunavut is not typically considered a basin jurisdiction because its share of the basin's land area is so small) (Figure 3.1). Under the Canadian Constitution, provinces have primary responsibility for the management of water resources, and thus each province has its own institutional framework for governing water.

The federal government is also a key basin jurisdiction because of its constitutionally defined responsibilities relating to concerns such as fisheries, national parks, transboundary waters, Aboriginal peoples and the territories. Yukon Territory has received province-like powers from the federal government through a devolution process. In the Northwest Territories, the devolution process is expected to be completed in 2013, at which point decision making and administrative responsibility for lands and waters will be transferred to the government of the Northwest Territories. Until that happens, many of the key functions undertaken by provincial governments will continue to be shared between the territorial government and a federal agency, the Department of Aboriginal Affairs and Northern Development Canada.

Indigenous peoples occupy a special place in Canada's society because of their historical, legal, treaty and constitutionally defined rights. Hence, the fact that the basin includes the traditional territories, treaty areas and land claim settlement regions of First Nations, Métis and Inuvialuit (or western Canadian Inuit) peoples is critical to any discussion of transboundary water governance in the basin. As noted in the next section, the river basin organisation that is the subject of this chapter, the MRB Board, was created in part to provide a coordinating function among these actors.

The basin is under pressure from a number of sources. Climate change is an important current and future stressor on the basin's social and ecological systems (Schindler and Donahue 2006). Numerous studies have already identified changes in hydrological and meteorological parameters linked to observed changes in climate (Abdul Aziz and Burn 2006; Schindler and Donahue 2006). Significant effects on aquatic and terrestrial ecosystems are anticipated from these kinds of changes in the future (MRBB 2003). Climate warming and increasing climatic variability will affect all sectors of society, but especially the basin's Indigenous peoples, who depend on its ecosystem for their lives and livelihoods.

While climate change is an extremely serious concern, it is not the only source of current and future environmental change in the basin. Fossil fuel development is an especially important economic activity that takes place throughout the MRB. Alberta's oil sands, located near the town of Fort McMurray (Figure 3.1), are well known around the world for the environmental impacts they create (Holroyd and Simieritsch 2009; Jordaan 2012). That oil sands development results in large-scale land disturbance and massive socio-economic change is uncontested. However, the precise nature of the impacts of emissions from oil sands projects on air and water quality is a matter of ongoing scientific debate. Some studies link mining activities to downstream pollution (Kelly et al. 2009; Kelly et al. 2010; Kurek et al. 2013), while others suggest that natural sources account for contaminants found in air and water that have been attributed to oil sands development activities (Hall et al. 2012; Wiklund et al. 2012).

Electricity generated from hydro-electric power stations in the basin is also extremely important to the economies of the major basin jurisdictions. Two stations located on the Peace River account for approximately 30 per cent of the generating capacity of British Columbia Hydro (MRBB 2003). Like fossil fuel developments, hydro-electric power generation has resulted in significant environmental and social impacts. British Columbia Hydro's Bennett Dam on the Peace River (Figure 3.1) has resulted in changes to seasonal patterns of stream flow that have long been thought to have negatively affected the ecology of the vital Peace–Athabasca Delta (Prowse and Conly 2000). Recent studies suggest that the hydrological changes observed in this critical delta may in fact be attributable primarily to climatic drivers, rather than to the Bennett Dam (Wolfe et al. 2012). Importantly, Wolfe et al. (2012) indicate that current water levels in the Peace–Athabasca Delta are anomalously high relative to the paleolimnological record. Thus, anticipated climate-driven changes have serious implications not only for the ecology of the basin, but also for proposed future hydroelectric power projects on the Peace River (BC Hydro 2010).

In summary, the MRB is a nationally significant transboundary resource. Major economic developments already exist, and many more are planned. These projects have resulted in significant environmental and social changes, but considerable uncertainty remains regarding their impacts on human health and ecosystem conditions. Climate change adds another layer of uncertainty. In light of pressures on the basin, divided jurisdiction over the basin, and the presence of multiple, competing interests, effective transboundary governance is essential.

3.3 POLITICS, STRATEGIES AND INSTITUTIONAL DESIGN

The jurisdictions that share the MRB have independent responsibilities for water resources. As a result, even though the basin is located entirely within Canada, the kinds of problems that are typical in shared international basins (Loures et al. 2008) also exist in this case. Examples include upstream–downstream effects, inadequate monitoring and uncoordinated water policies. Recognising this reality, the basin jurisdictions began working towards a more cooperative approach to water management in the early 1970s (Box 3.1). These efforts coincided with a period of national

BOX 3.1 TIMELINE OF COOPERATION BETWEEN BASIN JURISDICTIONS

1961	Construction of the Bennett Dam begins.
1967	First commercial oil sands project (Alberta portion of the MRB). Construction of Bennett Dam is completed.
1969	Canada, Alberta, Saskatchewan and Manitoba sign the Master Agreement on Apportionment that creates the Prairie Provinces Water Board.
1972	Canada, British Columbia (BC), Alberta, Saskatchewan, Yukon and the Northwest territories (NWT) meet in Inuvik, NWT, to create the Mackenzie River Basin Liaison Committee.
1977	Signing of Memorandum of Understanding among above governments to create the Mackenzie River Basin Committee and undertake research on basin hydrology and ecology.
1977	Berger Inquiry submits its report on the proposed Mackenzie River pipeline project.
1988	Basin jurisdictions begin work on a transboundary agreement.
1996	Northern River Basins Study Board submits its report to Canada, NWT and Alberta.
1997	Responsible ministers from Canada, BC, Alberta, Saskatchewan, Yukon and NWT sign the Mackenzie River Basin Transboundary Waters Master Agreement; creation of the Mackenzie River Basin Board under the Agreement.
2000	Bilateral agreement between Yukon and NWT (with Canada) is signed.
2005	BC and Alberta sign a Memorandum of Understanding regarding bilateral negotiations.
2007	Alberta, NWT and Canada sign a Memorandum of Understanding regarding bilateral negotiations.
2009	MRBB publishes Bilateral Water Management Agreements Guidance Document.
2012	Bilateral negotiations between Alberta and NWT (with Canada) ongoing.

leadership from the federal government, and a relatively high degree of cooperation among the provinces and the federal government on water issues in Canada, during the 1970s (Booth and Quinn 1995).

In 1972 the federal government and the governments of British Columbia, Alberta, Saskatchewan, Yukon and the Northwest Territories recognised a need to coordinate on issues related to the MRB. Creery (1979) suggests that the drying out of the Peace–Athabasca Delta, which was attributed to development of the Bennett Dam on the Peace River (Figure 3.1), was a key driver. At a 1972 meeting in Inuvik, Northwest Territories, the basin jurisdictions agreed to create the Mackenzie River Basin Liaison Committee, and to charge it with facilitating information exchange, especially in relation to proposed developments and their potential impacts on downstream jurisdictions (Creery 1979). Membership of this liaison committee was limited to government officials. These early efforts led to a new memorandum of understanding, signed in 1977, that committed the parties to a three-year programme of research focusing on the ecology and hydrology of the basin. Simultaneously, the parties to the memorandum agreed to create the Mackenzie River Basin Committee as a replacement for the former liaison committee; membership of this new institution continued to be limited to officials from the basin's governments.

The MRB Committee published the findings from its research in 1981 and made nine recommendations to the governments. One of these recommendations related to the need to establish a transboundary water management agreement for the basin:

> It is recommended that the jurisdictions at an early date conclude an agreement through which trans-boundary water management issues such as minimum flows, flow regulation, and water quality can be addressed at jurisdictional boundary crossing points in the MRB, and which establishes a permanent board to implement the provisions of the agreement. (MRB Committee 1981: x)

Industrial development in the MRB was increasing during the 1960s and 1970s, and numerous new developments were being proposed. The first commercially viable oil sands operation began production in 1967, although at a very small scale compared to contemporary levels (Humphries 2008). More important at the time were plans for a major pipeline project to carry natural gas from northern fields to southern Canada down the valley of the Mackenzie River. An inquiry commissioned by the Government of Canada to investigate the environmental, social and economic impacts of this project recommended a ten-year moratorium on its development (Berger 1977). This specific project did not proceed. Nonetheless, the MRB Committee recognised that as pressure for development increased in the basin, it would become increasingly difficult to negotiate agreements for joint management of its water resources among the basin jurisdictions. As a result, early work on such an agreement began in 1988 and proceeded through several iterations of negotiations and public comment. An agreement was drafted in the early 1990s and finalised in 1997 when the responsible ministers from each jurisdiction signed the Mackenzie River Basin Transboundary Waters Master Agreement (henceforth referred to as the Master Agreement) (Government of Canada et al. 1997b).[2]

As illustrated by Figure 3.1, the various jurisdictions that share the basin do not have equal stakes. Nunavut, for example, did not exist when the Master Agreement was being developed, and today is not usually considered a basin jurisdiction. Large portions of the basin lie within Yukon and Saskatchewan, but most of the transboundary issues that motivated the creation of the Master Agreement did not fall within their areas. Thus, the political jurisdictions with the most at stake were: British Columbia, due to its upstream hydro-electric power developments; Alberta, due its position downstream of British Columbia's hydro projects, and its own energy and industrial developments upstream of the Northwest Territories; the Northwest Territories, as the jurisdiction downstream of all the others; and the federal government, because of its overall responsibilities for transboundary concerns and its specific responsibilities for water in the territories.

Differential stakes among the basin jurisdictions account for the fact that a major parallel study that took place in the years leading up to the signing of the Master Agreement involved only Alberta, the Northwest Territories and the federal government. In response to concerns about pulp mill effluents generated in northern Alberta but flowing into the Northwest Territories, these three governments commissioned the Northern River Basins Study (NRBSB 1996). The Northern River Basins Study drew on extensive scientific research, traditional knowledge from Indigenous peoples, and public input. The report included specific recommendations to the governments relating to monitoring, ecosystem health, pollution prevention, future research, public participation and basin-wide management. In the context of this chapter, the Northern River Basins Study considered the roles that could be played by different kinds of river basin organisation with varying levels of responsibility and authority. Ultimately, the Northern River Basins Study endorsed an advisory board model, and called for the prompt signing of the Master Agreement, which at the time was awaiting the signatures of the responsible Ministers.

Recommendations in the Northern River Basins Study relating to the membership of the proposed advisory board provide an insight into the evolving landscape of water governance in Canada during the period under consideration here. As noted earlier, the membership of the original coordinating bodies in the basin, the MRB Liaison Committee and its replacement, the MRB Committee, was limited exclusively to government officials. In contrast, the Northern River Basins Study report recommended that the advisory board it called for should include government representatives, First Nations, industries, municipalities, environmental interests and basin residents, 'without dominance by any one constituency or interest group' (NRBSB 1996: 179). The authors of the Northern River Basins Study report envisioned a body that had the resources and mandate to be proactive:

> An advisory board, properly informed with power to direct research and monitoring activities, with a strong ability to influence governments, adequately supported and delivered by credible individuals in an open public arena could make major contributions to the well-being of the northern rivers and the people who live there. (NRBSB 1996, 179)

In a signal of the significance they assigned to the Northern River Basins Study, Canada, the Northwest Territories and Alberta published a formal, public response in which they committed to further research and to signing the Master Agreement that had existed in draft form since the early 1990s (Government of Canada et al. 1997a). Significantly, in their published response the governments endorsed the idea of an advisory board, but did not comment on the Northern River Basins Study Board's suggestion that it should have broad membership and a strong mandate supported by the necessary resources (Donihee et al. 2000). This omission would prove to be a telling indication of the unwillingness of the basin governments to share power with a new river basin organisation.

The Mackenzie River Basin Transboundary Waters Master Agreement that was signed in 1997 by Canada, Alberta, British Columbia, Saskatchewan, the Northwest Territories and Yukon reflects the preferred style of intergovernmental cooperation in Canada during the 1990s, as well as key principles that were shaping discussions around transboundary water resources in the international realm during this time, for example, information exchange and cooperation among watercourse states (McCaffrey 2008; see also Chapter 2 by Jaspers and Gupta, this volume). Table 3.1 summarises how these preferences and principles were expressed in terms of the five institutional design rules discussed by Huitema and Meijerink in Chapter 1 of this volume.

Key underlying principles in Part C of the Agreement (Government of Canada et al. 1997a) include the following:

1. Managing the Water Resources in a manner consistent with the maintenance of the Ecological Integrity of the Aquatic Ecosystem;

Table 3.1 Rules underlying the institutional design of the Master Agreement and the Mackenzie River Basin Board

Rule types	Rules in case study
Authority rules	The MRB Board has no independent authority of its own beyond the mandate it received from the parties to the Master Agreement. Instead, its mandate focuses on coordination, facilitation, dispute resolution and information gathering and sharing.
Aggregation rules	Decisions are made by supermajority of two-thirds of the members of the Board, although there are constraints on Aboriginal members (e.g. they are appointed at the pleasure of the respective ministers and they cannot pass resolutions or action items without the support of some government members).
Boundary rules	The boundaries of the Board's mandate are defined by the hydrologic basin, and the signatories to the Master Agreement have relatively well-accepted territorial and constitutional jurisdictions. Entry and exit rules are specified in the Master Agreement. Membership is limited to representatives of governments, and to Aboriginal peoples (who are not considered as representatives of the general public).
Information rules	The Board has a mandate to conduct research and to generate information that contributes to a whole-basin perspective on problems or issues; thus, it has some scope to generate its own information. Discussions around the board table are private, and minutes of these discussions are not publicly available.
Pay-off rules	The annual budget for the board that was established in 1997, \$280 000, is provided by the basin jurisdictions according to a clear formula specified in the Master Agreement. The budget has not been changed since 1997, meaning that the present value of the budget is significantly less than in 1997 due to inflation.

2. Managing the use of Water Resources in a sustainable manner for present and future generations;
3. The right of each to use or manage the use of the Water Resources within its jurisdiction provided such use does not unreasonably harm the Ecological Integrity of the Aquatic Ecosystem in any other jurisdiction;
4. Providing for early and effective consultation, notification and sharing of information on developments and activities that might affect the Ecological Integrity of the Aquatic Ecosystem in another jurisdiction; and
5. Resolving issues in a cooperative and harmonious manner.

The Master Agreement does not define ecological integrity, sustainable management, or what constitutes cooperative management. Furthermore, fulfilling these objectives depends on the substance of bilateral water management agreements that were to be negotiated between jurisdictions under the broad framework of the Master Agreement. As of the end of 2012, one bilateral agreement had been completed (between Yukon and the Northwest Territories), and active negotiations between Alberta and the Northwest Territories also were under way. Alberta and British Columbia had signed a Memorandum of Agreement regarding their bilateral negotiations, but had not made significant progress towards an agreement. However, by early 2013 the process appears to have been reinvigorated. Several new agreements (for example, between Alberta and British Columbia, and between British Columbia and Alberta), and an updated Yukon and the Northwest Territories agreement are expected by the end of 2013 (Miltenberger 2013).

Section 2 (D) of the Master Agreement created the river basin organisation that is the focus of this chapter: the MRB Board. The boundaries of the Master Agreement, and the jurisdiction of the Board, were hydrologically defined – and this boundary rule has not been contentious. For instance, both the MRB Committee report in 1981, and the Northern River Basins Study in 1996 and the formal response of the governments, endorsed a basin approach. When the agreement was negotiated the hydrological boundaries established in previous studies were accepted as constituting the MRB Board's jurisdiction. Establishing the boundaries was apolitical as it was viewed as 'a GIS [geographic information system] job and if somebody came and said the line needs to move a mile east or a mile west because they've now done more detailed GIS work . . . fine. That is not an issue.'[3]

The aggregation rules that define the Board's membership are considerably more limited in scope than those proposed by the Northern River Basins Study. As defined by the Master Agreement, the Board has up to three officials appointed by the federal government and one each appointed by the provinces and territories. If member jurisdictions wish to exit the agreement, the Master Agreement specifies that one year's notice to all of the other jurisdictions must be provided. A distinctive feature of the Board's membership, which does reflect the Northern River Basins Study recommendation, is the provision for five Aboriginal members. These members are elected by Aboriginal organisations in each of the provinces and territories, and are then appointed by the respective jurisdictions. It has been suggested that the inclusion of Aboriginal members brings a necessary perspective to the Board.[4] As mentioned previously, Aboriginal peoples in Canada have specific traditional, legal and treaty rights to water. However, the Board's aggregation rules reflect the notion that no single faction shall dominate the board and so decisions require a supermajority of two-thirds of members present to consent to a decision or undertaking. Therefore, in most cases Aboriginal members cannot pass resolutions or action items without the support of some government members.

The MRB Board's overall mandate is to support the principles of the Master Agreement and to provide a forum to coordinate and facilitate information exchange between each jurisdiction. Specific duties relating to the underlying information rules are spelled out in the Master Agreement; these include monitoring the progress of bilateral water management agreements that are to be completed between jurisdictions that share part of the MRB; creating State of the Aquatic Ecosystem Reports for the responsible ministers every five years; reviewing the terms of the agreement and making recommendations for revision to the ministers at least every three years; encouraging consistent monitoring programmes; and recommending uniform water quality and quantity objectives. The board is also charged with considering the needs of Aboriginal peoples, keeping them informed of the MRB in culturally relevant ways and incorporating traditional ecological knowledge and values. Pay-off rules are clear-cut. The Master Agreement specified that the total annual costs to be shared by the parties shall not exceed Canadian $280 000 (Part D, Section 3); an amount that has not been increased since 1997.

The MRB Board's mandate is broader than was considered ideal by the officials from Alberta and Saskatchewan who participated in the negotiation of the Master Agreement. These jurisdictions had prior experience with another Canadian internal transboundary agreement, the 1969 Master Agreement on Apportionment among Alberta, Saskatchewan, Manitoba, and Canada, which created the Prairie Provinces Water Board. The 1969 agreement apportions the waters of the Saskatchewan River system in southern Canada, which flows through Alberta, Saskatchewan and Manitoba (Percy 2012). It tasks the Prairie Provinces Water Board with ensuring that each jurisdiction receives its full allocation, facilitating information exchange and preventing and resolving disputes (Saunders and Wenig 2006). Board members from Alberta and Saskatchewan preferred the narrower scope of the Prairie Provinces Water Board.[5] However, with its focus on principles of ecological integrity and sustainable water management, the Mackenzie River Basin Transboundary Waters Master Agreement has a much broader scope than the 1969 Master Agreement on Apportionment for the Prairie provinces (Saunders 2012). This is mainly because the Government of the Northwest Territories felt that a Prairie Provinces Water Board-style arrangement that focused on apportionment and measurement of flows at the border would not have been appropriate in light of the more complex situation in the MRB.[6] In terms of authority rules, the MRB Board is an institution designed to facilitate and coordinate rather than to regulate. It does not have authority independent from the authority of the governments that created it. It does not issue licences or make resource management decisions. It can influence decisions by providing information and acting as a 'friend of the tribunal' in provincial, territorial and federal public hearings (Donihee et al. 2000). However, the MRB Board has no authority to intervene in the parties' jurisdiction; indeed, Part G of the Master Agreement explicitly protects the 'proprietary rights or interests of the Parties'. This institutional design was purposeful. None of the jurisdictions wanted to cede their autonomy to make resource management decisions or to influence their development trajectory to a multijurisdictional management board.[7]

Despite some disagreement about the Board's design and role, there was enough of a consensus that the Master Agreement was signed by the parties. Undoubtedly part of the reason for this agreement was the fact that the Master Agreement is non-binding, its language is vague and its principles rely upon the fulfilment of the bilateral water management agreements that were to be negotiated in future. As the years have passed, however, how the Master Agreement has been interpreted and implemented has become a cause for concern for the Northwest Territories. The Northwest Territories has been advocating for changes to the Board's membership and funding because it views the Board as ineffective, lacking independence from government, and missing needed authority.[8] Upstream hydroelectric developments, and especially oil sands development in Alberta, inspired the creation of the Northwest Territories water strategy (Box 3.2), the success of which depends strongly on the existence of bilateral agreements that respect the Northwest Territories's concerns. The Northwest Territories Minister of Environment and Natural Resources, Michael Miltenberger, has made several unsuccessful attempts to meet with the other ministers to discuss reforming the Board within the confines of the Master Agreement.[9]

As noted above, since the Master Agreement was signed in 1997, only one bilateral agreement had been completed: in 2000 between Yukon and the Northwest Territories. In this case, the issues were relatively straightforward and the stakes much lower than is the case between British Columbia and Alberta, or Alberta and the Northwest Territories (Saunders 2012). The current impetus for completion of agreements, which are expected to be completed in 2014 (Miltenberger 2013), strongly reflects the desire of the Northwest Territories to secure bilateral agreements. In an effort to reinvigorate the process of creating bilateral agreements, the *Bilateral Water Management Agreements Guidance Document* (hereafter: Bilateral Guidance Document) was created (MRBB 2009), largely under

BOX 3.2 THE NORTHWEST TERRITORIES (NWT) WATER STRATEGY

Aboriginal peoples residing in the NWT have a history of partnership with government in making decisions about land and water resources. This is demonstrated in the NWT's recent water strategy, Northern Voices, Northern Waters, which states, 'today, as in the past, the deeply held values of Aboriginal people have brought water issues to the forefront of the NWT' (Government of the Northwest Territories 2010a: 3). The water strategy, which was developed in partnership between the government of the Northwest Territories and Aboriginal governments, recognises the cultural, spiritual and historical importance of water to Aboriginal people and outlines a commitment to appropriately involve Aboriginals in implementing the water management strategy. In the words of Mark Warren (personal communication 2012), former Board member and former Assistant Deputy Minister of Environment and Natural Resources in the NWT, 'A lot of our perspective is driven by the people who were born and raised and lived in the NWT since time immemorial which is the Aboriginal population'.

the leadership of the Northwest Territories.[10] Although there was some resistance from the other parties to the creation of this document[11] it was completed and published by the MRB Board in 2009.

The Bilateral Guidance Document (MRBB 2009) reiterates the importance of including Aboriginal concerns in the bilateral water management agreements and establishes procedures and considerations for the jurisdictions as they negotiate bilateral water management agreements. Key components intended to be incorporated in the bilateral agreements include surface and groundwater quality and quantity, aquatic ecosystem health and jurisdiction-specific issues. The Bilateral Guidance Document also outlines a future role the Board may have that is specific to monitoring the implementation of the bilateral agreements (one of its specific responsibilities under the Master Agreement). Having an independent third party separate from the MRB Board monitor the progress of implementing the agreement is one of the options the document considers. Significantly, the MRB Board is not playing a role in the current bilateral negotiations. This reflects its mandate under the Master Agreement. However, as Donihee et al. (2000) argue, even within its current mandate the MRB Board could play a larger role in transboundary water management in the basin. Why this has not occurred is discussed in the next section.

3.4 PERFORMANCE OF THE MRB BOARD AS A COORDINATING RIVER BASIN ORGANISATION

3.4.1 Introduction

Evaluation of the performance of river basin organisations is a key goal in this book. In this section, we evaluate the MRB Board relative to four interrelated performance considerations identified in Chapter 1 (Huitema and Meijerink, this volume) (that is, coordination, accountability, legitimacy and environmental effectiveness). Except as noted below, we have adopted the definitions of these terms as provided in Chapter 1.

3.4.2 Coordination

Coordination is arguably one of the MRB Board's most important functions. However, despite the Master Agreement's promotion of a more holistic approach to management of the waters of the basin, the kind of coordination that the Board can undertake on its own is circumscribed by the authority rules contained in the Master Agreement (Table 3.1). To illustrate, nothing in the wording of the Master Agreement requires or permits the MRB Board to respond proactively to developments in any jurisdiction even if these developments have transboundary implications (Donihee et al. 2000). The Board could play a legitimate role as an intervener in provincial planning and regulatory processes to provide information or advance general principles. However, advancing specific positions supporting or opposing particular projects would run contrary to the agreement's clauses relating to cooperation and respect for the parties' jurisdictions (Donihee et al. 2000).

The Master Agreement does permit jurisdictions to strengthen the coordination function through bilateral agreements. For instance, the bilateral agreement between Yukon and the Northwest Territories contains strongly worded requirements for prior notification and consultation between the parties that Saunders (2012) suggests are typical of those found in international agreements. However, Saunders (2012) also correctly notes that achieving these requirements in negotiations between Alberta and the Northwest Territories will be much more difficult because the stakes are so much higher. For the MRB Board this is uncertain because the MRB Board's future coordinating role ultimately relies upon the successful negotiation and the actual content of the bilateral agreements.[12]

Dispute resolution is a common function of river basin organisations in other parts of the world (see Jaspers and Gupta, Chapter 2 in this volume). By helping to resolve disputes, river basin organisations can support coordinated approaches to the governance of shared basins. The MRB Board can play a role in resolving disputes when questions or disputes relating to the Master Agreement are brought directly to the Board by a member (Donihee et al. 2000). However, reflecting the limited and carefully constrained role that the parties to the Master Agreement wanted when it was negotiated, the Board can only issue conclusions and recommendations in such cases, and the parties involved need not respond. Furthermore, as Saunders (2012) notes, the Master Agreement does not permit either a judicial role or any form of binding arbitration – meaning that the dispute resolution procedures it contains are weak. Bilateral agreements could contain additional dispute resolution procedures, but the one that has been negotiated (between Yukon and the Northwest Territories) does not take this step (Saunders 2012).

The Board has a mandate to recommend objectives for water quality and quantity, which is another avenue for coordination that would support a whole-basin perspective. However, it does not have the authority to ensure or require that the basin jurisdictions have common policies.[13] As noted in section 3.3, the Master Agreement specifically protects jurisdictional autonomy. Importantly, as the sole successfully negotiated instance demonstrates (Yukon and the Northwest Territories), bilateral agreements negotiated under the Master Agreement can contain more specific water quality and quantity objectives if the parties desire them (Saunders 2012). This again reinforces how the MRB Board's ability to coordinate monitoring is hamstrung by the absence of bilateral agreements. The fact that it has not been formally involved in critical initiatives (for example, the Canada–Alberta oil sands monitoring programme) reinforces the institutional redundancy that exists in the basin.

The MRB Board has a mandate to conduct research and to generate information that contributes to a whole-basin perspective on water-related issues. In a thorough assessment of the Master Agreement from a legal perspective prepared for the Board, Donihee et al. (2000) suggested that the signing of bilateral agreements is not a precondition for the Board to address basin-wide issues relating to the sustainable use of water resources or aquatic ecosystems in a more proactive fashion. Despite this opinion, the Board has not pursued this opportunity. Instead, it has focused its energy on the production of State of the Environment reports;

submissions of these to the Ministers is required every five years under the terms of the Agreement. Two such reports have been submitted (MRBB 2003, 2012). The second report was several years overdue; it provided a brief overview and summary of key issues in the basin raised in the first report (MRBB 2012).

The Board has a special mandate to involve Aboriginal peoples in its deliberations, and to integrate traditional knowledge. Aboriginal Board members are explicitly not representatives of the general public. Instead, they are supposed to represent the interests of Aboriginal peoples in the provinces and territories from which they were appointed. This is a complex task because of the diversity of Aboriginal peoples in each jurisdiction; as noted in section 3.2, the basin is home to distinct First Nations, Métis and Inuvialuit peoples. At the same time, due to the existence of land claims agreements in the territories that assign specific governance functions to Aboriginal peoples, the legal position of Aboriginal peoples in those jurisdictions differs from the position of Aboriginal peoples in the provincial portions of the basin. Practical challenges also exist relating to the ability of the Aboriginal members of the Board to participate in its activities. For instance, resources to support capacity building have been insufficient.[14] Most significantly for the subject of coordination, the role that has been permitted to the Aboriginal Board members has been described as 'token' by a senior territorial official;[15] this is completely unacceptable from the perspective of the Northwest Territories, he suggested, where Aboriginal governments are viewed as partners with the territorial government (Box 3.2).

The Board's performance of its coordination functions should not be surprising in light of the constraints under which it operates. For its day-to-day work, the Board has a Secretariat with a staff of one person (MRBB 2013). Interviewees emphasised that the budget of the MRB Board, which has not been increased since 1997, is too limited to undertake significant original research. Additionally, as a former executive director noted, the different jurisdictions have struggled to come to agreement on what research initiative should be undertaken with the limited funds that are available.[16] He attributes this in part to the Board's composition. Along with several other interviewees, he noted that the Board is dominated by bureaucrats, and that this can work against a broader perspective because the government members are focused too strongly on the interests of their own jurisdictions. Funds also have been insufficient to coordinate the various monitoring programmes within the basin jurisdictions.[17]

The fact that the recent monitoring programme instituted in 2012 for the oil sands due to public and scientific concerns is a joint initiative between Canada and Alberta (Environment Canada 2012), and does not involve the MRB Board, raises questions about the relevance of the Board, and the strength of the political commitment to shared management of the basin. Despite its position as the downstream jurisdiction, and potentially a recipient of many of the contaminants of concern under the programme, the Northwest Territories also is not part of the plan. Recognising this concern, the Government of the Northwest Territories (2012) has requested (unsuccessfully) that it and its Aboriginal partners be involved in the implementation of the monitoring programme.

3.4.3 Accountability

How accountability should be evaluated in a case such as the MRB Board is complicated and contested. The MRB Board was created by ministers from the respective jurisdictions. It exists at the pleasure of these ministers and is formally accountable only to them; from this perspective, an Alberta Board member suggested, accountability is not a concern.[18] In contrast, other interviewees pointed to larger accountability questions. For example, the respective ministers are under no obligation to present reports from the Board to their respective legislatures, or to the general public.[19] Hence, the ministers themselves are not regularly accountable for the success or failure of the Board to achieve its mandate, or for the extent to which the terms of the Master Agreement are met.[20]

The Board's accountability to Aboriginal peoples in the basin is also an important question that deserves consideration.[21] In terms of the legal text of the agreement, the Board is not specifically accountable to Aboriginal peoples in the same way that it is accountable to the ministers. From the perspective of the Northwest Territories, however, the question of accountability to Aboriginal peoples is less clear-cut. As noted previously, Aboriginal governments are viewed as partners by the Government of the Northwest Territories (Box 3.2). This stands in contrast to the general perspective of the provincial and federal governments in Canada towards Aboriginal peoples (Phare 2009).

The key informants interviewed for this chapter reported varying levels of satisfaction regarding the extent to which the Board was fulfilling its mandate. Referring to the kinds of issues discussed above, some interviewees suggested that the Board was not particularly effective in fulfilling its mandate under the Master Agreement and, importantly, was not being held to account. For example, the Board is required to submit reports annually. However, it did not publish any annual reports during the years 2007–2011 and only recently managed to fulfil its overdue reporting requirements with an update in 2013 (MRBB 2013). As noted earlier, the MRB Board was late in delivering its second State of the Aquatic

Ecosystem reports. The Board also has a specific mandate to incorporate traditional ecological knowledge, but has struggled to do so according to interviewees. In a recent report, the Board itself acknowledged that, 'when compared with the availability of western science, Traditional Knowledge is underrepresented in all areas of the Mackenzie River Basin' (MRBB 2012: 10).

These kinds of concerns speak to accountability, in the sense that historically it does not seem to have mattered to all the signatories to the Master Agreement whether or not the Board fulfils its mandate. In fairness, however, it must be emphasised that it is entirely within the power of the basin jurisdictions (with the consent of all parties) to increase the Board's budget, to change the instructions given to government members regarding their role on the Board, to encourage the Board to expand its activities within its mandate, to provide it with the support it would need to integrate traditional ecological knowledge more effectively – or even to reconstitute it through a new agreement. Any of these activities would require the consent of all basin jurisdictions. Although most jurisdictions agree that modifying the budget is necessary, agreement on the precise nature of the changes has not been forthcoming (MRBB 2013).[22] According to interviewees from the Government of the Northwest Territories, theirs is the only jurisdiction that has advocated (albeit unsuccessfully) for a reformulation of the MRB Board.[23]

3.4.4 Legitimacy

As in the case of accountability, multiple perspectives exist regarding the legitimacy of the MRB Board. The concept of 'legitimacy' is conventionally defined as the approval of institutions by those subject to their actions (Gearey and Jeffrey 2006; Suchman 1995). The MRB Board has the legitimacy that flows to bodies constituted under agreements entered into by parties that themselves have legitimacy. It certainly has not exceeded its mandate, or even (arguably) reached the limits of its mandate (Donihee et al. 2000; Saunders 2012).

Legitimacy is often evaluated in terms of 'inputs' and 'outputs' (see Huitema and Meijerink, Chapter 1 in this volume). On the input side, the Board has provided a forum for the parties to share information. Discussions around the board table are private, and minutes of these discussions are not publicly available. Thus the MRB Board creates a safe environment in which dialogue can occur and contentious issues can be discussed.[24] The fact that it has a poor track record of engaging the general public within the basin, or indeed being known to residents of the basin – a problem identified by several respondents – also does not undermine its legitimacy from this perspective; this kind of engagement was not sought by the signatories to the Master Agreement. Where the MRB Board clearly falls short on the input side is in meeting the requirement to engage Aboriginal peoples and to make effective use of traditional ecological knowledge. In terms of making use of traditional ecological knowledge, as noted previously, the Board itself identified this as a deficiency (MRBB 2012). The Board's Traditional Knowledge and Strengthening Aboriginal Partners Committee, the membership of which includes the Aboriginal Board members and traditional-knowledge practitioners, has recently developed a best-practices report to support integration of traditional knowledge into environmental monitoring and assessment in the MRB (Stevenson 2012).

On the output side, the Board falls short relative to its specific mandate to provide annual reports and State of the Environment reports every five years. Numerous relevant concerns relating to output legitimacy were identified above. Importantly, some interviewees pointed to the fact that after 15 years only one bilateral agreement exists as evidence of the Board's ineffectiveness. This criticism is unfair, we suggest, because the Master Agreement assigns the Board no role in the negotiation of the bilateral agreements. The rate of progress in negotiating the bilateral agreements has been determined entirely by the signatories to the agreement, rather than by the Board.

3.4.5 Environmental Effectiveness

The Master Agreement includes protection of the ecological integrity of the MRB as a guiding principle, but the bilateral water management agreements that will operationalise this principle have not yet been completed. In its State of the Aquatic Ecosystem Reports, the MRB Board has noted that poor or limited baseline data and inadequate monitoring have made evaluating environmental performance within the basin difficult (MRBB 2012). As a result, assessing the environmental performance of the MRB Board is challenging because of gaps in knowledge regarding pre- and post-MRB Board environmental conditions. Furthermore, it is important to remember that the MRB Board does not have the authority to intervene directly in the environmental decision making of the parties. This means that any evaluation of the environmental performance of the Board must be conducted with reference to the constraints under which it functions. Ultimately, it is the provinces, territories and federal government that are responsible for environmental outcomes within the basin.

3.5 CONCLUSIONS

Why does water governance in the MRB matter to Canadians, let alone to people in other parts of the world? After all, the MRB is an internal basin in a country that most citizens of the world view as relatively water wealthy. At the same time, it is located in a sparsely populated part of the country that most Canadians may never visit. We suggest that the establishment of an effective regime for cooperative governance of water and related resources in the basin should matter a great deal; especially to Canadians, but also to scholars and practitioners of transboundary water management in other countries.

For Canadians, effective transboundary water governance in the MRB should be a national priority. An economically motivated rationale can be constructed relatively easily. The basin contains vast energy and mineral resources. In the face of increasing climatic variability and growing pressure from existing and planned developments, a cooperative approach to addressing the environmental and social impacts that result from developments is needed to create a more secure climate for investment. Public opposition to the Mackenzie Valley Pipeline in the 1970s is one factor that led to the shelving of that project (Berger 1977). A major pipeline currently proposed to take oil from Alberta's oil sands to key markets in Asia, the Northern Gateway project, appears to be facing the same fate. Recognising the breadth and depth of public opposition to this project, the government of British Columbia threatened to deny the pipeline developer the permits it would need to proceed (Simpson 2012). These examples illustrate that whether or not and how economic developments in the basin proceed depends in part on the extent to which the interests and needs of multiple basin jurisdictions can be satisfied.

Economic arguments for cooperation aside, we suggest two reasons why a moral imperative exists to move beyond a narrow focus on individual sectoral and jurisdictional interests in the basin. First, the basin contains nationally and internationally significant water-dependent ecosystems that have not yet been entirely despoiled. Experiences from around the world demonstrate that remediating damaged ecosystems is orders of magnitude more difficult than protecting them before they are compromised. Effective transboundary water governance based on deeper cooperation among basin states is therefore essential to avoid damaging the MRB's ecosystems. As Saunders (2012) suggests, such a shift is needed to give Canadians confidence that the basin's natural environment will be protected for future generations.

Second, the MRB is home to a large and diverse population of Indigenous peoples. Given their occupation of the lands of the basin for millennia and their ongoing dependence on its ecosystems for their lives and livelihoods, a ready moral argument exists for adopting a truly cooperative approach to governance in the basin. However, pragmatic and distinctly Canadian reasons also exist for a cooperative approach to governance in the basin where Aboriginal peoples are partners rather than stakeholders. Canada's Constitution recognises and affirms Aboriginal and treaty rights. Furthermore, in the Northwest Territories and Yukon portions of the basin, Aboriginal governments exist under land claims and settlement agreements (Isaac 2012a). This legal reality means that in Canada Aboriginal peoples are not simply 'minorities' or 'stakeholders' who should be treated fairly on moral grounds. Instead, given that Canada is one of the very few countries in the world where the rights of the state have been fettered at law (Isaac 2012b), Aboriginal peoples must be viewed as partners in the governance of the MRB; in this respect, the Northwest Territories provides an appropriate model for the relationship between Aboriginal peoples and the state (for example, see Box 3.2). How this legal reality fits with prevailing approaches to transboundary governance based on political jurisdictions treating with each other, remains to be seen – but it cannot be ignored. Constructive solutions developed in Canada's MRB may be relevant for the numerous other countries that share basins with Indigenous populations.

Fortunately, a solid foundation exists for designing a transboundary governance regime that addresses the concerns discussed here. To begin, it is important to re-emphasise that an institutional framework for cooperative governance in the basin already exists. It clearly is not without flaws. For example, Saunders (2012) correctly notes that the other Canadian example discussed in this chapter, the Master Agreement on Apportionment in the Prairie provinces, which was discussed above, is much stronger than the Master Agreement for the MRB because in the former bilateral agreements among the parties were negotiated at the same time as the main agreement, and included as schedules. In contrast, the Master Agreement for the MRB has been hobbled by the failure to negotiate the key bilateral agreements. Nonetheless, the jurisdiction most concerned about transboundary governance in the basin viewed the Master Agreement as a useful foundation. The Northwest Territories Minister for Environment and Natural Resources suggests that his concern is not with the language of the agreement; instead, the problems he identified related to implementation of the agreement, and to political will.[25] A regime based on principles of collaboration among the basin jurisdictions, Aboriginal governments and key non-state actors ultimately may be needed to achieve effective transboundary water governance in the MRB. However, as we argue in this chapter, the current regime has hardly been pushed to its limits.

It is also important to emphasise that a wealth of examples from around the globe in both domestic and international basins provide insights into how the gaps in the current design can be addressed. Increasing awareness of the challenge of transboundary governance in internal basins in the United States (motivated partly by cases such as the Tri-State Water Wars among Georgia, Alabama and Florida) has resulted in the publication of both general principles and guidelines (e.g. Draper 2006) and model compacts for interstate water governance (e.g. Muys et al. 2007). Principles and benchmarks from this vast literature provide a solid foundation for meeting the challenges of transboundary water governance in the MRB (see de Loë 2010 for an example of how these principles can be applied in the MRB). Indeed, the book containing this chapter is itself a contribution to the large and growing body of empirical knowledge.

Finally, we close by returning to one of the basic underlying themes of this book: the importance and limitations of institutional design. The constraints under which the MRB Board operates today exist by design. The parties to the Master Agreement desired a coordinating body with no authority to direct the actions of governments. Instead, they sought a river basin organisation that would provide a neutral forum for discussing issues and monitoring the implementation of the agreement. This reinforces a central argument that runs throughout the book: the design of river basin organisations is extremely important and must be suited to the context in which the organisation will function. The Master Agreement and the MRB Board were designed to be compatible with the preferred style of intergovernmental cooperation in Canada that existed in the 1980s and 1990s. With the benefit of hindsight, the issues identified in this chapter relating to the implementation of the Master Agreement were entirely predictable. Nonetheless, given Canada's historical, cultural, legal and political landscape, it is highly unlikely that one of the other three models for river basin organisations discussed in Chapter 1 of this volume would have been suitable when the agreement was created.

Paradoxically, however, this case study also illustrates that while good institutional design is necessary, it is far from sufficient. Up until the time that the Northwest Territories became concerned about the impacts of oil sands development, the parties to the Master Agreement seem to have been satisfied with the relatively low level of interaction and cooperation that was taking place, and did not desire a more active or effective MRB Board. Ultimately, this case demonstrates that no amount of institutional design can fully compensate for a lack of political will, or an inability to recognise the benefits of a cooperative approach. This is hardly a novel finding even just within the transboundary water governance field. However, it speaks again to the importance of designing institutions with context in mind (Ingram 2008). The Master Agreement was negotiated at the tail end of a period of relatively high interjurisdictional cooperation. The decades of the 1970s and 1980s in Canada were a time of leadership from the federal and the provincial and territorial governments on water issues. This is reflected in the spirit of collaboration that led to the precursors to the MRB Board, the Northern River Basins Study, and ultimately the Master Agreement. Unfortunately, as a former senior official in the federal government who was closely involved in the creation of the Master Agreement noted, both federal leadership and the collaborative spirit of previous decades are greatly diminished today.[26] The challenge facing those concerned about the MRB will be to ensure that the bilateral agreements currently being negotiated establish a durable regime that results in more effective shared governance in the basin. Success in this respect may well invigorate the MRB Board and clarify its role in transboundary water governance.

ACKNOWLEDGEMENTS

The authors are extremely grateful to the nine people who shared their insights during the interviews. During the time this chapter was being prepared, negotiations regarding bilateral agreements were under way. Many of the people to whom we spoke played roles, direct and indirect, in those negotiations. Hence, we are especially thankful that they were willing to share their insights during this busy time. As the authors of this chapter, we are fully responsible for errors and omissions, and for the ways in which we used the information provided to us by interviewees, including how we interpreted their comments.

NOTES

1. Canada's Constitution Act, 1982, uses the term 'Aboriginal' with reference to three groups of Indigenous peoples found in Canada: Indians (First Nations), Métis and Inuit. In this chapter, we use the internationally recognised term 'Indigenous' peoples, except in places where 'Aboriginal' is locally appropriate (for example, this is the term used by most people in the Northwest Territories).
2. The full text of the agreement is available at: http://www.mrbb.ca/uploads/files/general/19//mackenzie-river-basin-transboundary-waters-master-agreement.pdf (accessed 21 July 2013).
3. Robert Harrison, personal communication, 13 August 2012, telephone interview.
4. David Livingstone, personal communication, 7 August 2012; Harrison, personal communication, 13 August 2012, telephone interview.
5. Jack Van Camp, personal communication, 28 August 2012, telephone interview.
6. Gary Bohnet, personal communication, 5 September 2012, telephone interview.

7. Robert Harrison, personal communication, September 2012, telephone interview.
8. Gary Bohnet, personal communication, 2012, telephone interview.
9. Michael Miltenberger, personal communication, 13 September 2012, telephone interview.
10. Mark Warren, personal communication, 8 August 2012, telephone interview.
11. David Livingstone, personal communication, 7 August 2012, telephone interview.
12. Don Pittman, personal communication, 4 October 2012, telephone interview.
13. Robert Harrison, personal communication, 13 August 2012, telephone interview.
14. Jack Van Camp, personal communication, 28 August 2012, telephone interview.
15. Gary Bohnet, personal communication, 5 September 2012, telephone interview.
16. Jack Van Camp, personal communication, 28 August 2012, telephone interview.
17. Michael Miltenberger, personal communication, 13 September 2012, telephone interview.
18. Robert Harrison, personal communication, 13 August 2012, telephone interview.
19. Mark Warren, personal communication, 8 August 2012, telephone interview.
20. Michael Miltenberger, personal communication, 13 September 2012, telephone interview.
21. Michael Miltenberger, personal communication, 13 September 2012, telephone interview.
22. Mark Warren, personal communication, 8 August 2012, telephone interview.
23. Mark Warren, personal communication, 8 August 2012, telephone interview.
24. Robert Harrison, personal communication, 13 August 2012, telephone interview; Don Pittman, personal communication, 4 October 2012, telephone interview.
25. Michael Miltenberger, personal communication, 13 September 2012, telephone interview.
26. David Livingstone, personal communication, 7 August 2012, telephone interview.

REFERENCES

Abdul Aziz, O.I. and D.H. Burn (2006), 'Trends and variability in the hydrological regime of the Mackenzie River Basin', *Journal of Hydrology*, **319** (1–4), 282–294.

Berger, T.R. (1977), *Northern Frontier, Northern Homeland: The Report of the Mackenzie Valley Pipeline Inquiry*, Ottawa: Minister of Supply and Services Canada.

Bhat, A. (2009), 'The politics of model maintenance: the Murray Darling and Brantas River Basins compared', *Water Alternatives*, **1** (2), 201–218.

Booth, L. and F. Quinn (1995), 'Twenty-five years of the Canada Water Act', *Canadian Water Resources Journal*, **20** (2), 65–90.

Bowal, P. (2006), 'Canadian water: constitution, policy, and trade', *Michigan State Law Review*, **5** (Special Issue), 1141–1177.

British Columbia (BC) Hydro (2010), 'Site C clean energy project', available at http://www.bchydro.com/planning_regulatory/site_c.html (accessed 1 August 2013).

Creery, R.A. (1979), 'Avoiding further downstream crises: the Mackenzie River Basin Committee', *Canadian Water Resources Journal*, **4** (3), 60–66.

de Loë, R.C. (2010), 'Transboundary water governance in the Mackenzie River Basin, Canada', presentation at 7th Biennial Rosenberg International Forum on Water Policy, Buenos Aires, Republic of Argentina, 15–17 November.

Donihee, J., S.A. Kennett and J.O. Saunders (2000), *Defining a Role for the Mackenzie River Basin Board: A Review of Key Considerations and Options*, Calgary, AB: Canadian Institute of Resources Law.

Draper, S.E. (2006), *Sharing Water in Times of Scarcity: Guidelines and Procedures in the Development of Effective Agreements to Share Water Across Political Boundaries*, Reston, VA: American Society of Civil Engineers.

Environment Canada (2012), 'Backgrounder: Joint Canada–Alberta Implementation Plan for Oil Sands Monitoring', available at http://www.ec.gc.ca/default.asp?lang=En&n=757CF272-1&news=3705C71B-D87F-4577-AEBD-212C29F7448D (accessed 1 August 2013).

Fischhendler, I. and E. Feitelson (2005), 'The formation and viability of a non-basin water management: the US–Canada case', *Geoforum*, **36** (6), 792–804.

Gearey, M. and P. Jeffrey (2006), 'Concepts of legitimacy within the context of adaptive water management strategies', *Ecological Economics*, **60** (1), 129–137.

Government of Canada, Government of Alberta and Government of the Northwest Territories (1997a), 'Canada–Alberta–Northwest Territories Response to the Northern River Basins Study Report to the Ministers 1996', available at http://environment.gov.ab.ca/info/library/8701.pdf (accessed 17 September 2013).

Government of Canada, Government of British Columbia, Government of Alberta, Government of Saskatchewan, Government of the Yukon and Government of the Northwest Territories (1997b), 'Mackenzie River Basin Transboundary Waters Master Agreement', available at http://www.mrbb.ca/uploads/files/general/19//mackenzie-river-basin-transboundary-waters-master-agreement.pdf (accessed 17 September 2013).

Government of the Northwest Territories (2010), 'State of the Environment Report', Government of the Northwest Territories, Canada, available at http://www.enr.gov.nt.ca/_live/pages/wpPages/soe_conservation_sustainable_use.aspx#3 (accessed 1 August 2013).

Government of the Northwest Territories (2012), 'Joint Canada–Alberta Implementation Plan for Oils Sands Monitoring', Government of the Northwest Territories, Canada, available at http://news.exec.gov.nt.ca/joint-canada-alberta-implementation-plan-for-oils-sands-monitoring/ (accessed 27 July 2013).

Hall, R.I., B.B. Wolfe, J.A. Wiklund, T.W.D. Edwards, A.J. Farwell and D.G. Dixon (2012), 'Has Alberta oil sands development altered delivery of polycyclic aromatic compounds to the Peace–Athabasca Delta?', *PLoS ONE*, **7** (9), available at http://www.plosone.org/article/info%3Adoi%2F10.1371%2Fjournal.pone.0046089 (accessed 17 September 2013).

Holroyd, P. and T. Simieritsch (2009), *The Water That Binds Us: Transboundary Implications of Oil Sands Development*, Drayton Valley, AB: Pembina Institute.

Humphries, M. (2008), 'North American oil sands: history of development, prospects for the future', CRS Report for Congress, Vol. Order Code RL34258, Washington, DC: Congressional Research Service.

Ingram, H. (2008), 'Beyond universal remedies for good water governance: a political and contextual approach', paper presented at Water for Food: Quantity and Quality in a Changing World, Zaragoza, Spain, 24–27 June.

Isaac, T. (2012a), *Aboriginal Law: Commentary and Analysis*, Saskatoon, SK: Purich Publishing.

Isaac, T. (2012b), 'Aboriginal peoples and the law – legal certainty and the need for political leadership', Waterloo Aboriginal Education Centre, St Paul's University College, University of Waterloo, 2 October.

Jordaan, S.M. (2012), 'Land and water impacts of oil sands production in Alberta', *Environmental Science and Technology*, **46** (7), 3611–3617.

Jordan, J.L. and A.T. Wolf (eds) (2006), *Interstate Water Allocation in Alabama, Florida and Georgia: New Issues, New Methods, New Models*, Greinesville, FL: University Press of Florida.

Kelly, E.N., D.W. Schindler, P.V. Hodson, J.W. Short, R. Radmanovich and C.C. Nielsen (2010), 'Oil sands development contributes elements toxic at low concentrations to the Athabasca River and its tributaries', *Proceedings of the National Academy of Science*, **107** (37), 16178–16183.

Kelly, E.N., J.W. Short, D.W. Schindler, P.D. Hodson, M. Ma, A.K. Kwan and B.L. Fortin (2009), 'Oil sands development contributes polycyclic aromatic compounds to the Athabasca River and its tributaries', *Proceedings of the National Academy of Science*, **106** (52), 22346–22351.

Kurek, J., J.L. Kirk, D.C.G. Muir, X. Wang, M.S. Evans and J.P. Smol (2013), 'Legacy of half a century of Athabasca oil sands development recorded by lake ecosystems', *Proceedings of the National Academy of Sciences*, Early Edition.

Loures, F., A. Rieu-Clarke and M. Vercambre (2008), *Everything You Need to Know About the UN Watercourses Convention*, Gland: World Wildlife Fund.

Mackenzie River Basin Board (MRBB) (2003), *Highlights of the Mackenzie River Basin Board's State of the Aquatic Ecosystem Report*, Fort Smith, NT: Mackenzie River Basin Board.

Mackenzie River Basin Board (MRBB) (2009), *Bilateral Water Management Agreements Guidance Document*, Fort Smith, NT: Mackenzie River Basin Board.

Mackenzie River Basin Board (MRBB) (2012), *The Mackenzie River Basin Board's 2012 Issues Report: Oil Sands Development, Hydroelectric Development, and Climate Change in the Mackenzie River Basin*, Fort Smith, NWT: Mackenzie River Basin Board.

Mackenzie River Basin Board (MRBB) (2013), *Report to the Ministers April 1, 2007–March 31 2012*, Yellowknife, NWT: Mackenzie River Basin Board.

Mackenzie River Basin (MRB) Committee (1981), *Mackenzie River Basin Study Report: a Report Under the 1978–1981 Federal–Provincial Study Agreement Respecting the Water and Related Resources of the Mackenzie River Basin*, Regina, SK: Environment Canada, Inland Waters Directorate.

McCaffrey, S. (2008), 'Convention on the Law of the Non-Navigational Uses of International Watercourses, New York, 21 May 1997', available at http://untreaty.un.org/cod/avl/ha/clnuiw/clnuiw.html (accessed 1 August 2013).

Miltenberger, M. (2013), 'Northwest Territories Water 101: changing the paradigm', presentation at Changing the Water Paradigm, Canadian Water Network, Ottawa, Ontario, 19 March.

Muys, J.C., G.W. Sherk and M.C. O'Leary (2007), 'Utton transboundary resources center model interstate water compact', *Natural Resources Journal*, **47** (1), 17–115.

Northern River Basins Study Board (NRBSB) (1996), *Northern River Basins Study: Report to the Ministers 1996*, Edmonton, AB: Alberta Environmental Protection.

Percy, D.R. (2012), *Resolving Water-Use Conflicts: Insights From the Prairie Experience for the Mackenzie River Basin*, Toronto: C.D. Howe Institute.

Phare, M.A. (2009), *Denying the Source: The Crisis of First Nations Water Rights*, Surrey, BC: Rocky Mountain Books.

Prowse, T.D. and F.M. Conly (2000), 'Multiple-hydrologic stressors of a northern delta ecosystem', *Journal of Aquatic Ecosystem Stress and Recovery*, **8** (1), 17–26.

Prowse, T.D., C. Furgal, B.R. Bonsai and D.L. Peters (2009), 'Climate impacts on northern Canada: regional background', *Ambio*, **38** (5), 248–256.

Regional Municipality of Wood Buffalo (2012), *Municipal Census 2012*, Fort McMurray: Regional Municipality of Wood Buffalo.

Saunders, J.O. (2012), 'Managing the Mackenzie: negotiating a future with the basin in mind', *Resources*, **114**, 1–10.

Saunders, J.O. and M. Wenig (2006), 'Whose water? Canadian water management and the challenges of jurisdictional fragmentation', in K. Bakker (ed.), *Eau Canada: The Future of Canada's Water*, Vancouver: University of British Columbia Press, pp. 119–141.

Schindler, D.W. and W.F. Donahue (2006), 'An impending water crisis in Canada's western prairie provinces', *Proceedings of the National Academy of Science*, **103** (19), 7210–7216.

Simpson, J. (2012), 'You heard it here: Northern Gateway's dead', *Globe and Mail*, available at http://www.theglobeandmail.com/commentary/you-heard-it-here-northern-gateways-dead/article4589760/ (accessed 17 July 2013).

Statistics Canada (2012a), 'Population and dwelling counts, for Canada, provinces and territories, 2011 and 2006 censuses', available at http://Www12.Statcan.Gc.Ca/Census-Recensement/2011/Dp-Pd/Hlt-Fst/Pd-Pl/Table-Tableau.Cfm?LANG=Eng&T=101&S=50&O=A (accessed 1 August 2013).

Statistics Canada (2012b), 'Fort McMurray, Alberta (Code 0292) and Alberta (Code 48) (table), Census profile, 2011 Census', Statistics Canada Catalogue no. 98–316-XWE, Ottawa: Statistics Canada.

Stevenson, M.G. (2012), *Towards a New Current of Thought: Best Practices in Gathering and Incorporating Traditional Knowledge into Environmental Monitoring and Assessment*, Yellowknife, NWT: Mackenzie River Basin Board.

Suchman, M. (1995), 'Managing legitimacy: strategic and institutional approaches', *Academy of Management Review*, **20** (3), 571–610.

Wiklund, J.A., R.I. Hall, B.B. Wolfe, T.W.D. Edwards, A.J. Farwell and D.G. Dixon (2012), 'Has Alberta oil sands development increased far-field delivery of airborne contaminants to the Peace–Athabasca Delta?', *Science of the Total Environment*, **43**, 379–382.

Wolfe, B.B., R.I. Hall, T.W.D. Edwards and J.W. Johnston (2012), 'Developing temporal hydroecological perspectives to inform stewardship of a northern floodplain landscape subject to multiple stressors: paleolimnological investigations of the Peace–Athabasca Delta', *Environmental Reviews*, **20** (3), 191–210.

4. Designing an agency to manage a wicked water problem: the Oregon Watershed Enhancement Board

Denise Lach and Dan Calvert

4.1 INTRODUCTION

Water in the western United States (US) is and has historically been a contentious issue. While many of the country's largest rivers flow through the western half of the country, much of the landscape is desert or shrub steppe. Even those places with adequate rainfall enjoy a Mediterranean-like climate with dry summers, so almost all agriculture in the west is irrigated from rivers and through pumping groundwater. In an effort to encourage westward expansion in the nineteenth century, water policy was designed to favour the 'first' (that is, not Native Americans, Spanish or Mexican) settlers arriving from the east putting water to 'beneficial' (only to human) use. Through most of the nineteenth and twentieth centuries, water in the west was used to generate power, irrigate agriculture, provide municipal and industrial water, and transport goods and people across the country. Economic development in the west was tied explicitly to the management of water and these uses transformed the landscape. Regional as well as local organisational structures were developed to manage water and its many uses; state and federal agencies responsible for water management are fragmented across various interests including agriculture, energy production, and water quality.

The rivers and streams of the Pacific Northwest are also home to several species of anadromous fish – salmon and steelhead (salmonids) born in freshwater streams that spend most of their lives in the ocean and return to fresh water to breed. In Oregon, salmon life cycles begin and end in streams all along the coast and as far east as Idaho. As human populations grew and landscapes changed, salmon habitat across the region was negatively affected or destroyed. In Oregon, the status of salmon stocks is mixed but the absolute numbers appear to be a small fraction of that of a couple of centuries ago (Huntington et al. 1996; Lackey et al. 2006). By the 1990s, several species of Oregon salmonids were being considered for listing under the US Endangered Species Act. Any listing would require an extensive and expensive recovery plan overseen by several federal agencies and ultimately affecting all river uses currently in practice. There were several existing state agencies that could possibly take on the obligation of habitat restoration, although no single agency covered the myriad of landscapes, locations and land uses affected by a listing.

Born in controversy over endangered species conservation, the Oregon Watershed Enhancement Board (OWEB) was created to implement the restoration of water and riparian habitat, something difficult to do within existing water law and the fragmented state and federal agencies responsible for managing water, wildlife and landscapes. Rather than bringing existing agencies, organisations and interested parties together to address the issue of restoring aquatic habitat, the Governor and legislature created a new agency with a limited role (no encroachment on existing mandates), directions to work almost exclusively with local and place-based experts in a non-regulatory manner (devolution of responsibility from state to local landowners), access to increasingly adequate funds (plenty of resources to go around), and transparent reporting processes (overseen by citizens and interested stakeholders) to begin addressing what some consider an especially 'wicked problem' (Rittel and Weber 1973): the restoration of salmon habitat throughout Oregon.

Under the Endangered Species Act, when a species is listed as threatened or endangered, a recovery plan must be developed and approved by a federal agency (in this case, the National Marine Fisheries Service); the plans require habitat protection and/or restoration. In Oregon, protecting or restoring salmon habitat would affect all types of land owners from private to federal, types of activities from agriculture to power generation, and types of streams from the mighty Columbia to the backyard brook; all would be affected by any Endangered Species Act listing. Practices and land uses would need to be changed, potentially in drastic ways including removing dams that blocked migration routes, resettling human activities away from riparian areas, and reducing the amount of water available for irrigation or other uses. In 1995, the United States (US) government offered to pay US$100 million of the estimated US$160 million it would cost to restore the salmon (*Spokane Spokesman Review* 1995). There are no current and reputable estimates of the resources spent since 1995 on this wicked problem, although as of now, up to billions have been spent with no end in sight and still no more fish in the streams. A policy analyst recently reported that he did not know of a credible source about total expenditures, 'other than to say that it is a big number. Of course, the direct costs are the smallest component (but easiest to estimate). The foregone costs are the greatest, but most difficult to measure'.[1]

So, in a bold political move, the state of Oregon made a proposal to the federal government: let us capitalise on the nascent development of local, voluntary place-based organisations – watershed councils – to develop specific recovery responses for and from individual locales. This is an examination of how a poorly funded Governor's initiative – the Governor's Watershed Enhancement Board – was transformed into a state agency with a line-item budget and powerful reach across the state made possible through the creation, support and oversight of local watershed councils. Within the ideal types presented by the editors of the book, this is the story of an 'agency', which implements its mandate through 'partnership' organisations, each of them an autonomous, voluntary place-based organisation. OWEB is not a river basin organisation in the traditional sense, although its ecological focus and unique institutional design may portend the future of water resources management in the twenty-first century.

4.2 BACKGROUND AND HISTORY

Oregon is a relatively large state (255 026 km^2) in the northwest corner of the United States, divided by the Cascade Mountains into a water-rich 'west side' and a water-poor 'east side' (see Figure 4.1). Most of Oregon's nearly four million residents live west of the Cascades, with about two-thirds living in the Willamette Valley between Portland and Eugene. More than half of Oregon (53 per cent) is public land: national parks, forests, grasslands and monuments managed by federal agencies. The Columbia River, one of the largest rivers in the US, serves as the boundary between Oregon and Washington State to the north.

The Governor's Watershed Enhanced Board came into existence in 1987 with limited responsibilities and an equally limited biennial budget (US\$500 000–US\$1 000 000) to conduct training workshops for landowners, watershed councils, educators and other community members (Oregon Land Conservation and Development Commission 1990). About ten years later (1997), the Governor and Oregon legislature created the Oregon Plan for Salmon and Watershed (Oregon Plan) as a response to the Endangered Species Act listings of several Oregon salmon species. Four major principles underlie the Oregon Plan: (1) voluntary actions by landowners who have local knowledge and a sense of ownership; (2) coordination among state, federal and tribal restoration efforts; (3) ongoing monitoring of watershed health and salmon recovery; and (4) reliance on an independent multidisciplinary science team to evaluate the effectiveness and success of projects, and to guide investments.

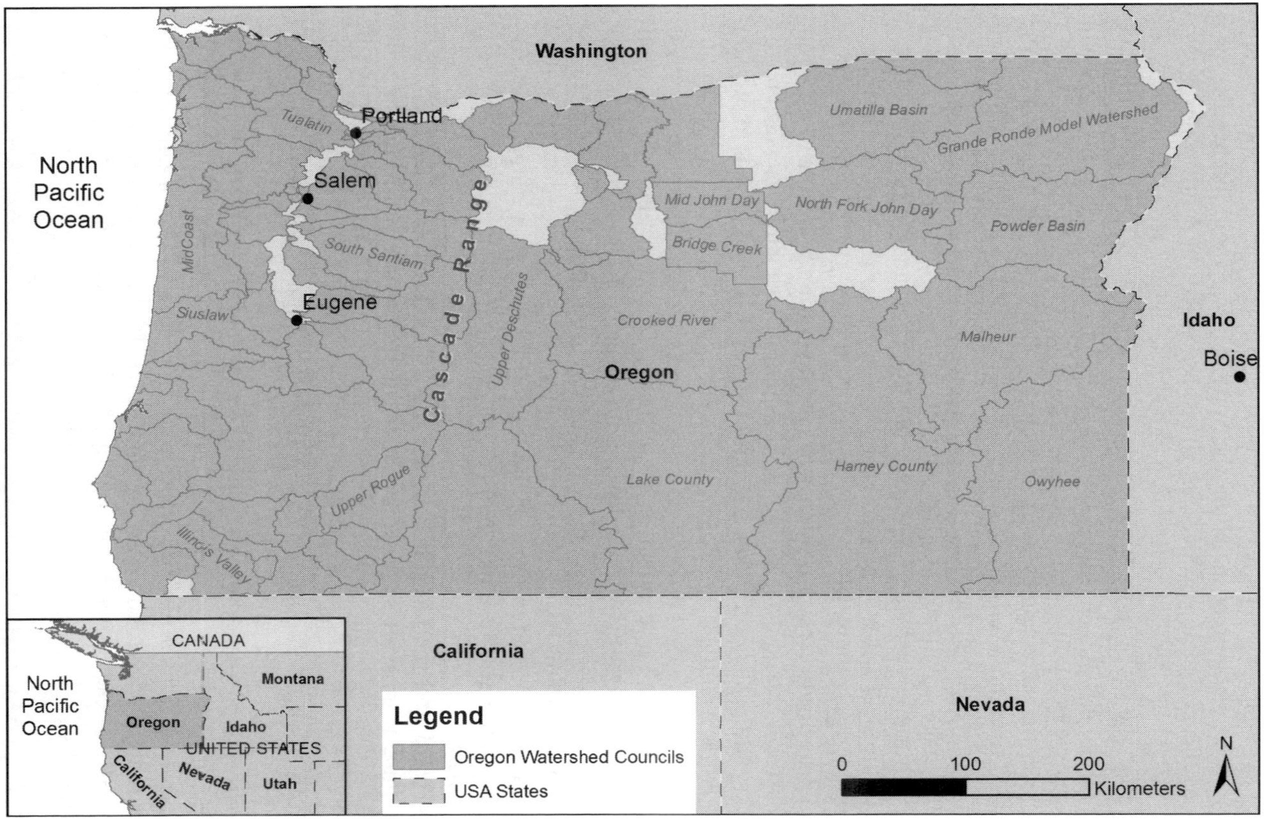

Figure 4.1 Location of Oregon and Oregon Watershed Councils

Funding for the Governor's Watershed Enhanced Board to administer the Oregon Plan was increased to about US$20 million with funds coming from multiple sources including the State General Fund, timber tax revenues, licence plate revenues, angling licence fees and other resources. At this point, federal restoration funds also began to be administered through the Governor's Watershed Enhanced Board. The budget funded watershed assessments and action plans, monitoring of watershed council projects, education and outreach, and project work. And watershed councils, not the state agency, were expected to carry out the major activities related to salmon habitat recovery across the state as a demonstration of the Oregon Plan principles.

In 1998, a citizen-generated Ballot Measure amended the Oregon Constitution to provide 7.5 per cent of lottery revenues to support salmon restoration as well as a single agency to administer the restoration funds. The new agency, OWEB, was created in 1999, expanding the coordination and prioritisation responsibilities of the Governor's Watershed Enhanced Board and extending responsibility for administering all watershed restoration and protection funds. As legislated, OWEB's objective is to provide funds and resources to support watershed councils, soil and water conservation districts in an effort to 'sustain an enduring, high capacity local infrastructure for conducting watershed restoration and conservation in Oregon'.[2] OWEB is tasked with and designed to implement a habitat restoration project based on the principles and practices suggested by the Oregon Plan. As of March 2013, more than 100 independent watershed councils covered most of the state of Oregon providing venues for local landowners to make deliberative decisions about habitat recovery (Figure 4.1).

OWEB and the Oregon Plan are tightly interconnected as OWEB is designed to provide support for citizen participation and influence over restoration decisions and outcomes (Rosenberg 2005: 102). While OWEB can be considered an agency type of organisation within the framework of this book, its connection to a policy that requires voluntary and local actions to implement its mission at the watershed level also contains seeds of a partnership organisation, suggesting that OWEB deviates in substantial ways from the ideal types presented by the framework. Do these deviations affect the creation and practice of the organisation in meaningful ways? And if so, in what ways do they affect the success of OWEB? These questions are explored after the design is considered in some detail.

4.3 INSTITUTIONAL DESIGN OF OREGON WATERSHED ENHANCEMENT BOARD

Table 4.1 gives the rule types of the Oregon Watershed Enhancement Board.

Table 4.1 Rule types of the Oregon Watershed Enhancement Board

Rule types	Rules in case study
Authority rules	State mandate for single issue (salmon stream restoration); coordinating and non-regulatory roles; expertise situated in managing grants and reporting results.
Aggregation rules	State coordinating body for stream restoration; appointed board where majority rules; local decisions made by consensus of involved stakeholders.
Boundary rules	The state of Oregon and local watershed councils officially approved by existing local government; membership rules defined by each watershed council.
Information rules	1. State level: biennial progress reports for legislature and citizens; narratives highlighting projects and individuals. 2. State and federal level: annual reports to funding and legislative agencies.
Pay-off rules	Distributes funds from state lottery and federal initiatives for project specific grants to both watershed councils and other state agencies.

4.3.1 Authority Rules

In preparation for creating OWEB, then-Governor John Kitzhaber engaged both private and public sector stakeholders to ensure that they understood and agreed to a voluntary watershed restoration approach. When the Governor brought the proposal to the legislature, he was supported by many credible stakeholders and proceeded to negotiate creation of a new state organisation to oversee implementation of the Oregon Plan through voluntary and local efforts. The state legislature determined the terms of authority for OWEB to implement the Oregon Plan, and the agency remains accountable to the legislature and the Governor for biennial reporting on the Oregon Plan.

One limit to the authority of the organisation is the insular process used to select and appoint a citizen Advisory Board (Board) (all Oregon state agencies have oversight boards or commissions, so this was not a unique situation at OWEB). Members are identified and appointed by the

Governor and then approved by the Senate for four-year, staggered terms. Neither OWEB nor the public is involved in the appointment process although legislative standards are in place to ensure that a wide variety of stakeholders will be represented on the Board.

OWEB's authority over state budget allocation and appropriation is also limited. Oregon citizens vote regularly on distribution of Oregon Lottery funds and currently about 7.5 per cent of all revenues are targeted for OWEB, which must use at least 65 per cent of allocated revenues for capital expenditures (for example, culvert replacements, equipment).[3] In practice, the legislature also exercises control over how the other 35 per cent of lottery funds are utilised and there have been instances when restoration funds have been directed to other agencies and away from OWEB priority projects.

OWEB's primary mechanism for accomplishing its mandate involves administration of a grant programme funded through the Oregon Lottery, games of chance that generate revenues in excess of US$1 billion a year meant to support state government initiatives such as OWEB (Esteve 2013). OWEB funds are distributed through competitive grants for projects that support voluntary efforts by community members and landowners to restore streams and other water courses (springs, ponds, and so on) in their local watersheds. OWEB is responsible for managing project applications, application reviews and funding recommendations to the OWEB Board, which makes the final decisions on grant awards.

In this aspect of institutional design, OWEB is most clearly an agency. The organisation has a very specific and narrow mandate to implement the Oregon Plan through local and voluntary watershed councils. The direction to implement through units not under their control, however, suggests OWEB might also be considered a coordinating institution, expected to work with citizens, as well as other state and federal agencies responsible for water and fish management to ensure that efforts are coordinated across those working on the issues. This concept of a coordinating institution is particularly important in light of the fact that OWEB has no actual regulatory enforcement power and can only accomplish its objectives via voluntary actions by local stakeholders facilitated by a watershed council. All these groups are represented on the Advisory Board, which sets priorities and evaluates the progress of OWEB in meeting its goals. While the functions of OWEB are limited to funding and training others to carry out activities likely to restore salmon habitat in Oregon, they have control over the framing of problems, the criteria for choosing projects and the distribution of funds.

4.3.2 Aggregation Rules

OWEB is the single state agency 'that provides grants to help Oregonians take care of local streams, rivers, wetlands and natural areas' through dedicated lottery funds.[4] However, there are distinct implementation structures under which it meets its responsibilities including oversight by both its own Board and an Independent Multidisciplinary Science Team, budgetary authority and the independence of the organisations through which the mandate is carried out.

OWEB is overseen by a Board comprised of citizens as well as tribal and agency representatives, all of whom have some stake in seeing salmon habitat managed. Eleven of the 17 members are voting members, one is a tribal representative and the other five are citizen representatives. Together, these six constitute a majority of the Board's voting members. These six individuals are selected through a specific department within the governor's office and approved by the state senate. The other five voting members are drawn from the citizen boards or commissions (each of which has a separate vetting process) of five state agencies, all of which have some responsibility in implementing the Oregon Plan (Board of Forestry, Board of Agriculture, Environmental Quality Commission, Fish and Wildlife Commission and Water Resources Commission). These five representatives are selected based upon evaluations between the director of OWEB and the director of the specific state agency, their personal interest in serving and ability to make the necessary time commitments. The five non-voting members represent federal natural resource agencies and the Oregon State University Extension Service, which provides outreach expertise and venues across the state. The board is the primary decision-making body in the allocation of funds and resources to support OWEB objectives. Board decisions are made through a simple majority vote. While potentially a set-up for politicisation, the current process for appointing members and conducting meetings has resulted in OWEB Boards that appear to work together even on contentious issues, providing a process and venue that allows contradictory positions to be heard and considered. In reading through the minutes of OWEB Board meetings over the years, it is possible to see how different perspectives are elicited and considered through Board discussion and public comment periods.

As part of the initiating legislation for OWEB, an Independent Multidisciplinary Science Team was created to advise OWEB, and the Governor and legislature on implementation of the Oregon Plan. Up to seven scientists with 'recognized expertise in fisheries, artificial propagation, stream ecology, forestry, range, watershed and agricultural management'[5] were to be jointly appointed by the Governor, President of the

Senate and Speaker of the House of Representatives. Appointments of the Independent Multidisciplinary Science Team require a unanimous decision by appointing authorities in an attempt to depoliticise scientist appointments. The Independent Multidisciplinary Science Team provides an annual report to the Governor and Legislature including any recommendations for changes or adjustments to OWEB and others responsible for implementing the Oregon Plan.

OWEB has control over decisions primarily through the allocation of funds to projects it deems to be worthwhile. OWEB has relatively wide latitude in the way it chooses to meet its mandate of providing resources to 'improve the conditions of Oregon's streams and rivers as well as wetlands and other natural areas'.[6] And the legislative statute gives OWEB authority to approve all standards and guidelines for grant applications, evaluation criteria and distribution of funds (interagency agreements, contracts for services, and so on). These two authorities – what to fund and how to decide what to fund – create a space where OWEB can exert state-wide priorities on salmon restoration efforts.

However, OWEB must rely on other organisations, particularly watershed councils, to carry out its mandate. Local citizens and landowners are empowered to define the specific environmental concerns in their watershed and ultimately to find ways to solve the problems as they characterise them, not follow state or federal agency templates. This reliance on devolved authority is central to the Oregon Plan, described by Governor Kitzhaber as 'designed to involve, empower, and incent private landowners to make voluntary commitments to watershed restoration and habitat restoration' (State of Oregon, Office of the Governor 1998).

In this institutional design feature, OWEB also has aspects of an agency organisation. While it is designated as the single state unit responsible for managing watersheds, it has active oversight from citizens and state and federal agencies who serve on the Board, scientists who serve on the Independent Multidisciplinary Science Team, and the Governor and legislature who appoint Board and Independent Multidisciplinary Science Team members. While legislation provides OWEB with a relatively stable funding base, the revenues from the State Lottery are variable and subject to capture by other state agencies. And, finally, while OWEB relies on independent units – watershed councils and others – to implement activities to meet its mandates, it exercises control over selection of projects and allocation of funds to manage the organisations implementing the activities.

4.3.3 Boundary Rules

While the geographical boundary of OWEB is the entire state of Oregon, its ability to get its mandate implemented relies on partnership organisations who voluntarily work in specifically characterised watersheds. The definition of a watershed is left to the local community organising the council. As seen in Figure 4.1, these watersheds range from relatively small basins to quite large sweeps of land.

Watershed council formation is initiated by local actors at a grass-roots level. In order for a watershed council to be recognised it must initially be approved by a governing body at the county level. Local government groups (that is, municipal, county) are tasked to bring residents and landowners together to form a watershed council or to designate an existing group as a watershed council. This local recognition is the first step to becoming an 'official' watershed council with access to OWEB funds and services. Oregon legislation directs the types of representatives to be involved with the watershed council, but does not designate exactly what constitutes a watershed council or area to be covered.[7] Those decisions are for council members to determine along with any operational, membership and decision rules. There are distinct incentives for local governments to encourage formation of a council, including access to funds that can be used locally to support restoration as well as economic development through jobs and infrastructure repair.

The idea of OWEB as a state-wide organisation is embedded in the design of the Board, with members who are expected to represent different areas of the state. Currently, however, eight of the 11 voting Board members are from the west side of Oregon. Board memberships shift regularly, so as recently as 2011 the membership was more evenly divided geographically with six members from western Oregon and five from eastern Oregon. In a recent review of the watershed project directory,[8] grants for restoration projects are distributed across the state with some exceptions for the mostly unpopulated areas in southeastern Oregon. There are definite concentrations of funded projects in urban areas around Portland and throughout the Willamette Valley but there are also concentrations in the northeastern corner of Oregon in the Grand Ronde Basin and along the south coast, both places with long histories of active watershed councils.

Currently, OWEB itself consists of 27 employees located in Salem, Oregon, with six additional staff at regional offices around the state. The expertise of employees reflects the mandate of OWEB with eight employees devoted to the grants programme, another three responsible for elements of the monitoring programme, and then two or more responsible for activities such as partnership relationships, reporting and data management. The expertise of employees clearly reflects the methods that OWEB uses to meet its goals through funding other units to perform the work. The field representatives typically have some type of technical expertise (for example, hydrology, wildlife conservation) although they are also responsible for working with individual watershed councils on grant writing, monitoring and reporting functions.

What makes OWEB less clearly an agency in this institutional rule set is the mandate that requires stakeholder and citizen involvement in its day-to-day business. This is amplified by direct oversight from private and public agency stakeholders through the entity of the Board. The skill set necessary for implementing mandates through other agencies is reflected not in the technical expertise usual at most water management organisations, but instead in skills related to organising and monitoring resource allocations as well as managing partnership relationships.

4.3.4 Information Rules

OWEB has an annual reporting responsibility to the Oregon legislature, the legislative fiscal office and the federal government (regarding federal funds). The reports to the state are based on 11 performance measures developed by OWEB in conjunction with the Oregon Progress Board (responsible for monitoring progress of all state agencies). Six of the performance measures relate directly to OWEB's mandated activities including four related to administration of grants and allocation of funds and another two related to customer service. The other five measures relate to the quality and quantity of fish and plant populations and habitat – all outcomes from activities not directly performed by OWEB.

Biennial reports are prepared for the legislature and citizens of the state to 'provide an update on the accomplishments and continuing efforts of people throughout Oregon to improve and protect clean water and recover and maintain healthy populations of fish and wildlife' (Oregon Plan for Salmon and Watersheds 2011). The report documents the amount of funding provided for various projects as well as outcomes for those projects, usually in the context of the performance measures (for example, miles of riparian habitat restored). One of the techniques OWEB uses to communicate its efforts is through 'watershed stories' that provide examples of local efforts, usually with a local 'hero'. There is a web page just for stories and they feature prominently in each biennial report alongside the technical and quantitative details. The use of narrative to integrate local knowledge and expertise of landowners suggests the division between OWEB's technical expertise (usually associated with monitoring both spending and outcomes) and their partners' experiential expertise (usually associated with on-the-ground restoration initiatives).

OWEB again displays institutional features of both an agency and a partnership. They provide annual reporting on indicators of success determined by another entity (the Oregon Progress Board) to both state and federal funding sources. But OWEB also gathers and disseminates information about local projects, experience and lessons learned to its funders and partners, as well as others interested in watershed council activity. To demonstrate success in ongoing engagement with locals, OWEB uses a narrative form with mostly non-rational appeals to define success.

4.3.5 Pay-Off Rules

As described above, the operating budget for OWEB primarily comes from Oregon Lottery revenues although there are also increasing sources of funding through federal programmes. For example, during the 2001–2003 biennium, OWEB secured US$74.6 million in private and federal funds for watershed restoration to be used in addition to the state-appropriated funds. In recent years, the economic downturn has affected both lottery sales and federal funding with a resulting decline in total funds available. OWEB's annual budget varies due to its reliance on the fluctuating revenues from the lottery, federal appropriations and grant funding, making it difficult to make (and sometimes keep) long-term commitments.

In 2004, OWEB set and continues to meet a limit of 6 per cent of total budget for administrative costs. While there is an expectation that goals are regularly met or exceeded, OWEB is limited by state compensation rules, so performance bonuses are typically not available for employee incentives.

As described above, the amount of money OWEB receives from the state is limited by law as to its sources and its relative amount (that is, 7.5 per cent of lottery revenues). For the most part, these funds are moved out to local watershed councils or others working on specific projects. Funds have been allocated to every watershed council that has requested them. For continued funding, watershed councils have to demonstrate that they have been effective in using earlier funding, especially in helping OWEB meet performance goals related to restoration outcomes. There has been some predation of OWEB funding by other state agencies to backfill budget shortfalls, which makes it difficult for OWEB to fully plan for or commit to long-term projects. It also contributes to the search for external funding described above.

In this aspect of institutional design, OWEB is clearly an agency with a committed, although variable, budget to meet its mandate. OWEB has the capacity to seek additional funding outside state sources to meet its performance measures, which were set by the state of Oregon with no input from stakeholders. Performance measures reflect OWEB's success in working with local councils to implement restoration projects in both process (for example, allocation of funds) and outcome (for example, restorations completed).

As described, OWEB is definitely an agency type of organisation, with an obligation to provide funds for stream restoration services across the state. It has a distinct mandate that separates it from other state water agencies, a committed budget to meet its objectives, and extensive oversight, monitoring and reporting obligations. Expertise in the agency is specific to its mandate, with most people involved in grant management, education and outreach to partners, and monitoring and data management. However, because OWEB is the state agency responsible for restoration of salmon habitat it must be aware of and sometimes fund the efforts of other state resource agencies, including water resources, to meet the demands of the federal listing of Oregon salmon species under the Endangered Species Act. The interdependence between the Oregon Plan and OWEB puts it in a coordinating role, which is reflected most clearly in its Board members representing relevant state and federal agencies. And finally, OWEB implements its goals through partnerships with local and voluntary watershed councils, each an independent and autonomous entity. The culture of a partnership organisation permeates OWEB in its open hearings and extended outreach activities, funding of outside initiatives over which it exercises limited control, and representative decision bodies (the Board and the Independent Multidisciplinary Science Team).

4.4 THE POLITICS OF DESIGNING AND IMPLEMENTING OWEB

A window of opportunity opened in the mid-1990s for those in Oregon interested in addressing water management for ecological purposes, in this case for restoring habitat for threatened and endangered salmon species. Prior to the creation of OWEB, a governor's initiative (the Governor's Watershed Enhanced Board) hobbled along with a very small budget trying to introduce and spread ideas about local and voluntary decisions in spaces defined as watersheds, new ideas in water governance in the state and the western US. The window was shoved open with the proposal to list several species under the Endangered Species Act, an experience that Oregonians had lived through recently with the Endangered Species Act listing of the Northern Spotted Owl (*Strix occidentalis caurina*) in old-growth forests of the Pacific Northwest. The resulting conservation plan, the Northwest Forest Plan of 1994, was controversial in calling for steeply decreased logging in National Forests (Thomas et al. 2006). The listing and resulting plan changed timber practices, lost jobs and reduced federal timber taxes for local governments, dealing the economies of many Oregon communities a blow they have yet to recover from. Oregonians were not particularly interested in reliving the polarisation between those relying on natural resources for jobs and those such as environmentalists and scientists who saw an indicator species of a failing ecosystem.

To say that the Endangered Species Act has been controversial puts it mildly; the Endangered Species Act, the Clean Water Act and other environmental regulations were the result of a unique political time and their implementation has had many unanticipated outcomes and consequences. It is undeniable that the Endangered Species Act and similar legislation serve as powerful tools for the conservation and management of streams and rivers, but they have come with a price. The structure and format of US environmental regulatory policies such as the Endangered Species Act have resulted in a seemingly endless stream of litigation and 'analysis paralysis' in which some claim that the only real winners have been lawyers (Noss et al. 1997). Under the Endangered Species Act, each listed species must have its own unique recovery plan entailing a specific set of actions and conditions. Salmon populations have been further broken down into Evolutionary Significant Units, each of which requires a distinct recovery plan (McElhany et al. 2000). The branch of the federal government primarily tasked with salmon recovery, the National Marine Fisheries, for example, has been working nearly 20 years to issue what is called a Biological Opinion to implement a salmon recovery plan for multiple Evolutionary Significant Units in the Columbia River Basin. The draft Biological Opinions for all salmon recovery plans have faced constant challenge by groups spanning the entirety of the political spectrum. With 17 Pacific Northwest salmon Evolutionary Significant Units currently listed as threatened or endangered under the Endangered Species Act, only five recovery plans have been approved and put into place (NOAA 2013).

So, when it increasingly looked as if the Endangered Species Act listings of salmon would occur in the mid-1990s, then-Governor Kitzhaber brought together the people he would need to convince that there was a different path to species recovery than the years of legal wrangling needed to put a recovery plan in place. Governor Kitzhaber proposed that the state take the lead in recovery efforts by employing the Endangered Species Act § (4)(b)(1)(A), which says that listings may be deferred 'after taking into account those efforts, if any, being made by any state'. This represented the first time a state had spearheaded an Endangered Species Act recovery effort without federal intervention (Smith et al. 1997). After crafting a recovery plan with like-minded stakeholders based on local and voluntary actions by watershed councils, the Governor took it out on the road to the coastal communities most likely to be affected by the listing. With their support, the plan was approved and funded by the legislature. The National Marine Fisheries initially accepted the plan and did not list coastal salmon as threatened or endangered. Then, just days before the National Marine Fisheries had to rule on the decision, they sent a revised version of the agreement; the Governor just had to sign and the Oregon Plan could be launched.

However, Kitzhaber couldn't accept the terms of the revised agreement and drafted another version, 'pretend[ing he] never got the first one' (Collette 1998). He also sent a letter to then-Vice-President Gore describing the new vision, and the White House sent a team of negotiators to meet with the Governor and the National Marine Fisheries. In one week, a new deal was struck. In response, several groups including fishing associations and environmental groups sued the National Marine Fisheries, claiming there was no evidence the strategy would lead to recovered salmon habitat. A US magistrate ruled in the litigants' favour six months later.

Environmental groups who sued the National Marine Fisheries did not accept the premise that landowners would undertake voluntary actions to restore habitat. In the ruling, the Judge agreed that voluntary actions are speculative and therefore not a guarantee for successful restoration. Oregon and the National Marine Fisheries appealed against the decision and both lost their appeals. On 3 August 1998 Oregon coastal Coho were listed as threatened under the Endangered Species Act. (Since then, an additional ten Oregon species have been listed as threatened or endangered (Oregon Department of Fish and Wildlife 2013).) Surprisingly, several timber companies announced they would voluntarily comply with the strategy and even provide resources to support the effort. Backed by the federal decision that habitat recovery plans must be filed and approved by the federal government, not a local watershed council, most environmental groups agreed to come back on board to support the efforts of OWEB in implementing the Oregon Plan.

Even though the state lost its bid to demonstrate a new way to restore habitat in the face of the Endangered Species Act, its efforts created a grass-roots network of local community members, the private sector and environmental groups willing to work together to manage the recovery process. As quoted in Collette (1998), one salmon advocate who supported the use of OWEB to restore habitat describes the Governor as '[a] person who . . . understands that the government's not going to save salmon'. This is echoed by one of the Governor's staff:

> You don't recover salmon by issuing orders from Salem (Oregon's capital). But you don't recover them with orders from D.C. either. People on the ground have to want to do it. You can't make them do it. You get it done through long, hard work, stream reach by stream reach and gravel bar by gravel bar.

This commitment to local and voluntary decision making led volunteers to complete more than 1200 watershed restoration projects in the year between adoption and rejection of the Oregon Plan. They fenced off about 300 miles of stream banks as protection from livestock; planted thousands of trees along shorelines; and opened access to roughly 400 miles of streams for salmon spawning.

The outcomes of these processes indicated a local willingness to give the voluntary and local organisations a chance to address the wicked problem of salmon habitat restoration. These new governance ideas related to devolution and participatory democracy had been floating around the academic policy realm for a while, especially in relation to natural resource management (e.g. Fischer 2000, 2003, 2009; Guttman and Thompson 2004; Ostrom 1990, 2007). Early on, Governor John Kitzhaber used the regional platform of the Western Governors' Association to introduce ideas about participatory governance. He and a fellow western Governor (Mike Levitt, Republican, Utah) proposed a series of principles for collaborative governance which they called 'EnLibra' (Western Governors' Association 2009). Not everyone is convinced that devolved and collaborative processes are a good idea, and some are convinced this is another manifestation of a new world order ready to take over and destroy all that is good about the US (e.g. Stubbs 2003).

More seriously, and in further demonstration of the partisan politics associated with natural resource management, salmon recovery and the Endangered Species Act, even while OWEB continued its efforts to allocate resources to local councils for salmon recovery Republication legislators in the Oregon House of Representatives blocked the 2004 confirmation vote for the OWEB Director's second term. On his offer to step down, they threatened to withhold permission for the agency to apply for federal grants or even to use federal funds already secured (Steves 2004). The argument appeared to be about the use of OWEB funds for land acquisition, something that is within the legal authority for OWEB but appeared unclear to some local landowners. It was just around this time that new administrative rules for land acquisition were adopted after a formal development process and public review. It is very possible that there was both public and agency confusion over application of the new rules. Since 2004, OWEB has conducted extensive discussions with applicants, grantees and other landowners about the constraints and inefficiencies in the existing policy. New rules were created and implemented in 2010 to clarify the process, expanding eligible costs that can be covered by OWEB funds, and priority setting that includes inputs from local residents.

The design manifest in OWEB definitely brought ecological goals to the forefront of water management and did so in a way that supplemented rather than replaced the mandates of existing organisations for water quality, allocation, power generation, irrigation, and so on. The fact that OWEB is a non-regulatory agency with no actual enforcement power further separates it from other agencies. The decision power of the new agency was limited to a specific topic (salmonid habitat restoration) and to specific activities (funding others to meet the mandate), creating and pushing into a space in state and federal structures not already occupied. The bureaucracy of OWEB is small, its technical expertise is concentrated in allocating resources through competitive processes, and its actions and outcomes are transparently recorded in regular public meetings, reports and conferences. OWEB did not replace the existing water governance structure in Oregon; instead it became a resource that basins and others responsible for managing water could access for funds and/or assistance with community engagement.

Faced with an expensive and potentially impossible problem to solve, the Oregon Governor and legislature sidestepped political infighting and ideological battles by creating a new state agency that took on the mandate of salmon recovery through an institutional design that limited OWEB's power to direct other organisations, a large and mostly stable budget to which others had direct access, and a responsibility for monitoring and reporting the success of others. And this was done primarily through a focus on ecological goals set by a federal agency and operational implementation through local and voluntary councils who represented the people most likely to be affected by any changes in land use, practices or infrastructure needed to restore habitat across the state. Instead of creating a super-agency that served to direct state-identified and funded projects, OWEB acts as a well-funded pass-through organisation with funds for potential agency competitors, environmental litigants, and local landowners to conduct projects they characterise as important.

While OWEB has been successful in managing the polarisation over endangered species recovery plans through this design, it is not as clear that the institutional strategy has been as successful for salmon recovery.

4.5 EVALUATING OWEB

4.5.1 Introduction

As OWEB gains experience in administering its initiatives and watershed councils gain expertise in habitat restoration, ongoing assessment and information dissemination provide feedback and learning mechanisms for those involved, interested and/or just watching. As issues arise or conditions require, the legislature makes revisions to the OWEB mandate and frequent reporting allows everyone to quickly see the outcome of changes.

4.5.2 Coordination

OWEB was born in the crisis of an endangered species listing in a political environment of years of rancour over other listing issues (that is, the Spotted Owl). A Governor convinced that local solutions were the only feasible solutions to this wicked problem opened a window for framing a new organisation reliant on non-governmental partners for success. While many existing state and federal organisations were responsible for managing different aspects of water resources in the state, none was explicitly mandated to manage the restoration of salmon habitat, which affects almost all coastal waters as well as streams throughout Oregon. Although some existing water organisations were organised around hydrological units (for example, regional water basins responsible for managing water allocation rights), most were organised around sectors (for example, energy, agriculture, forestry).

Enter an organisation designed only to administer funds, build capacity in partners and report progress on the single topic of habitat restoration for salmon recovery. Primarily focused on building capacity of local watershed councils, OWEB also provides access to both state and federal funds as well as expertise and venues for engaging with a public increasingly trained in water resource issues and collective decision making.

While OWEB is designed as and meets many of the criteria of an agency type of organisation, it also operates as a partnership organisation with a resulting culture of outreach and engagement with everyone aligned with the single mandate of habitat restoration. It provides a visible location for all queries and initiatives relevant to salmon recovery, bringing together governmental and non-governmental organisations as well as individual landowners and citizens around a collective (although imposed from the federal government) enterprise. It provides resources to monitor required systems and expertise to write required reports, complementing (and maybe even replacing) existing obligations of other agencies. In this case, the authority of an agency is paired with the culture of a partnership to create a structure for addressing a wicked water management issue in Oregon.

4.5.3 Accountability

OWEB reframed salmon recovery, habitat restoration and water management as a local and voluntary responsibility. While there are agencies such as Oregon Fish and Wildlife responsible for solutions such as salmon hatcheries, there is agreement among all parties that technical solutions are only supplementary to the concerted effort needed to be made on the ground, on people's property, to restore miles of habitat. This includes not only a change of practice, but also a change of heart about personal commitment to change (one of the explicit EnLibra Principles).

As described above, OWEB is overseen by an Advisory Board consisting of governmental, non-governmental and individual stakeholders who are appointed by the Governor and state Senate. Non-voting members from relevant federal agencies also serve. This Board makes ultimate decisions about which projects get funded, general policy directions, and hears recommendations from other organisations (for example, the Independent Multidisciplinary Science Team) about changes needed to OWEB programmes. The Independent Multidisciplinary Science Team also serves as an independent body, reviewing both the relevant science of recovery and restoration as well as the progress of OWEB projects.

The Board meetings are open to the public with meeting agendas and minutes posted on the OWEB website.[9] The website also contains archives of all documents including annual and biennial reports, project trackers, links to the Oregon Plan, and supporting documentation for the creation and operation of OWEB (for example, relevant legislation).

While OWEB is responsible as an agency for reporting to the Oregon Governor and Legislature on its progress, the website is designed to provide watershed councils and citizens with access to information, networks and resources regarding salmon recovery and habitat restoration. Again, the culture of a partnership type of organisation is reflected in the way OWEB opens meetings, books and information to the wider community, especially the watershed councils.

4.5.4 Legitimacy

OWEB was created when the 'extreme event' of an endangered species listing for Oregon salmon was proposed. In this case, the state had recently been involved in protracted and polarised conversations about a different listing that resulted in a major recovery strategy that affected jobs and communities across the state (that is, the Northwest Forest Plan of 1994). Given the animosity between environmentalists and others dependent on natural resources, the Governor began development of the new approach with extensive counsel and input from public and private sectors as well as individual landowners. He approached development with a set of principles related to local and voluntary responses to wicked problems that he had been promoting as a guide for managing natural resources in settings with multiple perspectives, agendas and needs (the EnLibra Principles).

These principles are enacted in the way that OWEB interacts with its constituents (both public and private) with requirements for and practices of open meetings, extensive consultation and inclusion of multiple perspectives on the Advisory Board. The principles are also enacted in the recommendation for watershed decision making through deliberative processes and consensus. For the most part, OWEB is not separated or distinguished from this idea of deliberative practice; the success of OWEB is seen as analogous to the success of watershed councils and especially to their use of deliberative processes.

OWEB and its Board have designed (and continue to refine) methods for allocating resources to both public and private parties, criteria for assessing the use of those resources and methods for communicating with all interested parties. From a close reading of Board meeting minutes, it is clear that parties to habitat restoration regularly bring concerns about OWEB policies and practices to the Board. It is equally clear that there are usually multiple perspectives to each of these concerns, and the meetings are structured to provide opportunities for all to be heard. The Board makes regular recommendations for changes to policies and practices, some of which are then codified in statute.

The partnership culture and organisational design of OWEB have created both input and output legitimacy through the development of multiple venues for providing feedback, learning about results and opening the agency to a wide spectrum of perspectives, all of them working to restore the coasts and streams of Oregon.

4.5.5 Environmental Effectiveness

A recent review of the salmon recovery tracker maintained by the Oregon Department of Fish and Wildlife[10] provides no evidence that salmon populations are recovering. While there is some upward trending of a few species in particular places, for the most part numbers are not increasing. There may be many reasons for this, including ocean conditions and predation. However, while federal landowners are in the process of developing or have completed recovery plans for their properties (as required by the Endangered Species Act) and watershed councils are working on local restoration projects (as required by OWEB), there is no evidence that local and voluntary watershed councils are the right organisational strategy for the recovery of salmon in Oregon waters. Even more interestingly, there has been no significant outcry or demand by legislators or landowners to try something else. There is still pending litigation associated with the recovery plans and potential controversial restoration and management strategies such as dam removal; but for now, the state is sticking with the ideas behind the Oregon Plan (including OWEB) as the way forward in managing stream restoration for salmon recovery.

4.6 CONCLUSIONS

OWEB is clearly an agency type of organisation although it relies on a partnership structure to get its responsibilities accomplished. And while it has state-wide geographical responsibility, its work is done in specific watershed locations. The agency structure is limited to the skills needed to allocate resources for local projects, monitor progress of those projects and report to its constituents, although its mandate is the broader recovery of salmon habitat and recovery of salmon in Oregon.

These tensions appear to be ameliorated, at least in part, by the underlying principles used to create, implement and maintain the agency to treat its constituents as partners: with the ideas that large problems need both grass-roots engagement and top-down commitment, especially in funding; that local solutions can be additive and maybe even cumulative in solving regional problems; and that expertise resides not just in skills but also in different perspectives brought to a project. No one expected salmon recovery to be quick but many expected it to be as painful as other experiences with Endangered Species Act listings, with lawsuits and protests, job losses and business shut downs. Instead, OWEB experimented with the concepts of devolved authority and participatory governance to extend decision making and resources to landowners, companies and individuals working together in watersheds which they define themselves on problems they decide are important to habitat restoration. The practice of monitoring and reporting embedded in the OWEB design allows all who are interested to track the success or failure of both the salmon and the watershed councils.

NOTES

1. R. Lackey, personal communication, 1 April 2013.
2. Available at http://www.oregon.gov/OWEB/Pages/wsheds_councils_main.aspx.
3. Available at http://www.oregonlottery.org/About/Lottery101/HowareFundsAllocated.aspx.
4. Available at http://www.oregon.gov/OWEB/GRANTS/Pages/index.aspx.
5. Available at http://www.oregonlaws.org/ors/541.914.
6. Available at http://www.oregon.gov/OWEB/Pages/about_us.aspx.
7. Available at http://oregonwatersheds.org/ors-541.388.
8. Available at http://oe.oregonexplorer.info/RestorationTool/.
9. Available at http://www.oregon.gov/OWEB/Pages/index.aspx.
10. Available at http://www.odfwrecoverytracker.org/.

REFERENCES

Collette, C. (1998), 'The Oregon way', *High Country News*, available at http://www.hcn.org/issues/141/4548/print_view (accessed 1 March 2013).

Esteve, H. (2013), 'Lawmakers like Oregon Lottery's money, just not how it's made', *Oregonian*, 4 March, available at http://www.oregonlive.com/politics/index.ssf/2013/03/lawmakers_like_oregon_lotterys.html (accessed 1 March 2013).

Fischer, F. (2000), *Citizens, Experts, and the Environment: The Politics of Local Knowledge*, Durham, NC: Duke University Press.

Fischer, F. (2003), *Reframing Public Policy: Discursive Politics and Deliberative Practices*, Oxford: Oxford University Press.

Fischer, F. (2009), *Democracy and Expertise: Reorienting Policy Inquiry*, Oxford: Oxford University Press.

Guttmann, A. and D. Thompson (2004), *Why Deliberative Democracy?*, Princeton, NJ: Princeton University Press.

Huntington, G., W. Nehlsen and J. Bowers (1996), 'A survey of healthy native stocks of anadromous salmonids in the Pacific Northwest and California', *Fisheries*, **21** (3), 6–14.

Lackey, R., D. Lach and S. Duncan (2006), 'Wild salmon in western North America: the historical and policy context', in R. Lackey, D. Lach and S. Duncan (eds), *Salmon 2100: The Future of Wild Pacific Salmon*. Bethesda, MD: American Fisheries Society, pp. 13–55.

McElhany, P., M.H. Ruckelshaus, M.J. Ford, T.C. Wainwright and E.P. Bjorkstedt (2000), 'Viable salmonid populations and the recovery of evolutionarily significant units', US Department of Commerce, NOAA Technical Memorandum NMFS–NWFSC–42.

National Oceanic and Atmospheric Administration (NOAA) (2013), 'Salmon and steelhead listings', NOAA Fisheries and PNW Regional Office, available at http://www.nwr.noaa.gov/protected_species/salmon_steelhead/salmon_and_steelhead_listings/salmon_and_steelhead_listings.html (accessed 21 September 2013).

Noss, R.F., M.A. O'Connell and D.D. Murphy (1997), *The Science of Conservation Planning, Habitat Conservation under the Endangered Species Act*, Washington, DC: Island Press.

Oregon Department of Fish and Wildlife (2013), 'Threatened, endangered, and candidate fish and wildlife species', available at http://www.dfw.state.or.us/wildlife/diversity/species/threatened_endangered_candidate_list.asp (accessed March 2013).

Oregon Land Conservation and Development Commission (1990), 'Governor's Watershed Enhancement Board state agency coordination procedures guide', available at http://www.oregon.gov/LCD/docs/sac/gweb_sac.pdf (accessed April 2014).

Oregon Plan for Salmon and Watersheds (2011), '2009–2011 biennial report', available at http://www.oregon.gov/OWEB/Pages/biennialreport2011.aspx (accessed March 2013).

Ostrom, E. (1990), *Governing the Commons*, New York: Cambridge University Press.

Ostrom, E. (2007), 'A diagnostic approach for going beyond panaceas', *Proceedings of the National Academy of Sciences*, **104** (39), 15181–15187.

Rittel, H. and M. Weber (1973), 'Dilemmas in a general theory of planning', *Policy Sciences*, **4**, 155–169.

Rosenberg, S. (2005), *Watershed Restoration in Western Oregon: Landowners, Watershed Groups, and Community Dynamics*, Eugene, OR: University of Oregon.

Smith, C.L., J.D. Gilden, J.S. Cone and B.S. Steel (1997), 'Contrasting views of coastal residents and coastal Coho restoration plans', *Fisheries*, **22** (12), 8–15.

Spokane Spokesman Review (1995), 'US to pay for salmon recovery taxpayers, not ratepayers, to cover most of cost', 15 March, available at http://www.spokesman.com/stories/1995/mar/15/us-to-pay-for-salmon-recovery-taxpayers-not/ (accessed March 2013).

State of Oregon, Office of the Governor (1998), 'WGA Enlibra Speech', speech by Governor Kitzhaber, 4 December, available at http://archivedwebsites.sos.state.or.us/Governor_Kitzhaber_2003/governor/speeches/s981204.htm (accessed 22 September 2013).

Steves, D. (2004), 'Head of environmental agency steps down', *Eugene Register Guard*, 10 July, available at http://news.google.com/newspapers?nid=1310&dat=20040410&id=GFFWAAAAIBAJ&sjid=uusDAAAAIBAJ&pg=3392,2298660 (accessed 1 March 2013).

Stubbs, R. (2003), 'EnLibra: a plan to destroy America from the inside out', *News with Views*, available at http://www.newswithviews.com/BreakingNews/breaking17.htm (accessed 1 March 2013).

Thomas, J., J. Franklin, J. Gordon and N. Johnson (2006), 'The Northwest Forest Plan: origins, components, implementation, experience and suggestions for change', *Conservation Biology*, **20** (2), 277–287.

Western Governors' Association (2009), 'EnLibra', available at http://www.colorado.edu/AmStudies/lewis/west/enlibra.pdf (accessed 1 March 2013).

5. Partnering for success in England: the Westcountry Rivers Trust

Hadrian Cook, David Benson and Alex Inman

5.1 INTRODUCTION

An official government document from the United Kingdom (UK) proclaims that:

> A vibrant participatory democracy should strengthen our representative democracy. The third sector – through charities, voluntary organisations and social enterprises – has much to offer from its traditions of purposeful altruism and selfless volunteering. Equally, we believe that political activity is a worthwhile and essential part of British life, and we want to restore people's faith in politics. (Department of Communities and Local Government 2008)

Changes in strategic political thinking in the UK have consequently witnessed official government policy moving towards the engagement of non-statutory non-governmental organisations (NGOs) in environmental management. Although these changes predate the general election of May 2010, that replaced the Labour government of Gordon Brown, they have become more salient under the Coalition led by David Cameron (Conservative) and Nick Clegg (Liberal Democrat). Under the Prime Minister's 'Big Society' agenda in particular, the perceived 'virtues' of the voluntary (or third) sector include direct and positive citizen participation, reducing ties with bureaucratic regulation and promoting more cost-effective, local service delivery. While these priorities may clearly be viewed as part of a 'neoliberal agenda' the long-standing role of voluntary action in the UK (Cook and Inman 2012) has shown it to be both effective and trusted when working with all kinds of economic actors. The key word is 'trust', and provided that factors such as staff continuity, institutional culture and, naturally, adequate financial and other resources are well maintained, NGOs are strongly positioning themselves as a major player in public participation, for example, in river basin governance.

For many decades prior to the introduction of the European Union (EU) Water Framework Directive 2000 (European Communities 2000),

UK water management was characterised by its closed, consensual and agency-led approach to governance (Cook 1998). Ironically, prior to the 1940s, water governance was actually predicated on strong local control with most management tasks, primarily water supply and sewage treatment, allocated to local government authorities (Bell and McGillivray 2000). Other tasks such as pollution control, drainage and fisheries then came to be institutionalised at the catchment scale within a plethora of rather technocratic agencies. Rationalisation of this approach began with the introduction of the Rivers Board Act 1948 that established a nationally consistent system of river boards, based on catchments. River authorities then assumed control over most water tasks, apart from sewage treatment and drinking supply (ibid.). In the early 1970s, water governance in England and Wales underwent a process of regionalisation, whereby tasks were reallocated to regional water authorities with jurisdiction over specific river basins (Haigh 2005: 4.1). Significant issues arising from the inability of these bodies to tackle chronic water pollution precipitated the Water Management Act 1989. The Act created multiple privatised water companies to provide services (Cook 1998), while a new national government agency, the National Rivers Authority (subsequently incorporated into the Environment Agency in 1996), was established to enforce water pollution legislation. Thus, water governance had by this point become strongly associated with more agency-controlled forms of consensual management with limited roles for the 'public' and non-industry actors. Introduction of the Water Framework Directive in 2000 in some ways began to change this approach, through its provisions for public participation in river basin management planning (Benson et al. 2012), for Article 14 of the Directive requires that member states encourage the involvement of all stakeholders in the implementation process. However, the process has been widely perceived as too top-down in England and Wales, with the Environment Agency controlling it via limited collaboration between actors (Fritsch and Benson 2013). Regionalisation under the Water Framework Directive process has, however, occurred in parallel to rising activity at the sub-regional level, with the widespread growth of partnership-type organisations that often have little or no connection to river basin planning (Cook et al. 2012; Benson et al. 2013). Although reasons for this growth are multifaceted, one driver has been a local-level reaction to the perceived historic inadequacy of agency led water management in solving pollution and habitat loss issues (Benson et al. 2013). The work of one such river basin organisation, the Westcountry Rivers Trust, is the subject of this chapter. We can understand its characteristics through the typology of institutional arrangements, based on those devised by Elinor Ostrom (see Huitema and Meijerink, Chapter 1 in this volume).

5.2 THE WESTCOUNTRY RIVERS TRUST

One successful example of a partnership river basin organisation (RBO), as described in Chapter 1 of this book, which has emerged from this paradigm shift is the Westcountry Rivers Trust (WRT) that serves the South West of England. The WRT is an environmental charity[1] established in 1995 to 'secure the preservation, protection, development and improvement of the rivers, streams, watercourses and water impoundments in the Westcountry and to advance the education of the public in the management of water' (WRT 2012a).

The WRT provides a good case study of a partnership RBO for several reasons. Firstly, it is one of the longest-established partnership organisations of its type in the UK, so its impacts can be studied over a long temporal scale. Secondly, over this period it has been highly successful in mobilising resources, building collaborative relations and addressing environmental problems. Finally, the WRT provides a model of partnership development for 'lesson drawing' (Benson and Jordan 2011) by other contexts, both in the UK and internationally. Indeed, the WRT's experience was reported in a UK Government House of Lords report as an exemplar of 'good practice' (House of Lords 2012). In these respects, the WRT provides a useful heuristic device for learning on how to initiate, develop and sustain a successful partnership RBO.

As we go on to discuss throughout this chapter, the WRT performs advisory and brokerage roles based around integrated catchment management principles. Primarily, the Trust delivers advice to farmers and riparian owners in implementing management change, ideally so as to improve profitability and diversification and to restore ecosystem function in catchments in South West England. Although the WRT has proved successful in leveraging resources from multiple sources, including the EU, to support this approach it is essentially a bottom-up initiative based on consensus decision making between diverse partners. A strong emphasis is placed on securing the voluntary collaboration of such stakeholders alongside the utilisation of local knowledge in decision support. Another approach increasingly utilised for achieving its aims for integrated catchment management is through facilitating payments for ecosystem services (PES). This approach is complementary to statutory regulation (URS/Scott Wilson 2011). Under the polluter-pays principle, there is regulation or market compliance (permit trading being an example), whereas PES involves direct payments to the 'service supplier' from a beneficiary. The WRT, as we discuss below, is proving adept at broking such 'win–win' opportunities to support more effective catchment management.

In order to better analyse how this success has been achieved, the rest of this chapter adopts the following structure. Section 5.3 provides an overview of the case via its physical, economic and social context. Section 5.4 gives an overview of the historical evolution of the WRT. Section 5.5 analyses the institutional characteristics of the organisation to understand its design. Discussion is provided on the actors propagating its development in respect of their 'political' strategies. Consideration is given to how well the WRT functions, not only in a democratic sense but also in terms of coordinating between multiple partners to generate positive environmental, economic and social outcomes. The analysis is mainly based on primary and secondary data sources such as websites, original documents, official reports and personal correspondence. Some interview data with key actors is also included. Section 5.6 provides our conclusions.

5.3 CONTEXT

The Trust's operations are focused on Devon and Cornwall in South West England, where multiple threats exist to water resources, hence providing significant drivers for integrated catchment management. Agriculture and sewage pollution, alongside industrial pressures present major obstacles to effective resource management. However, it is diffuse pollution of rivers and other water bodies that remains an apparently intractable problem. Agriculture employs a higher proportion of the local workforce than the national average across the South West region of England, with both counties economically reliant on the industry and associated sectors such as food and beverage production. The region has a mild, largely frost-free climate, high rainfall and variable land quality, and includes areas suitable for dairy, meat and arable production. As a result, non-point-source pollution from fertiliser application (both artificial and organic) in addition to high levels of animal stocking provide a particular threat to river and coastal water quality in some catchments; most specifically, the 975 km^2 catchment of the Tamar which contains some 139 km of river (see Figure 5.1). Another factor is the historical legacy of continuing contamination from intensive metalliferous mining (including zinc and arsenic) and china clay production (Cornwall Rivers Project 2008) which poses ongoing problems for water managers. Point-source pollution from wastewater treatment plants and modification of waters from industrial activity also provide challenges.

Drivers for water quality improvement in South West England are legal and regulatory, economic and cultural. Improvements to the ecological status of South West waters are mandated under the EU Water Framework Directive (European Communities 2000). The South West River Basin District covers 21 000 km^2 that includes Cornwall, Dorset, Devon and parts

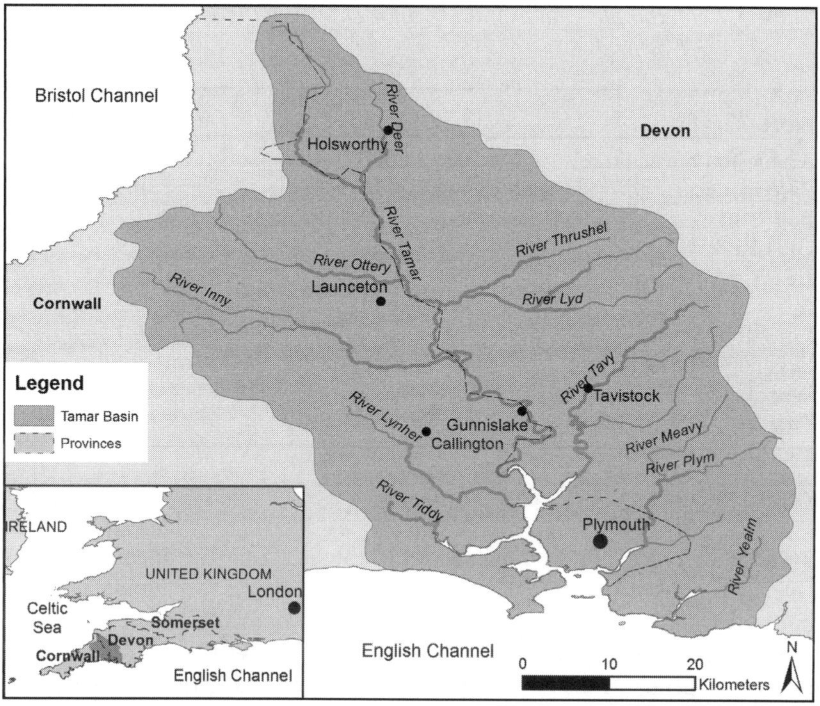

Figure 5.1 *The river channel network in the Tamar catchment in the South West of England with main towns*

of Hampshire, Somerset, Wiltshire and the Isles of Scilly. Environment Agency estimates suggest that only 42 per cent of surface water bodies and 57 per cent of groundwater will reach 'good ecological status' by 2015 (Environment Agency 2009). Devon and Cornwall are also reliant economically on the tourism industry, with nearly 15 per cent of total employment in Cornwall and the Scilly Isles within this sector (ONS 2012). With leisure activities such as fishing, surfing, sailing and visiting beaches highly popular for tourists, there remains a strong incentive to improve bathing water quality. Another critical driver in this respect is drinking water quality. In the Tamar, companies such as South West Water are beginning to take a lead in source drinking water protection through supporting better 'upstream' on-farm management practices to reduce their 'downstream' processing costs under PES (South West Water n.d.). In addition, as we describe below, generating ecosystems services through land management practices has been encouraged through EU and government funding as a means of supporting the local economy in Devon and Cornwall, where

Table 5.1 Example land use: the Tamar catchment in Devon and Cornwall

Land use type	Area (km^2)	% of catchment
Arable and horticulture	372	20.9
Bare ground	16	0.9
Bog	20	1.1
Grassland	1026	56.9
Heath	31	1.7
Inland water	4	0.2
Littoral sediment	2	0.1
Urban and gardens	118	6.5
Woodlands	223	12.6
Total	1812	100.0

Source: WRT.

average incomes are well below both English and European averages (Devon County Council 2012). Finally, another driver for change has been cultural with local actors taking the decision to collaborate in protecting their natural environment. As we later discuss, the WRT was formed by a concerned group of locals in response to declining water quality, thereafter emerging to become genuinely bottom up and community focused.

The Trust concentrates its activities on specific catchments within South West England, where water quality and habitat issues are significant. Although the WRT's activities cover a range of rivers in the region (including the Dart, Taw, Teign and Exe) its activities started in the Tamar Catchment in the 1990s.

The land use is a mixture of arable agriculture, grazing and forestry although overall it is grassland cover that predominates (Table 5.1). To the east is Dartmoor, to the west Bodmin Moor. Both include considerable areas of upland acidic peat soil. There are significant alluvial flats (wetlands) in the middle reaches of the catchment and the inclusion of 'littoral sediment' serves to underline the importance of coastal habitat in the region; the bulk of the urban area is the city of Plymouth.

5.4 EVOLUTION

River basin partnership organisations were relatively rare in the UK until quite recently (Cook et al. 2012; Benson et al. 2013). Hitherto, catchment management had been left to the devices of publicly owned regional water authorities who also provided water services. The Water Act of 1989 privatised the utilities and established statutory bodies that became regulators in the public sector: first the National Rivers Authority, then the all-embracing environmental regulatory body, the Environment Agency, in 1996. One of the first recognised examples of a partnership in England and Wales, the WRT, was established in 1995.

The origins of the Trust can be traced back to 1993 when a group of friends and conservationists met to express concerns over the ecological health of Westcountry rivers which, despite repeated government regulation and agency intervention, was declining (WRT 2012b). From this initial meeting, a decision was taken to form the Trust with Arlin Rickard appointed as the inaugural director. Several guiding principles were chosen to inform the Trust's main aim of improving water quality, along with enhancing habitats and biodiversity, namely:

- Good catchment management should result in benefit to the whole of the community as well as the natural environment;
- That this is best achieved through positive engagement in the community and by forming meaningful partnerships with all those who have an interest in, or an influence over, the quality of our rivers; and that
- The work of the Trust should always be backed up by sound scientific research. (Ibid.)

Scientific support for the integrated catchment management approach was secured via a partnership with the Royal Holloway Institute for Environmental Research. Once charitable status was secured, the trustees also then began to seek funding sources.

In this respect, the Trust has been successful in obtaining funding from the EU and UK government for three significant projects that have underpinned its early development in successive phases. Firstly, the Tamar 2000 SUPPORT project was a pilot scheme run by the Trust between 1996 and 2000 aimed at reducing diffuse pollution in the catchment. Finance was provided under Objective 5b EU regional structural funds, by the former Ministry of Agriculture, Fisheries and Food and various private sources. Under the project, around 300 land managers were supplied with individually tailored advice on combining environmental improvements with the attainment of economic benefits, for example, through reducing fertiliser use or employing soil conservation measures. A bespoke plan was produced for each farm incorporating this advice into a programme of best management practices, including the fencing of river banks subject to erosion, establishing riparian buffer zones and improved ditch management. Several strategic partners were engaged with, most notably the National Rivers Authority (1989–96) (and subsequently its successor the

Environment Agency), the Wetland Ecosystem Research Group and BDB Associates, who provided technical guidance to farmers.

The Trust then secured Phase I funding for the Westcountry Rivers Project, again from the EU and Ministry of Agriculture, Fisheries and Food under Objective 5b structural funds. Beginning in 2000, the 18-month project extended the integrated catchment management approach to the rivers Taw and Torridge in North Devon. Another 300 farmers and water managers were provided with advice on best management practices and landscape rehabilitation.

Phase II funding, for the Westcountry Rivers Project and the Cornwall Rivers Project, was secured in 2002. The Trust initially received an Objective 1 grant for £744 500 from the EU, which was match funded by the Department of the Environment, Food and Rural Affairs (successor to Ministry of Agriculture, Fisheries and Food) and supplemented by a further £364 870 from the private sector – a total of £1.86 million (approximately €2.2 million). This funding was utilised to extend the Phase I programme to Cornwall between 2002 and 2004. Initial catchments targeted under the project were the rivers Lynher, Fowey, Camel, Ottery, Inny, East and West Looe, Seaton, Neet, Strat, Fal and Cober. Further funding, which pushed total expenditures to £2.6 million (approximately €3 million), allowed extension of the programme until the end of 2005, to include additional rivers such as the White, Menalhyl, Gannel, Mevagissey and Caerhays. In total, the project covered 2455 km^2 of mostly agricultural land (Defra 2006). Around 870 farmers and riparian owners were contacted by the Trust and offered an individual site assessment, a free integrated management plan and ongoing advice. 'Best farming practice' information sheets were produced in order to disseminate advice, along with the use of 'champion farmers' demonstration sites as practical case studies. In its evaluation of the project, the Department of the Environment, Food and Rural Affairs (Defra 2006: 9) noted multiple benefits most notably that, on average, the advice given generated '£1369[2] per farm in efficiency savings, soil retention, [and] improved nutrient management'.

According to the WRT (2012b), the projects combined generated multiple outputs and physical outcomes. Between 1996 and 2006, *inter alia*, more than 1000 farmers and landowners were visited and given advice; multiple integrated river basin resource management plans were generated; over 100 km of riverbank was fenced to prevent erosion; over 260 km of river was surveyed with management actions taken; and 75 buffer zones were introduced.

With the initial objectives now complete, the Trust has expanded its operations to include several parallel projects that build on its earlier success (WRT 2012c). For example, the River Improvement Project, funded by the Catchment Restoration Fund, aims to enhance the ecological status of waters along the rivers Exe, Dart, Teign and Axe, and in the South Hams (Devon) and South Cornwall. A joint initiative with the River Exe and Tributaries Association is now undertaking work to promote salmon habitats across the Exe catchment. In addition, funding from INTERREG supports three further projects: the WATER (Wetted Land: The Assessment, Techniques and Economics of Preservation) project, the AARC (Atlantic Aquatic Resource Conservation) project, and the COLLABOR8 project (promoting regional economic prosperity and sustainability in North West Europe).

The UK agricultural industry is regulated *inter alia* through advice, health and safety legislation, incentives and codes of good practice aimed to minimise environmental damage through negative externalities. While demands to provide cheap, secure food production are less frequently acknowledged today, the water industry increasingly looks to farmers as environmental service providers.[3] A significant feature of current integrated catchment approaches is therefore PES, reflected in two further WRT projects. The Upstream Thinking project funded by South West Water (South West Water n.d.) provides a good example of how the Trust can act as an 'ethical broker' in bringing together a 'willingness to pay' for an environmental good, in this case by the water company, with a provider of these goods, namely farmers. Another example is the Westcountry Angling Passport Scheme which provides revenue to farmers and riparian owners from anglers while providing incentives to protect water resources. In this respect, the Trust can play an active role in identifying 'win–win' gains for stakeholders so that protecting the environment can contribute to profitability through creating new markets. Several other PES-type mechanisms are currently in development for carbon sequestration, country sports recreation (via the Country Sports Project, funded partly by the Rural Development Programme for England) and working with supermarkets on sustainable food production.

Bioregional planning is also promoted by the WRT. For example, any current analysis of sectoral interests would produce serious distortions towards the food production sector. The planning role is therefore provided through the adoption of 'aspirational maps' that produce a range of different land use and land management scenarios. These might focus on water quality (for habitat and consumption purposes), on habitat conservation per se and enhancement, on carbon sequestration, on recreation, and so on. Weighted geographic information system (GIS) overlays utilising environmental information in 'intelligent catchment design' are being employed to aid map production (WRT n.d.). These GIS layers comprise those for: clean and fresh water ecology; clean and fresh drinking water; water regulation for flooding; water regulation for drought; a climate layer (greenhouse gas regulation and air quality); recreation and tourism; and provision of habitat. These are combined as an environmental services model that is compared with the intensity of agriculture. As an outcome, there is a 6 per cent catchment area where land use in intensive agriculture conflicts with conservation requiring land targeting of potential 'hotspots' (ibid.).

The WRT also played host to the Tamar Catchment Area Partnership (T-CAP), which consisted of over 100 stakeholders representing around 30 organisations and groups. The Tamar catchment was chosen on the basis of its size and ability to operate with the Environment Agency's 83 English management catchments, whilst being small enough to allow stakeholder participation. The aim was to address failures in communication with stakeholders and its methodology was derived through creating stakeholder working groups that looked at water quality and quantity (both flooding and drought), carbon sequestration, habitats for wildlife, and recreation and culture. The groups consisted of service providers, including farmers and landowners, and customers of the services (groups and businesses), as well as experts. The methodology used GIS-generated maps to create multifunctional areas that are important in the provision of ecosystem services. Partners were then able to better appreciate the value of the catchment, and identify areas where complementary actions could occur, leading to cost–benefit improvements (Couldrick 2013).

5.5 INSTITUTIONAL DESIGN, POLITICAL STRATEGIES AND FUNCTIONAL PERFORMANCE

5.5.1 River Basin Organisation Institutional Design: From Agency to Partnership Working

In Chapter 1 (this volume), Huitema and Meijerink described how RBOs can be understood through the typology of institutional arrangements, based on Ostrom's rules (Table 5.2). The institutional approach to river basin management in England and Wales since the early 2000s has been primarily of an agency form due to the over-riding requirements of EU policy.[4] In England and Wales, the Environment Agency, as the government-designated competent authority, has determined river basin management planning activities under the Water Framework Directive. Ten regional-scale river basin districts[5] have been established to implement

Table 5.2 The different institutional features evident in the agency type river basin districts and partnership Westcountry River Trust

Rule types	River basin districts	Westcountry Rivers Trust
Authority rules	Centralised decision making	Collaborative decision making
Aggregation rules	Delegated central control, weak coordination	Consensus and strong coordination between actors
Boundary rules	Centrally determined participation	Voluntary and locally inclusive participation
Information rules	Centralised reporting, technocratic knowledge generation	Local knowledge and scientific evidence
Pay-off rules	Centralised objective achievement	Costs and benefits shared between actors ('win–win' solutions)

the Water Framework Directive, with authority rules centrally imposed by the Environment Agency. Mandates are limited to ensuring river basin management planning within the overall context of EU objectives for good ecological status of waters. In terms of aggregation rules, the Environment Agency has delegated some control over planning activities to the regional scale, but coordination measures with other organisations are weak. Planning in each river basin district is nominally subject to advisory direction by so-called River Basin Liaison Panels comprised of multiple stakeholders such as the Agency, government bodies, local governments, businesses and regional authorities (Benson et al. 2013). However, panel members are appointed by the Environment Agency which also chairs meetings and coordinates responses, meaning that their actual influence on decision making remains questionable (Fritsch and Benson 2013). Boundary rules also conform to an agency-type RBO. Geographical boundaries of river basin districts were centrally determined by the UK government in conjunction with the EU Commission, while the Environment Agency primarily adopts a technocratic management approach based on the application of engineering and scientific expertise. Stakeholder and public participation in planning is largely limited to obligations under the Water Framework Directive (ibid.). Similarly, information provided on the planning process by the Environment Agency is generally in the form of reporting to central authorities and the EU as part of the Water Framework Directive requirements, although this data is made publicly available. Information rules therefore reflect another critical aspect of an agency-type RBO (see Chapter 1, this volume). Finally, pay-off rules in this context are related to the ability of the Environment Agency to meet water quality objectives within centrally determined budgets, which are currently constrained by a fiscal crisis in central government. While this approach remains the current dominant paradigm, a combination of mistrust of officialdom by certain stakeholders and lack of faith in such agency forms of RBO to address chronic water management issues has led to the recent emergence of novel institutional forms that together map out an increasingly polycentric governance landscape at the sub-regional scale.

These novel institutions, we observe, more readily conform to the partnership form of RBO identified in Chapter 1. In their 2011 survey of English and Welsh 'watershed partnerships', Cook et al. (2012; see also Benson et al. 2013) note several defining features of this institutional shift away from the agency RBO paradigm. Although the number of groups detected was relatively small – only 39 in England and Wales – it was increasing. Here, very different forms of authority rules are evident. Organisations were primarily bottom-up collaborative initiatives with membership drawn from different sectors: 18 per cent of group participation was derived from private individuals, with public institutions accounting for 22 per cent. Many groups, moreover, had evolved in response to complex, 'wicked' problems such as diffuse pollution or habitat loss. In terms of aggregation rules, consensus-based forms of decision making were strongly promoted amongst these organisations, resulting in high degrees of trust and reciprocity amongst partners. Boundary rules were generally not centrally imposed amongst these partnerships, reflected in the mainly voluntary and localised nature of membership. Indeed, a majority were constituted as charities (46 per cent) or limited companies (36 per cent), with local stakeholder ownership highly prevalent. Information rules and pay-off rules also tended to put an emphasis on locally indigenous and scientific knowledge in decision making, while more evenly spreading the costs and benefits of collaboration through seeking 'win–win' outcomes for different actors. An example of the latter was the role of some groups in helping farmers reduce nutrient run-off through on-farm best management practices, thereby increasing farm profitability, but also improving water quality.

The WRT is perhaps an example par excellence of this new institutional form of governance. Firstly, in an authority rules sense, the Trust is non-hierarchical, collaborative and largely institutionally self-organising, that is, independent of state agencies. As discussed above, its focus is overtly on the sustainable development of ecological resources and local economies within the context of integrated catchment management. Secondly, aggregation rules of the Trust have evolved to develop consensual decision making. A variety of stakeholders, from farmers to industry actors, are now included in project design and delivery, for example, in the Upstream Thinking and Angling Passport initiatives. In this respect, the Trust may be considered highly representative of the constituency in which it operates. By being closer to the ground compared to agency organisations, and including stakeholders in deciding management solutions, it promotes a participatory form of democracy that can result in positive environmental, economic and social outcomes. Thirdly, partnership membership is maintained by boundary rules that incorporate full-time and voluntary members. The organisation employs a staff of 25 who collectively possess a range of expertise in:

- freshwater, marine, estuarine and terrestrial ecology and conservation;
- spatial data, mapping and modelling using GIS;
- catchment modelling and risk assessment;
- agricultural, land and environmental management.

This team is split into several sections in order to manage the operations of the Trust: a management division headed by the current Trust Director Dr Dylan Bright, a fisheries team, a catchment team and a finance team. Their work is supplemented by volunteers and scientific experts, who help in the implementation of the various projects outlined above. Partnership-type boundary rules are also embedded by the WRT's charitable status and ability to work effectively as an NGO. Under UK law, a charity is a voluntary organisation that provides benefit to the public. In institutional terms, charities are governed by a board of trustees responsible for management decisions, including expenditure. While some charity staff can be employed, overwhelmingly they are run by volunteers who donate their time. Fourthly, Trust information rules place an emphasis on local knowledge, although the integrated catchment management approach adopted provides a requirement for sound science as a basis for decision making. Consequently, while local information from farmers, businesses and the public is sought in project implementation, the WRT works closely with scientific experts and academic institutions to generate evidence. Finally, as the WRT is strongly embedded within the local communities in Cornwall and Devon, and includes these stakeholders in decision making, its pay-off rules ensure that their interests are prioritised.

One example of such a pay-off is the Trust's PES strategy. To summarise from Couldrick[6] and Bright,[7] PES as realised by the WRT is founded on five principles:

1. 'Bioregional planning': taking a holistic approach to catchment management, including economic dimensions and statutory regulation.
2. Community conservation: nothing can progress without local leadership and support; achieving 'win–win' situations is central to the philosophy.
3. Payment is via the market: a major source of funding for ecosystem services comes not through public expenditure (that is from European, UK or local revenues) but from the direct beneficiaries of environmental gain.
4. Working sequentially to engage the public.
5. An 'honest broker' is sought who mediates the flow of cash from beneficiary to 'service provider', particularly farmers and landowners. The WRT has fulfilled this role in terms of payment from the UK Government, the EU and South West Water.

Under this approach, costs and benefits are therefore distributed more widely amongst partnership members. This institutional model has proved inspirational for other partnerships groups across the UK, providing significant scope for transfer (Benson and Jordan 2011) to other contexts. The WRT has been the first and, to date, arguably the most successful of the affiliated bodies within the nationally organised Association of Rivers Trusts, now called The Rivers Trust. Due to its success in leveraging funding and extending the integrated catchment management concept across a significant area of the South West, the Trust has also since received recognition by government. Evidence provided by the Trust to the House of Lords Committee inquiry into EU water policy was incorporated into their Lordships' report to the UK Government (House of Lords 2012). Here, they state that:

> The WRT works in partnership with external individuals and organisations to share expertise and facilitate better information transfer. By collaborating with a whole range of stakeholders – ranging from individual businesses through to academic institutions, NGOs and government departments – the Trust aims to circumvent sectoral interests and encourages joint solutions to the complex environmental problems our society currently faces. (House of Lords 2012: 32)

5.5.2 The Politics of Organisational Design

Huitema and Meijerink (Chapter 1, this volume) also argued that organisational design and form could not be separated from the political nature of environmental management. Here, 'architecture' is understood in terms of the product of evolving strategic interaction between actors, the strategies of parties and individuals, natural and political windows of opportunity, and venue shopping. Some of these elements are evident in the successful development of the WRT.

Most significantly, development of the Trust has reflected windows of opportunity, in addition to the principles and political strategies of its founders; although arguably it is the 'non-political' nature of its operations that have promoted its success. The founders of the WRT identified that, despite the best efforts of statutory bodies and government departments, the integrity of the rivers was diminishing and impacting on fish stocks. This natural crisis, caused by governance failure, in turn determined the founding principles of good catchment management, community engagement and sound scientific research stated above. The founders also influenced the decision to seek charitable status for the organisation. It was obtained not only because this would provide tax-efficiency, but also because NGO status would engender trust among stakeholders (WRT 2012b). As an NGO with charitable status, the WRT is also well placed to mobilise funding from different sources for delivery of catchment management while working with multiple stakeholders in a non-political way. It is this apolitical status, gradually developed by the founders through strategic interaction between actors, which has helped to reduce conflict. In a context such as UK water governance, government agencies are often in regulatory opposition with groups such as farmers. But through learning to incorporate often competing values in collaborative decision making, the scope for conflict has been reduced through organisational design. Simply stated, the English word 'trust' is used in both senses, as a technical description for a charity or NGO, and as a noun that stresses virtues such as reliability, truth and strength. All are essential in stakeholder engagement.

One means of incorporating values has been through facilitation techniques such as workshops and farm visits. As mentioned above, to date over 1000 farmers and landowners have been engaged with (WRT and Rickard 2013). Particular use is made of expert facilitators to reconcile stakeholder views and: 'in that way not only "opinion" farmers, within the farming community, but also other key stakeholders (water companies, regulatory bodies and local authorities) have been involved' (ibid.).

This philosophy reflects the views of the WRT's first director and founding member, Arlin Rickard. Now Chief Executive of The Rivers Trust, he suggests that the WRT will fail where there is an unwillingness to engage with stakeholders and where insufficient funding stifles development, notably at a difficult time for the economy.[8] This situation will self-evidently compromise political relationships or operational frameworks, including a reduction in informal understandings, of channels of communication including educative and transformative processes, of

Memoranda of Understanding and, naturally, formal financial relationships manifest in a contract. Likewise, engagement with state agencies, as well as the public, is considered central.

5.5.3 Functional Performance

Chapter 1 in this volume also laid out a set of evaluation criteria to assess functional performance of RBOs. Criteria include the ability of organisations to coordinate between competing policy objectives, in addition to their accountability and democratic legitimacy. Using these indicators, it is possible to show that the WRT is a highly successful example not only of a partnership organisation but also of RBOs more generally that complements, rather than replaces, the functions of agency-led regional river basin districts.

RBO coordination, in Chapter 1, is related to the ability to reconcile public decisions. One significant feature of the WRT approach is that it is sufficiently inclusive, adaptive and flexible to overcome issues such as contradictory politics and spatial fit while exploiting synergies between water management and cognate policy areas. The ethical broker status and task-specific spatial functions of the Trust means that it provides an effective mechanism for coordinating between often competing policy sectors, that is, farming, regional economic development and environmental protection. In this respect, it appears to partially reconcile the promotion of intensive farming, driven by commercial and political demands, with the interests of the environment and other users of environmental services. A good example of this coordinating role is the PES schemes. The Upstream Thinking model brings together the consumers of the environmental good, namely South West Water and its customers, with 'producers' of clean water such as dairy farmers, to coordinate across previously divergent (and competing) sectors. Agency river basin districts, as discussed above, are bound by authority rules to focus on achieving good ecological status of waters. This obligation can place the Environment Agency in conflict with farming and industrial sectors, making such coordination problematic. Indeed, the UK government has recognised this fact through stating its intention to promote the subregional integrated catchment management model across the country in support of the Water Framework Directive process in its future policy (Fritsch and Benson 2013).

One criticism of partnership RBOs discussed in Chapter 1 of this volume is that they can, ironically, lead to reduced accountability to the public through becoming dominated by sectional interests such as environmentalists or landowners. This view, however, presents a rather narrow conception of accountability since, in reality in England and Wales, more 'public' forms of RBO such as the river basin districts are perhaps less accessible to ordinary citizens due to their dominance by agency actors (Fritsch and Benson 2013). While the Trust originally only included a small, exclusive number of local actors, it has gradually evolved to increase its public accountability. Opportunities for the public to become involved in the Trust's activities are now greater, while care is taken to include private actors such as farmers in decision processes. The T-CAP project is just one example of how this is achieved. Consequently, although not an entirely new form of deliberative democracy, the Trust has provided innovative water management mechanisms for engaging actors who may otherwise have been marginalised in a more top-down approach.

Success is also evident if we consider issues of legitimacy. As Chapter 1 argues, dialogue infers the need for bilateral communication with the public and use of local knowledge. As frequently stressed above, these aspects are placed at the heart of the Trust's operations. Stakeholders, ranging from individual businesses to farmers, schools, academic institutions, other NGOs and government departments, all share knowledge to better aid decision support. The Trust aims to circumvent sectoral interests by encouraging joint solutions. There is up-front stakeholder engagement through building networks and elucidating novel mechanisms (including economic measures) that link researchers and research users in policy formulation. In practical terms, the operation of the WRT has long involved a range of commonplace practices, including farm visits to engage with farmers and to advise on practical matters, stakeholder meetings, educational initiatives and the collation and interpretation of technical data, such as water quality. It was also hypothesised in Chapter 1 that partnerships may generate greater input legitimacy than agency RBOs, but that in contrast, due to the complexity of multi-actor decision making, they may provide lower output legitimacy. Depending on how output legitimacy is measured, however, the Trust would tend to disprove this proposition. Input legitimacy is high due to the inclusive nature of decision making. But, critically, outputs of goods and services are potentially more than those attainable under the higher-level river basin district process. Indicators of output 'success' – such as advice provided for land managers, plans produced, best management practices introduced, landscape restoration achieved, farm efficiency savings made and funds channelled to local actors through the various projects and innovative PES schemes – all show a positive trend over time.

5.6 CONCLUSIONS

In conclusion, the WRT approach has so far pioneered stakeholder engagement and it is an operational (and economically efficient) model based in a workable contractual framework, poised for extension and adoption elsewhere in UK catchment management. The institutional design, strategic principles, democratic make-up and flexibility of this partnership organisation may be considered key to its success. In institutional terms, the WRT's status as a charity has enabled it to successfully access a range of funding sources, primarily from the EU, that have driven the development of the organisation through a series of project stages. By functioning as a largely self-organising, non-hierarchical collaborative with a full-time staff supported by dedicated local volunteers, the Trust has also been able to mobilise flexibly in response to the complex management issues it faces. Including different stakeholders, from farmers to business interests and government agencies, in the decision making has also helped the effectiveness of this institutional form. As noted above, underlying philosophies have also driven the partnership success. Right from its initiation, the principles of community-based, scientifically founded integrated catchment management have served as key guiding objectives, determining strategic actions. More latterly, PES has opened up new opportunities to pursue management objectives. As a partnership organisation, the Trust has also provided avenues for democratic engagement by increasingly locating decision making close to stakeholders on the ground. In contrast to top-down agency models evident in other areas of UK water management, most notably the Water Framework Directive implementation process, it is able to better coordinate the often competing demands of multiple water users for a common purpose. Again, by searching out 'win–win' management solutions, beneficial to all stakeholders, the Trust performs a coordinating role that enhances the accountability, transparency and legitimacy of its operations.

Multiple features of the WRT approach therefore lend themselves to drawing lessons. Firstly, the WRT cannot be divorced from its long-standing community programme and, by UK standards, represents a flagship initiative in stakeholder engagement. Secondly, in respect of this point, it provides salient lessons for other partnerships on the value of generating trust between stakeholders and utilising it to affect outcomes. Finally, the Trust provides an exemplar on how to mobilise different forms of capital – financial, social and human – to address critical sustainable development issues and achieve positive outcomes. There remains no practical or theoretical reason why other kinds of organisations cannot emulate this experience. Elected local government organisations and wildlife trusts, or the

Farming and Wildlife Advisory Group, would seem positioned politically to occupy the same niche in the UK, although it is arguable as to whether they have the ready capacity to operate as an honest broker in catchment management. Perhaps the key issue after all is trust in any formal or informal contractual understanding. Legitimacy for NGOs lies in their mission, not-for-profit status and sometimes open membership schemes (see Cook and Inman 2012). In local authorities there is a presumed electoral mandate. New governance arrangements would have to embrace both, although this aspect is outside the scope of this chapter.

ACKNOWLEDGEMENTS

We would like to thank Arlin Rickard (The Rivers Trust), Dr Dylan Bright and Dr Laurence Couldrick of WRT, and Martin Ross of South West Water plc for their advice and information. Opinions expressed remain those of the authors.

NOTES

1. Charity no. 1135007, Company no. 06545646.
2. Around €1600.
3. Interview with Martin Ross, South West Water plc (2010).
4. The UK government has promoted Catchment Sensitive Farming which aims at encouraging farm best management practice through financial and technical assistance, although these schemes cannot be considered genuine RBOs (see Cook et al. 2012 or Benson et al. 2013 for further discussion).
5. One more river basin district, the Solway Tweed, is jointly administered with Scottish authorities.
6. Interview with Laurence Couldrick, WRT (n.d.).
7. Interviews with Dylan Bright, Trust Director (2010, 2012).
8. Interview with Arlin Rickard, founding member of The Rivers Trust (2012).

REFERENCES

Bell, S. and D. McGillivray (2000), *Environmental Law*, 5th edn, London: Blackstone.

Benson, D. and A. Jordan (2011), 'What have we learnt from policy transfer research? Dolowitz and Marsh revisited', *Political Studies Review*, **9** (3), 366–378.

Benson, D., A. Jordan, H.F. Cook and L. Smith, (2013), 'Collaborative environmental governance: are watershed partnerships swimming or are they sinking?', *Land Use Policy*, **30** (1), 748–757.

Benson, D., A. Jordan and D. Huitema (2012), 'Involving the public in catchment management: an analysis of the scope for learning lessons from abroad', *Environmental Policy and Governance*, **22** (1), 42–54.

Cook, H.F. (1998), *The Protection and Conservation of Water Resources: A British Perspective*, London, UK and New York: Wiley International.

Cook, H.F., D. Benson, A. Inman, A. Jordan and L. Smith (2012), 'Catchment management groups in England and Wales: extent, roles and influence', *Water and Environment Journal*, **26** (1), 47–55.

Cook, H.F. and A. Inman (2012), 'The voluntary sector and conservation for England: achievements, expanding roles and uncertain future', *Journal of Environmental Management*, **112**, 170–177.

Cornwall Rivers Project (2008), 'The Cornwall Rivers Project', available at http://www.cornwallriversproject.org.uk/ (accessed 1 March 2013).

Couldrick, L. (2013), 'Tamar catchment pilot – creating a shared understanding. The catchment based approach, from pilot to wider adoption', paper presented at CIWEM/ DEFRA Conference, 14 February, SOAS, London.

Department for Environment, Food and Rural Affairs (2006), *Case Studies Aimed at Reducing Diffuse Water Pollution from Agriculture in England*, London: Department for Environment, Food and Rural Affairs.

Department of Communities and Local Government (2008), *Real People: Real Power. Communities in Control*, London: Department of Communities and Local Government, available at http://www.communities.gov.uk/documents/communities/pdf/886045.pdf (accessed June 2009).

Devon County Council (2012), 'EU transition status for Devon, Plymouth and Torbay', Exeter: Devon County Council, available at http://www.devon.gov.uk/transition_oct12.pdf (accessed 1 March 2013).

Environment Agency (2009), *Water for Life and Livelihoods: River Basin Management Plan – South West River Basin District*, Bristol: Environment Agency.

European Communities (2000), 'Directive 2000/60/EC of the European Parliament and of the Council of 23 October 2000 Establishing a Framework for Community Action in the Field of Water Policy', *Official Journal of the European Communities*, L 327/1 of 22 December, pp. 1–72.

Fritsch, O. and D. Benson (2013), 'Integrating the principles of IWRM? River basin planning in England and Wales', *International Journal of Water Governance*, **1** (3–4), 265–284.

Haigh, N. (2005), *Manual of Environmental Policy: The EU and Britain*, London: Maney Publishing and Institute for European Environmental Policy.

House of Lords (2012), *European Union Committee – Thirty-Third Report. An Indispensable Resource: EU Freshwater Policy*, London: House of Lords, available at http://www.publications.parliament.uk/pa/ld201012/ldselect/ldeucom/296/29607.htm#a14 (accessed 1 March 2013).

Office for National Statistics (ONS) (2012), *The Geography of Tourism Employment*, Cardiff: Office for National Statistics, available at http://www.ons.gov.uk/ons/dcp171776_287932.pdf (accessed 1 March 2013).

South West Water (n.d.), 'Upstream thinking', Exeter: South West Water, available at http://www.southwestwater.co.uk/index.cfm?articleid=8329 (accessed 1 March 2013).

URS/Scott Wilson (2011), 'Barriers and opportunities to the use of payments for ecosystem services', Final Report, London: URS/Scott Wilson.

Westcountry Rivers Trust (WRT) (2012a), 'Homes', Launceston: Westcountry Rivers Trust, available at http://www.wrt.org.uk/ (accessed 1 March 2013).

Westcountry Rivers Trust (WRT) (2012b), 'About the Trust: history', Launceston: Westcountry Rivers Trust, available at http://wrt.org.uk/wrtHistory.html (accessed 1 March 2013).

Westcountry Rivers Trust (WRT) (2012c) 'Trust projects', Launceston: Westcountry Rivers Trust, available at http://www.wrt.org.uk/projects.html (accessed 1 March 2013).

Westcountry Rivers Trust (WRT) (n.d.) 'Intelligent catchment design?' Launceston: Westcountry Rivers Trust, available at http://www.groundwateruk.org/model/Downloads/Paling_N(WRT).pdf (accessed 1 March 2013).

Westcountry Rivers Trust (WRT) and A. Rickard (2013), 'An ecosystems approach and rivers trusts: the Westcountry Rivers Trust', available at http://www.theriverstrust.org/environment/ecosystem/eco_approach.htm (accessed 1 March 2013).

6. State-founded water boards in industrialised Western Germany

Frank Hüesker and Christoph Bernhardt

6.1 INTRODUCTION

With the implementation of the European Union (EU) Water Framework Directive in 2000 (European Communities 2000), the institutional design, politics and performance of river basin organisations (RBOs) in Germany reappeared on the agenda of both researchers and practitioners (e.g. Moss 2003; von Keitz and Kessler 2008; Fichter-Wolf 2009; Thiel 2009). In North Rhine-Westphalia (NRW), the most populous state of Germany, RBOs had already existed for decades in the form of nine so-called Special Law Water Boards (Sondergesetzliche Wasserverbände). Each of these specific RBOs is rooted in its own state law, giving it the authority to regulate problems related to flood protection as well as water quantity and quality.

Beyond the Water Framework Directive's implementation, these existing Special Law Water Boards might also address novel challenges that come close to, but are not formally within their scope of responsibility. Examples include the regional decline of water consumption, the consequences of climate change, and, in an institutional sense, the implementation of new governance models, for example, participatory procedures. As a result of these developments, NRW's specific type of RBO has once again become relevant for advocates of integrated water resource management in environmental politics (for example, Green Party representatives in government, environmental stakeholders and managers of RBOs) and sciences (e.g. Moss 2003; Bernhardt et al. 2009). These institutional arrangements were established several decades ago, and certainly represent a special case within Germany. More importantly for this chapter, they are perceived as effective in terms of achieving goals, especially regarding the improvement of water quality and flood protection (e.g. Monsees 2008). Therefore this type of RBO might also serve as a model for mastering the above-mentioned new challenges for the rest of Germany, and even beyond.

This chapter investigates the assumption that the institutional design of water boards has led to successful problem-solving and output legitimacy in relation to the perceptions of water policy stakeholders in NRW. Following this edited volume's theoretical outline, and in a slight reordering of its central questions, we first examine the ways in which water politics in one specific catchment translated historically water-related interests and power distribution into the institutional design of a Special Law Water Board. What kind of aggregation rules were developed by the state of NRW (or the state of Prussia in earlier times) when the newly developed boundary rules forced public and private water users involuntarily to become members of an RBO? How did the water board members at that time jointly negotiate interests and solve problems, given that they had largely been brought together in a single organisation by the power of a state law and the pressure of environmental problems caused by other actors? In section 6.4 we turn to the first key question of this book, describing the present institutional design of our case: primarily an agency-type RBO with important elements of the autonomy type. To clarify aggregation rules, we show how the representation of interests in the water boards' committees is organised according to the detailed and sophisticated formal regulation and in the daily life of managing the water board. In this context, special emphasis is also placed on the pay-off rules and their effects on the power constellations within the aggregation rules of the water board.

To answer this edited volume's third central question regarding the water board's performance, we then discuss the water problems that led to the establishment of the water board about 100 years ago and were resolved in the context of legitimate and transparent coordination processes. Moreover, we will examine whether the institutional design has proven itself valuable in the face of currently pressing and decisively altered challenges. To do this, we discuss the extent to which this performance is interconnected with geographical conditions.

Historical research on NRW has already produced some relevant information regarding these guiding questions. With particular reference to the highly industrialised Ruhr area, Büschenfeld (1997), Olmer (1998), Eiden (2006), Paul (1995) and others have illustrated how (the level of) water consumption and water pollution caused regional water crises in the early twentieth century and thereby encouraged far-reaching countermeasures such as the foundation of the Special Law Water Boards. Until now, however, research in Germany has hardly focused on analysing the contemporary performance of these water boards. As the history of a water board's development and its founding history provide fundamental insights into the questions posed above, the chapter adopts both a historical and a contemporary perspective.

From a German perspective, these specific state-founded water boards

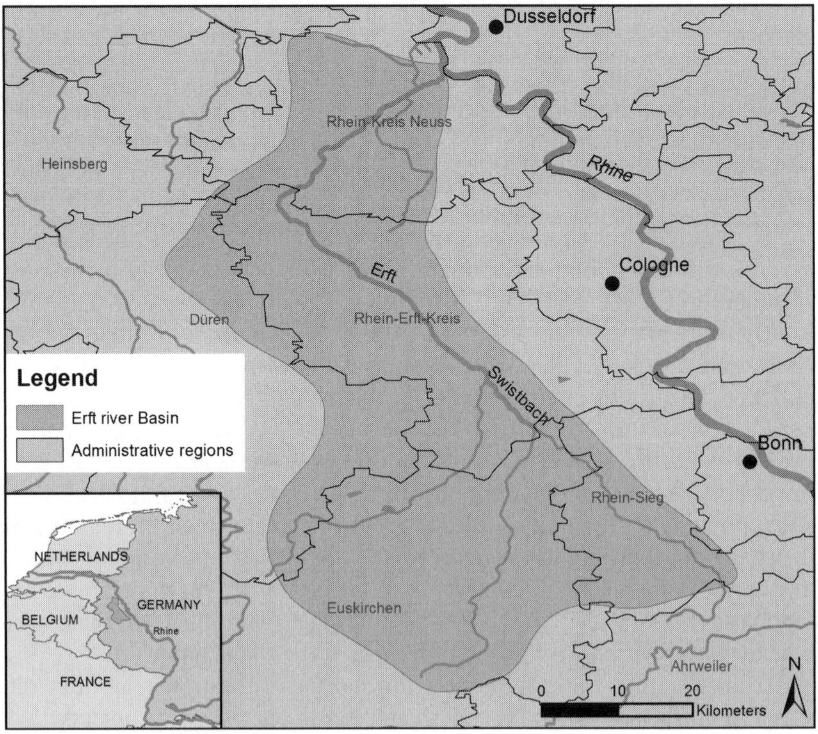

Figure 6.1 The Erft River Basin

constitute a singular type of RBO that can only be found in the state of NRW. At present, there are nine water boards with this particular institutional design. Some of them have been in operation for many decades, and many water policy-related stakeholders think of this model as highly successful. Here we examine the Erftverband, which operates in the Rhineland, west of Cologne (see Figure 6.1).

The Erftverband case is a particularly useful means of examining the above-mentioned three central questions. Our preliminary empirical findings in 2009 showed that this RBO had managed water resources legitimately and effectively for decades according to practitioners and decision makers interviewed, even though it works for the public good in a difficult setting, with powerful actors such as mining companies. The Erftverband's overall area of operation comprises 4216 km². It supplies water to 2.7 million people and provides wastewater treatment to approximately 1.2 million people. The annual water supply discharge capacity is around 611 m³ (Erftverband 2012).

As regards methodology, we analysed a number of historical and contemporary written sources, such as archival documents, contemporary literature, legal Acts, regulations and water board reports. In the period 2009–2011, the authors also conducted a number of interviews and two workshops in cooperation with regional decision makers from water boards, regional districts, municipalities, district governments, the Federal State Ministry of the Environment and representatives from politics and other stakeholder interests. Our empirical research rested on the interviews with practitioners and decision makers, examining the functioning, the politics and the performance of Special Law Water Boards in NRW.

6.2 FROM LOCAL 'MELIORATION' ASSOCIATIONS TO RIVER BASIN ORGANISATIONS IN THE RHINE'S SUB-BASINS AROUND COLOGNE

When the Prussian legislator established the Emscher Association as the first (there are now nine) Special Law Water Board in 1904, it did not break entirely new legal ground. Instead, it upheld the older legislation for a more coordinating type of RBO. This point is essential for an adequate understanding of today's Special Law Water Boards and their peculiarities. In this section we introduce the case and context by looking into the history of different kinds of water boards in the sub-basins around the Rhine. These territories were part of the Prussian state for decades in the nineteenth and twentieth centuries and nowadays lie in the German state of NRW. By describing the historical context in which these water boards developed we will focus on the Erft basin and the institutional steps prior to creating the RBO itself.

In this context, the much older 'melioration associations' are of particular relevance. These had been built in large numbers from the early nineteenth century so as to combat water-related problems of the pre-industrial age, for example, in the Erft basin. In one of the main texts on this issue, the regional administration of Cologne underlines that the agricultural meliorations constitute the nucleus of today's public water management authorities (Bezirksregierung Köln 1967: 195). The problems of a pre-industrial water economy most notably included inadequate maintenance of waters, disturbances due to the operation of numerous mills, insufficient flood control, morphological degradation, as well as an increasing pre-industrial commercial pollution. For an evocative depiction of the situation in the Rhine province and the area of Euskirchen see Dix (1997: 138ff, 169ff).

To solve these problems, several associations (more a coordinating

RBO in our typology) were established, such as the Erftverband (1859) and the Niers Association (1867) (Niersverband 1978; Paul 1999: 139ff). After a few decades, however, it became apparent that these associations were increasingly less capable of performing adequately in the face of industrialisation and urbanisation, which had become ever more challenging since the middle of the nineteenth century (Paul 1999: 139ff). Several amendments to the law were needed to extend the permitted justification (or authority rules) for establishing a state-imposed 'compulsory association' for agricultural meliorations to the regulation of reservoirs, for example, dams (1891), and, moreover, to the regulation of wastewater discharge and treatment (1904). These changes in law and additional tasks marked turning points, from a water board paradigm of 'river regulation' towards the paradigm of 'water pollution control'.

As a consequence, aggregation rules developed from the primacy of agricultural interests towards industrial and urban interests. Owing to the formulation of the Emscher Association, the extension to wastewater treatment found its way into the law. Subsequently, it was transferred to other associations (Bochalli 1966: 2f) and, finally, laid down comprehensively in the 1913 Prussian Water Law (Büschenfeld 1997). In fact, the newly established associations now comprised entire river basins rather than mere individual river sections. In the long-term history of water boards, they therefore also represent the development from a local 'single-purpose' association with limited assertiveness towards a regional and multifunctional association with comprehensive responsibilities and a statutory right to intervene in all residents' affairs. By looking at the increasing authority rules (more legal competences) and the changing boundary rules (larger territory and number of members) we can follow the development of the RBOs in Western Germany from coordinating agency to autonomous type.

Turning to the Erft basin, the basin's water policy agenda is shaped by mining activities and therefore surface water regulation is not as important as outlined in the introduction to this edited volume. In principle, the distribution of groundwater resources between municipalities and mining companies – the two principal user groups – was and has remained the prevailing challenge. In recent decades, however, wastewater treatment has become extremely important for the water board, partly as a result of the RBO's policies and strategies, which we will describe now and in section 6.3.[1]

The greater Cologne area has centuries of experience in dealing with coordinating corporate water boards institutionalised by water laws. As elsewhere in Europe throughout the pre-industrial age, the activities of early water boards were mainly focused on river regulations and agricultural meliorations. This already involved dealing with the multiple consequences of even earlier interventions. For instance, riverbed shifting was not the only reason for the River Erft's much-lamented increased degradation at the edge of the brown coal mining area. Instead, accumulation of sediments and ground elevations played a significant role as well (Paul 1999: 29f). Subsequent to the provision of a legal basis by the 1843 Prussian Private River Law and the 1848 Dike Law, the first Erftverband was established on 3 January 1859 (Bezirksregierung Köln 1967: 193). In the first decade of its existence, the association already raised enormous amounts (around 1.5 million gold mark), rendering 4275 hectares of former marsh or flooded land suitable for arable farming. Moreover, it extended 216 km of minor watercourses and built flood protection dams as well as numerous bridges, weirs and locks (Bezirksregierung Köln 1967: 193).

At that time, however, the Erftverband's territorial boundary rules were still limited to the river section located between the district of Bergheim and the confluence of the rivers Erft and Rhine, and within these limits, remained confined to the imminent vicinity of the river's shore area. Regarding performance, like many similar associations at that time the Erftverband increasingly lost its financial capacity to adequately maintain its water management facilities from the early twentieth century onward.

At the same time, other parts of the early industrialised Cologne region (for example, along the rivers Sieg and Ruhr) were facing a number of challenges beyond the additional and spatially extensive 'melioration ventures' (Dix 1997: 141). Since the 1920s, several associations and water boards with large catchment areas were established in order to ensure flood protection and drinking water supply, most notably the Agger Association on the right bank of the Rhine. For the Erft region, however, a modern river basin-centred water board was not established until 1959. In this case, the impetus was provided by the strongly intensified brown coal mining in the 1950s, which was accompanied by large-scale pumping of groundwater. This, in consequence, threatened the groundwater level and the supply of water within this region (Bezirksregierung Köln 1967: 198).

6.3 POLITICS AND ACTORS' STRATEGIES WHILE INSTITUTIONALISING RIVER BASIN ORGANISATION ERFTVERBAND IN 1959

6.3.1 Conflict Over the Representation of Stakeholders during the Establishment of Erftverband

Prior to the foundation of Großer Erftverband as a Special Law Water Board, several possible forms of coordinating, agency and autonomous types of RBOs had been discussed for more than 20 years. For instance, both a 'normal' coordinating water board according to the water boards' regulations as well as an autonomous institution under public law had been regarded as conceivable options. To be able to respond to continuing disagreement, a Public Water Management Authority Erft was established just in case in 1955 (Linckelmann 1957: 8; Bezirksregierung Köln 1967: 198).

Regarding politics and actors' strategies, a range of specific interests lay behind the different institutional designs, as Huitema and Meijerink pointed out in Chapter 1 of this edited volume. For example, the Registered Society of Rhenish Brown Coal Plants considered the establishment of a water board unnecessary. Instead, it argued in favour of a 'private agreement', for example, private law contracts between the mining sector and citizens concerned. Industrial representatives, however, argued in favour of establishing a State Institute for Safeguarding Water Supply in the Erft Region (Verein Rheinischer Braukohlenwerke e.V. 1956; Linckelmann 1957: 34).

Fundamental water policy controversies arose over authority and aggregation rules during the legislative procedure. Differences in opinion concerned the representation of different stakeholders in the water board's committees as well as the selection criteria, which differed substantially from those applied by prior types of water boards. According to the classical model of agricultural cooperatives, the particular benefit of a certain action or the property size defined the contributions payable. The contribution, in turn, determined the individual member's share of the vote. Restrictions to this principle were made in a selective way: namely by means of the regulation according to which an individual large landowner was allowed to hold no more than 40 per cent of all votes. Obviously, this clause was meant to avert a concentration of voting power as well as to undermine the principle of self-government within the co-operative. Moreover, those boards for water pollution control which also granted a vote and seat to 'aggrieved' persons imposed further restrictions on the classic principle of 'level of contributions defines voting power' (Linckelmann 1957: 8.).

During the pre-war period, the most pertinent legislation on the Ruhr, Wupper and Lippe water boards thus foresaw either a minimum voting share of 30–40 per cent for such aggrieved municipalities and districts, or restricted the maximum percentage of votes for mining companies to 40 per cent. Such regulation of aggregation rules was considered a workable compromise between the competing principles (Linckelmann 1957: 10f).

In its rationale concerning the distribution of votes in the water board committee, the draft bill categorically argued that the size of financial contribution must not be understood as an appropriate means for the distribution of votes. Instead, it was regarded as a mere correlative of the damage caused or anticipated; a criterion deemed to be counterbalanced and compensated by the interests of the aggrieved parties (Linckelmann 1957: 12). According to this draft bill, the water board was also entrusted with tasks for which the 'user-aggrieved party' dualism is irrelevant. Taking into account the protection of minorities and the potential for various combinations of interests, it thus claimed to assess the voting weight as objectively as possible.

It proposed dividing up the 36 members of the water board committee as follows: mining companies should be represented with eight delegates (22 per cent), supplemented by ten closely linked members of the electricity industry (28 per cent), whose contribution liability was estimated at 80 to 90 per cent of the overall contributions received (§ 3 of the draft bill on the establishment of Großer Erftverband, cf. Erftverband Foundation Act 1957, see Landtag NRW 2007).

This considerable deviation from the principle of 'level of contributions defines voting power' was modified later on in that payments from the mining companies were accounted for separately as 'special payments' beside the basic tariff calculation (see section 7.4). Doubtless, the mining companies had a minority position in the discussion on the water board committee: represented public water supply authorities and municipalities (three members); industry (six members); administrative and urban districts (seven members); agriculture (three members). And these other interest groups without exception had to expect disadvantages for their respective water supply in the case of planned groundwater extraction.

Apart from this incomparable decoupling of contribution level and distribution of votes, critics also found fault with the fixed setting of voting shares as an 'inflexible mechanism' (Linckelmann 1957: 13). In the end they were, however, unable to alter the distribution of votes – and hence power – in the water board committee.

Apart from the decision to raise the votes of drinking water resources management from three to four, the Erftverband Foundation Act (3 June 1958) confirmed the above-mentioned aggregation rules (Großer Erftverband n.d.: 2). The way votes were distributed among various interest groups also found expression in the design of the management board. In 1970, Cologne's city director Josef Baumann chaired the board as the representative of municipalities and administrative districts. The representative of industry acted as the first deputy and the mine director of Rheinische Braunkohlewerke AG served as the second deputy (Großer Erftverband 1971: 10).

The supreme water authority controlled and steered the water economy within the administrative district of Cologne. Within these confines, the public water authorities regarded Großer Erftverband as an instrument capable of mastering all problems arising from the mining sector's extraordinary undertaking and of performing in the interests of the public (Bezirksregierung Köln 1967: 204). However, in light of the considerable administrative interventions at that time and the resulting groundwater level reduction, it appears more appropriate to speak of mitigation and administration rather than of 'mastering' water policy problems from a present-day perspective. Directly after its legal institutionalisation we can now see Erftverband's characteristics as an agency and an autonomous type of RBO.

6.3.2 On the Changing Tasks of Großer Erftverband

In the decade after its establishment, both the structure of the water board's territory (regarding the boundary rules) and the water-related challenges apparent underwent substantial change. Between 1959 and 1969, the population increased by more than 200 000 and rose further to 1.5 million inhabitants in 1974 (Großer Erftverband 1971: 8; Großer Erftverband n.d.: 1). As a result, water demand increased steadily on the part of public water utilities. Due to the initiation of 'rationalisation measures', this increase remained relatively moderate until 1970, but began to accelerate soon afterwards (Großer Erftverband n.d.: 4).

In the years 1969 and 1970, mining and electricity industries jointly consumed approximately the same amount of water as public utilities, and were responsible for about half of the water consumed by the industrial sector (Großer Erftverband 1971: 22f). Moving away from the initial motivation for establishing the Großer Erftverband, the construction of sewage treatment plants soon became its most important field of activity. As early as 1963, the water board founded its own Operational Group for Sewage Treatment Plants, which took over 31 previously municipal sewage treatment plants within a period of seven years. Parallel to this, and at high financial cost, the number of inhabitants connected to sewage treatment plants was increased. As a consequence of the newly founded water board's strategy, aggregation rules (namely distribution of votes) changed markedly as it generally still had to follow the 'level of contributions defines voting power' principle.

For the water board, extensive monitoring programmes constituted one of its major remits. Each year, approximately 100 000 monitoring data records from 5000 measuring points were collected and evaluated solely for the purpose of developing groundwater levels. Once we add the comprehensive measurements to assess water quality and the chemical and biological quality of waters, it is evident that monitoring became one of the water board's most important functions.

In 1969, the adoption of the old Erftverband's tasks entailed a sudden and significant expansion of responsibilities, in particular with regard to personnel and the maintenance of the river course (Großer Erftverband 1971: 16, 35f, 48–53). The total annual revenue of almost 15.5 million Deutschmarks (1970 budget) doubled to approximately 30 million Deutschmarks until the mid-1970s. The lignite mining sector had to contribute another special payment amounting to one-third of the budget (10 million Deutschmarks) in 1970.

Conflicts within the water board remained at a relatively low level, at least if one considers the number of objections entered against membership or demands for payment. With 479 documented members by around 1970, less than ten conflicts per year were registered for the individual branches (Großer Erftverband n.d.: 3; Großer Erftverband 1971: 46–50). With regard to the ability to achieve consensus amongst often conflicting water-related interests, these figures appear remarkable from today's perspective.

6.4 ERFTVERBAND'S INSTITUTIONAL DESIGN AS AN AGENCY AND AN AUTONOMOUS TYPE OF RBO TODAY

6.4.1 Introduction

Considering the historical and geographical context described in section 6.2, as well as the politics and strategies described in section 6.3, we have seen the many interests involved and key decisions which ultimately shaped the design of the institutions of Erftverband. Looking at Erftverband from the 1970s up until today, this institutional design is relatively stable in terms of boundary, authority, information and pay-off rules. We can describe many aspects of Erftverband's institutional design as being in line with the agency and the autonomous types of RBOs. Table 6.1 gives an overview following the typology of RBOs developed in the introduction (Chapter 1 in this volume).

6.4.2 Aggregation Rules: Legal Features

According to the Erftverband Foundation Act (1957; see Landtag NRW 2007) and the related Erftverband Statutes (7 October 1993; see Erftverband 2010), all municipalities and administrative districts are legally obliged

Table 6.1 Institutional design of the Erftverband today

Rule types/ RBO types	Rules of the Erftverband
Authority rules/ Agency RBO	The Erftverband has the legal mandate to observe water conditions in the coal district. This mandate includes 11 tasks, for example, the management of groundwater, the protection of wetlands, maintenance of surface waters, flood protection and sewerage networks. The legal mandate is issued by the state of NRW with no time or other specific limits. Within this legal framework the Erftverband is free and independent regarding its self-government. The water policy competences and resources taken away from the municipalities are integrated into the RBO; the term 'Special Law Water Board' shows that this disempowerment has to be justified by extraordinary water policy problems. If not satisfied with Erftverband's performance NRW's parliament can change or determine the special law.
Aggregation rules/Agency RBO	Formal decision-making rules give large consumers and polluters (today mainly municipalities) the majority. The central principle is that water board members who pay higher tariffs get more votes in the water board assembly and committees ('level of contributions defines voting power' principle). In practise, cost-effective, consensual and rational decision making dominates the daily water management tasks of the board. The Erftverband has full control over the issues delegated. Control over decisions stays within the appointed leadership. The Erftverband has no veto power over administration, coordination is needed there. Key veto-players are members of the water board.
Boundary rules/ In-between agency and autonomous RBO	Core region (basin) and entire region (groundwater body) as defined in the special law by the state of NRW (most fundamentally in 1959); these boundaries fit to the perceived geographical boundaries of the basin (and the groundwater body). Water users and polluters within these boundaries in the sub-basin (private and public actors) are forced by special law into membership of the water board (the exception is for low consumption). This model of membership can be seen as organised stakeholder participation. Staff working at the RBO are mainly from engineering and natural science. Further participatory integration of citizens only if necessary (for example, the Water Framework Directive).
Information rules/ Autonomous RBO	The water board understands itself as an organisation collecting water management knowledge and data and providing it to its members (RBO takes over state functions). Arguments from natural science, engineering and economics are dominant; debates and decisions made are, in that sense, depoliticised. Accountability for the decisions rests within the leadership elected by the members.
Pay-off rules/ Autonomous RBO	Tariffs are the Erftverband's own source of income. It can specify the rules for contributing to it ('Veranlagungsregeln'). Members in general pay tariffs according to the amount of water used or polluted. Exceptions exist, but today not as rewards, rather as additional tariffs, for example, for mining companies causing extra damages.

to be part of the Erftverband. The same applies to mining companies, electricity companies and other water-using and water-polluting industries. According to the law, the meeting of delegates now comprises 102 members. The allocation of seats is determined by the amount of contributions. This is to say that the pay-off rules for calculating the member's contribution again plays a decisive role in defining a certain water board member's position of power and the aggregation rules. Today, the law provides one seat in the meeting of delegates for each of the two societal interest groups, for example, agriculture and fishing.[2]

In the Erftverband's meeting of delegates and the water board council, the voting power of municipal members has increased continuously since the 1970s. In that decade, the above-mentioned change of tasks (following the extension of the water board's competence in the sewage sector) engendered this new formal distribution of power and led to rising sewage-related concerns within the board. This, in turn, strengthened the voting weight of municipalities. As the operators of sewage works, they collect the respective fees from their citizens and pass them on to the Erftverband. With these changes in the distribution of votes in the meeting of delegates, the mining companies relinquished some of their power to the municipalities there.[3]

This shows the flexibility and adaptability of a mechanism of distributing votes according to the level of contributions. It also proves that – at least in structural terms – this mechanism does not systematically privilege any member group within the water board. It is open to political and strategic manoeuvring, not presently according to our empirical findings, but in the past as shown in section 6.3, and possibly in the future as well.

In order to protect the interests of the water board members, the Erftverband Foundation Act (§ 15) now foresees an exception to the basic principle of 'share of contributions/share of votes', according to which each municipality (or member group) may be represented by no more than 66 seats in the meeting of delegates. As the Erftverband's area clearly contains less than 66 municipalities, this also implies that each municipality and each county is represented directly in the meeting of delegates (notably, this does not apply to other Special Law Water Boards). In this way, each municipality within the Erftverband can decide for itself when it comes to choosing its delegates for the water board meeting.[4] In light of the compulsory membership of the water board, the direct participation of democratically elected municipal actors from all municipalities within the Erft River Basin can be regarded as a big advantage in terms of input legitimacy (accountability).

6.4.3 Pay-Off Rules as a Foundation and a Driver for the Aggregation Rules of the Erftverband

For the distribution of votes within the Erftverband, the principle of 'level of contributions defines voting power' and the so-called 'cooperative principle' have remained the most essential pay-off rules. The Erftverband's Assessment Regulations, which were agreed by the majority of its members, lay down the pay-off rules for calculating the tariffs as the sources for Erftverband's own income; and thus eventually determine the overall amount of contributions and the share of votes derived from it.

Overall, three criteria which are crucial to the pay-off rules within the Erftverband need to be considered and balanced in order to be able to make decisions on the assessment regulations within the water board: the user-pays principle, the avoidance of individual expenses, and individual benefits. While municipalities and districts covered 95 per cent of all contributions in 2011, mining and electricity industries covered 2.5 per cent, and water supply likewise covered another 2.5 per cent. Whereas 88 per cent of all contributions relate to wastewater disposal, tasks related to surface waters account for 9 per cent. Another 2 per cent can be assigned to hydrology-related tasks, while water supply accounts for 1 per cent.[5] These numbers prove that the portrayed change of tasks in the water board has entailed a significant shift in contribution shares and aggregation rules within the Erftverband since its foundation.[6]

The cooperative principle of the pay-off rules outlines the distribution of costs among board members according to the degree of utilisation (for example, based on a member's consumption of water). Considering that not all board members are affected by all 11 tasks of the water board, thematic subgroups have evolved, each of which is subject to particular pay-off rules (so-called assessment regulation).[7] Regarding pay-off rules for the groundwater member group, for instance, this implies that all those extracting and utilising groundwater are regarded as members (including mining companies). The contributions are calculated on the basis of the amount of extracted or utilised groundwater. According to the assessment regulations (which themselves have been decided on by the majority of members), the contributions drop progressively whenever consumption or utilisation figures are high. This is justified with the argument that the proportionate effort required for the water board also decreases with growing quantity.[8]

Pay-off rules strongly oriented towards the consumption or utilisation of water resources, as well as towards the creation or avoidance of costs, appear to be both structurally fair and flexible. As already illustrated by the debates at the time the water board was founded, this is also true for the water board's prospective capability to perform novel tasks, with regards to the implementation of the Water Framework Directive, for example. As already stated, an increasing use of resources also entails an increase in the level of contributions and influences aggregation rules. Especially with respect to the protection of resources, this nexus, however, provokes rather ambivalent consequences. Depending on whether a water board's member intends to save costs or seeks to extend its influence, it will or will not be interested in an economic or ecologic utilisation of resources. The element of quantity discount, however, runs counter to ecological performance criteria.

In addition to the aspects already mentioned in section 6.2, we also need to consider that according to the Erftverband Foundation Act (§§ 37–39), mining companies are subject to special pay-off rules. More precisely, they are obliged to pay an additional fixed price for particular mining-induced water board tasks which are taken into account neither in the assessment regulations nor in calculating the distribution of votes. This is because the contributions of mining companies would otherwise collapse altogether with massive cut-backs in coal mining. This would, in turn, contradict an appropriate application of the 'polluter-pays principle', since the mining industry is responsible for a great deal of the lowering of the groundwater table. These special payments by mining companies are thus not included in the aggregation rules.[9] From the perspective of input legitimacy, these 'polluter-pays principle'-based special pay-off rules appear justifiable. At the same time, however, a more comprehensive evaluation of output legitimacy would require comparing the absolute sum of fixed prices according to the Erftverband Foundation Act with the past profits of mining companies, which could be generated solely on the basis of water-related measures (such as water table extractions).

Regarding information rules, the water board employees see themselves as service providers for their members (municipalities, industries, mining companies, and so on) and as such aim to convince them of the management's lines of action. A well-functioning RBO, according to the employees' philosophy, depends on members being satisfied with its work; for this purpose the water board provides them with professional expertise and data. So offering transparent and sophisticated water policy monitoring data characterises the type of information rule in the Erftverband. This kind of expertise is unique in the basin and cannot be offered by the state authorities. This part of the water board's performance – turning to section 6.5 – is supposed to make it strategically quite difficult or impossible for other water policy actors to do anything without the active support of the water board as well.[10]

6.5 PERFORMANCE OF THE ERFTVERBAND AND OTHER WATER BOARDS IN NRW TODAY

6.5.1 Self-Perception of the Erftverband's Procedural and Managerial Performance

Beyond the above-mentioned legal documents, interviews conducted with members of the Erftverband and water policy decision makers in NRW showed unanimous agreement that the Erftverband currently follows consensus-based aggregation rules that parallel and somehow disregard formal details related to the majority-based distribution of votes described. The water board's management regards consensual decisions as an important element of the water board's accountability and procedural output legitimacy. In turn, consensual votes give the management accountability and the members recognise the advantages they themselves draw from the water board's overall performance. At an early stage of the policy process, on the management's initiative the committees of the water board test out the overlaps in members' interests.[11]

Moreover, regarding the performance criteria of coordination, the water board adopts the role of referee, as it brings together members from the entire river basin with sometimes diametrically opposed interests, and has the power to coerce these members to implement certain measures. For exactly this reason, the consensus principle in aggregation rules is considered essential. Only by this rule type will water board members be able to learn how to overcome conflicting interests and approve joint solutions over the years – even if they are not always entirely congruent with their own interest.[12]

Regarding the performance criterion of coordination, the ability to effectively solve water policy-related conflicts helps water boards to lessen the burden on the state and the public administration in the implementation of water-related policy goals. When compared to tax-financed state action, contributory funding, which is directly linked to the financing of specific tasks, would also raise the degree of acceptance among those affected by some measurements, or those responsible for paying.[13] So this perception connects the performance in financing water policy successfully and logically to the accountability and output legitimacy of the RBO.

Regarding the performance criteria dialogue, legitimacy and support, in our view it appears legitimate to solve the problems arising from conflicting interests and uses of water within the river basin by pursuing a consensus-based water board policy. In light of strongly divergent interests within the river basin, and the correspondingly high costs this can entail, it seems strategically wise to establish aggregation rules to achieve consensual solutions, if professionally accepted water engineering-related solutions are available and dissenting opinions can be softened. In this case, formal decision-making mechanisms following majority rules serve as a 'sheet anchor' for cases of conflict. In the following section we analyse how other water policy-related decision makers in NRW perceive the performance of this agency and autonomous type of RBO.

6.5.2 Perceptions of the Performance of Water Boards in North Rhine-Westphalia

Interviews with water policy-related actors from the state level of NRW (federal parliament representative; employee at the environment ministry; stakeholders from an industrial association, environment association and water board association) and on the level of district governments have shown that the board's effectiveness in reaching goals is hardly ever questioned. Surprisingly, in view of the fact that we are dealing with a very broad range of actors surveyed, we still come to the conclusion that the relevant policy actors dealing with water-related policies regard the overall performance of water boards as oriented towards sustainable development (including ecological aims), as cost-efficient and effective in problem-solving, and as legitimate. Exceptions included some stakeholder criticism of details on price levels or particular local measures (for example, water level of an individual reservoir).[14]

In principle, this generally positive feedback was also apparent with regards to the question as to whether water boards can be considered legitimate and effective institutional designs for tackling new challenges such as implementing the Water Framework Directive. For these purposes, the interviewees generally deem the existing state-founded water boards both appropriate and capable of performing in an accountable and legitimate fashion.

Our findings are even more remarkable if we bear in mind that these draw their procedural (input) legitimacy from the legal conviction that their *raison d'être* consists in solving exceptional water-policy challenges; that only such exceptional (for example, water quality or quantity) problems render such a marked legislative intervention by the state in the constitutionally delegated affairs of municipal self-governance (Tettinger et al. 2000; Köller 2009). In terms of Germany and NRW's constitutional law, normal practice would be to eliminate existing RBOs as soon as this extraordinary water policy problem had been resolved.

However, only a very few actors argued in support of such a move.[15] One frequently raised point is the comparison of the water board's performances with those regions in NRW where no comparable RBOs can be found (for example, Münsterland or East Westphalia). Here water policy problem-solving capacities of existing territorial, administrative institutions were assessed much more pessimistically by the interviewees.[16]

The enquiry focused in particular on how responsible decision makers at the state level and in district administrations assessed Special Law Water Boards as instruments for implementing their water policy agenda. The interviewees pointed out that the state of NRW may legally compel these water boards to accomplish certain tasks. Afterwards performance lies, in terms of authority rules, within the scope of their board's self-governance. The interviewees stated that the water boards would be completely free to decide on how to achieve their legal tasks (similar to municipalities). According to these considerations regarding the performance criterion coordination, the state's role would be limited to the task of functional and legal supervision. Its power to intervene would be of a merely provisional nature, hardly ever applied in practice.[17]

When compared to the municipalities, the water board's performance regarding the criteria coordination was described as more pronounced, especially since municipalities are unable to exhibit comparable professional knowledge and, moreover, are no longer endowed with sufficient financial resources. Although these state-founded water boards are disempowering the municipalities, none of the interviewees was critical regarding the institution of the water board in general. The positive perception of the water board's performance is crucial for this output-oriented legitimacy mentioned by the stakeholders interviewed.[18]

Another important finding regarding the accountability of the RBO highlighted by interviewees is that membership in the water board frees local politicians from the need to strive for political majorities in 'their' municipal parliament. Instead, it is easier to implement measures if it is possible to justify them with a decision taken by the board and if the costs involved disappear in an annual item labelled as 'payment to the water board'.[19]

Notably, the interview partners declared that compulsory membership was a core element regarding boundary rules (and the performance criteria of input legitimacy and accountability). According to this view, models of voluntary membership in water boards were bound to fail, as we described with regards to the historical institutional settings on the Erft. Interviewees stated that a certain political will at the state level is necessary to enforce such compulsory membership. This, in turn, would only be realistic in situations characterised by a certain set of water policy problems or under conditions where a noteworthy pressure to take action became manifest, requiring a long period of preparation in the case of NRW.[20]

6.6 CONCLUSIONS

In our opinion, the actors' positions outlined in section 6.5 also indicate the limits of the agency-type RBO as a model for institutional design. Due to the municipal right to self-administration (which is anchored in the German constitution), this model is transferable only in limited periods of extraordinary water policy-related problems, and to extend the authority rules in the direction of an autonomous RBO remains even more debateable in terms of constitutional law.

Moreover, once the institutions of this type of RBO have been designed, the general governance functions of these water boards can be regarded as relatively stable and difficult to reform. As a basic principle, any top-down implementation of new authority rules is only feasible by the complex means of federal state law revisions and, thus, does not appear viable in the short term.

The change of government in 2010 showed that these water boards in NRW are not purposefully used as a political instrument of water policy, at least at the time of conducting the research. The new green environment minister explicitly pursues an ecologically oriented water policy agenda. For this purpose, he should be politically able to make use of the water boards subordinated to him. The ministry staff has, however, noted the fact that their responsibility within water boards is confined to the task of legal supervision, and that a revision of authority rules within the special laws related to water boards is politically not realistic for the time being.[21]

The existing institutional design performs both legitimately and effectively within the scope of the legal duties (authority rules). The design is flexible within the framework of these authority rules, as demonstrated by Erftverband's pay-off and aggregation rules. However, at the same time, NRW's water boards do not seem to be forced by state authorities to solve novel water policy problems; at least not in the short term and outside the scope of the legally defined authority rules. This may be regrettable from the perspective of an ecologically effective performance, but might prevent agency and autonomous RBOs from being manipulated by short-term political strategies.

We have to bear in mind that it was Prussian legislation that established water boards of a new institutional design and endowed them with new authority rules. On this basis, the Prussian state succeeded in adjusting the formerly pre-industrial type of RBO in such a way as to improve its performance to tackle new industrial and urban water-related problems of hitherto unknown dimensions. As our analyses of aggregation rules have shown, the balancing of interests in the Erftverband has helped to commit the most powerful stakeholders (especially mining companies) to making a substantial contribution as regards the financing of water board tasks. In contrast to this (and partly by way of derogation from the principle of 'level of contributions defines voting power'), the municipalities acted as the most powerful stakeholders in water board committees and, as such, were able to exert a considerable influence on water board policies. Particularly in recent times, the broad acceptance of these water boards (from their members as well as from water management actors generally) and the assignment of new tasks testify to their performance and high output legitimacy. Even now, complex aggregation, pay-off, information and boundary rules have succeeded in keeping the conflict-laden relations between mining companies, municipalities and other stakeholders in a particular equilibrium.

The way pay-off rules are arranged to organise the assessment of contributions can be regarded as decisive to understanding the aggregation rules of negotiating interest within the water boards, the resulting distribution of power and the capacity to adapt to new tasks and challenges. The water boards' performance did, however, result in the resolution of individual problems, such as, for instance, the shortcomings in drinking water supply. Additionally, they also evoked new environmental problems, such as a spatial externalisation of water pollution and large-scale regional water-table extractions. However, the formation of efficient and differentiated aggregation and pay-off rules for tackling diverse water problems and interests under one shared public umbrella demonstrates that the balancing of partial interests and principles of solidarity has been a success.

Ultimately, it was the particular pressure exerted by water-related challenges that led NRW to experience such wide-ranging and unparalleled (at least in Germany) legal intervention in its system of municipal self-government. When the former German Democratic Republic was experiencing all-encompassing institutional transformation in 1990, the newly established *Länder* were considering whether or not to establish similar water boards to solve the water policy problems that existed in the open-cast mining regions of Lower Lusatia in the state of Brandenburg. Notwithstanding that these regions are comparable to those in NRW and, moreover, that these particular challenges were further exacerbated by climate change affecting the arid regions in the eastern parts of Germany, the new federal states finally decided not to establish Special Law Water Boards.[22] Other federal states, such as Saxony or Bavaria, entrusted their administrations with finding solutions to water-policy related problems. In this regard, they opted to a much lesser extent to delegate competences to institutions of autonomy or agency type. For these reasons, the state-founded water boards of NRW can be described as a success model of limited scope, which arose from one particular context and problem constellation. As such, it remains a special case within the German context for the time being.

ACKNOWLEDGEMENTS

The authors of this chapter have published a similar article in German in the Journal *Historical Social Research*, based partly on similar case studies and interviews, but differing conceptually: 'Von der Wasserkrise des Industriezeitalters zur EU-Umweltpolitik. Interessenaushandlung und Problemlösungsfähigkeit von Sondergesetzlichen Wasserverbänden in Deutschland seit 1900', *Historical Social Research*, **38** (2) (2013), 288–314.

NOTES

1. Interview with Erftverband, 21 December 2009.
2. We see here that the influence of agricultural interests and actors is very low compared to the nineteenth century around the Erft, as well as the mining sector there today and the powerful position of farmers in river basins in general. In the overall context of state-founded water boards in NRW, such a formal upgrading of particular societal interests does not constitute a standard. On the one hand, the representation of fishermen and farmers can be regarded as positive in that it strengthens participatory procedures by means of integrating non-state actors into the involuntary community of water boards. On the other hand, it might as well be interpreted as an unduly favourable treatment of certain interest groups to the disadvantage of excluded environmental groups or other groups using the river basin (for example, for fishing or sailing).
3. Interview with Erftverband, 18 March 2011.
4. Interview with Erftverband, 21 December 2009.
5. Interview with Erftverband, 18 March 2011.
6. See the website of the Erftverband for further details, www.erftverband.de (accessed 31 May 2013).
7. Interview with Erftverband, 21 December 2009.
8. Interview with Erftverband, 18 March 2011.
9. Interview with Erftverband, 18 March 2011.
10. Interview with Erftverband, 18 March 2011.
11. Interview with Erftverband, 18 March 2011.
12. Interview with Erftverband, 18 March 2011.
13. Interviews with Erftverband, 21 December 2009 and 18 March 2011.
14. Interview with stakeholder water boards, 12 May 2009. Interview with stakeholder environment, 4 February 2011. Interview with stakeholder economy, 10 May 2011.
15. Interview with parliamentarian, 13 May 2009.
16. Interview with stakeholder water boards, 12 May 2009. Interview with stakeholder environment, 4 February 2011.
17. Interview with Ministerium für Umwelt, Natur, Landwirtschaft und Verbraucherschutz NRW, 7 September 2010. Interview with district government Cologne, 7 September 2010. Interview with district government Düsseldorf, 6 September 2010.
18. Interview with Ministerium für Umwelt, Natur, Landwirtschaft und Verbraucherschutz NRW, 7 September 2010. Interview with district government Cologne, 7 September 2010. Interview with district government Düsseldorf, 6 September 2010.

19. Interview with Ministerium für Umwelt, Natur, Landwirtschaft und Verbraucherschutz NRW, 7 September 2010. Interview with district government Cologne, 7 September 2010. Interview with district government Düsseldorf, 6 September 2010.
20. Interview with Ministerium für Umwelt, Natur, Landwirtschaft und Verbraucherschutz NRW, 7 September 2010. Interview with district government Cologne, 7 September 2010. Interview with district government Düsseldorf, 6 September 2010. Interview with parliamentarian, 13 May 2009.
21. Interview with Ministerium für Umwelt, Natur, Landwirtschaft und Verbraucherschutz NRW, 7 September 2010.
22. Interview with water board association Brandenburg, 26 September 2011.

REFERENCES

Bernhardt, C., H. Kilper and T. Moss (eds) (2009), *Im Interesse des Gemeinwohls. Regionale Gemeinschaftsgüter in Geschichte, Politik und Planung*, Campus: Frankfurt am Main.

Bezirksregierung Köln (ed.) (1967), *150 Jahre Regierungsbezirk Köln*, Köln.

Bochalli, A. (1966), *Das Wasser- und Bodenverbandrecht mit der Neufassung des Wasserverbandsgesetzes und der ersten Wasserverbandverordnung*, 3rd edn, Köln: Heymann.

Büschenfeld, J. (1997), *Flüsse und Kloaken. Umweltfragen im Zeitalter der Industrialisierung (1870–1918)*, Stuttgart: Klett-Cotta.

Dix, A. (1997), *Industrialisierung und Wassernutzung*, Köln: Rheinland-Verlag.

Eiden, C. (2006), *Versorgungswirtschaft als regionale Organization. Die Wasserversorgung Berlins und des Ruhrgebietes zwischen 1850 und 1930*, Essen: Klartext-Verlag.

Erftverband (2010), 'Satzung des Erftverbandes vom 7. Oktober 1993 (GV. NRW., S. 732), geändert durch Beschluss der Delegiertenversammlung vom 6. Dezember 2010', available at www.erftverband.de (accessed 31 May 2013), [Erftverband Statutes].

Erftverband (2012), 'The Erftverband at a glance', available at http://www.erftverband.de/fileadmin/Erftverband/Presse/ev-blick-2012-englisch_internet.pdf (accessed 31 May 2013).

European Communities (2000), 'Directive 2000/60/EC of the European Parliament and of the Council of 23 October 2000 Establishing a Framework for Community Action in the Field of Water Policy', *Official Journal of the European Communities*, L 327/1 of 22 December, pp. 1–73.

Fichter-Wolf, H. (2009), 'Neue Entwicklungspfade in der deutschen Gewässerschutzpolitik? Implementierung der Europäischen Wasserrahmenrichtlinie (WRRL) und der Umgang mit "problems of spatial fit"', *ZAU Zeitschrift für angewandte Umweltforschung*, **19** (1–2), 173–189.

Großer Erftverband (1971), *1970. Zehn Jahre Verbandstätigkeit*, Bergheim/Erft.

Großer Erftverband (n.d.), *No Title*, undated historical document, Bergheim/Erft.

Keitz, S. von and P. Kessler (2008), 'Grenzen des Flussgebietsmanagements. Folgt die Wasserwirtschaft dem falschen Ansatz?', *Korrespondenz der Wasserwirtschaft*, **1** (7), 354–360.

Köller, S. (2009), *Funktionale Selbstverwaltung und ihre demokratische Legitimation. Eine Untersuchung am Beispiel der Wasserverbände Lipeverband und Emschergenossenschaft*, Berlin: Duncker und Humblot.

Landtag North Rhine-Westphalia (NRW) (2007), 'Gesetz über den Erftverband' (ErftVG) (GV. NRW 2007, p. 716), Düsseldorf, [Erftverband Foundation Act 1957].

Linckelmann, K. (1957), *Erwägungen zum Erftverband und zur Kerngestalt der Wasserverbände*, Bielefeld: Kirschbaum.

Monsees, J. (2008), *Governancestrukturen von Fließgewässern. Eine vergleichende Institutionenanalyse gewässerunterhaltender Verbände und Behörden*, Baden-Baden: Nomos.

Moss, T. (ed.) (2003), *Das Flussgebiet als Handlungsraum. Institutionenwandel durch die EU-Wasserrahmenrichtlinie aus raumwissenschaftlicher Sicht*, Münster: LIT Verlag.

Niersverband (ed.) (1978), *50 Jahre Niersverband*, Viersen.

Olmer, B. (1998), *Wasser. Historisch. Zur Bedeutung und Belastung des Umweltmediums im Ruhrgebiet 1870–1930*, Frankfurt am Main: Peter Lang.

Paul, J. (1995), 'Der Rhein als Klärbecken für städtische Abwässer: Abwasserpolitik in Köln und Düsseldorf seit dem 19. Jahrhundert', *Geschichte im Westen*, **10** (2), 135–148.

Paul, J. (1999), *Die Erft im Spannungsfeld zwischen Bergbau und Landwirtschaft*, Rheinlandia Verlag: Köln.

Tettinger, P., T. Mann and J. Salzwedel (eds) (2000), *Wasserverbände und demokratische Legitimation*, München: Vahlen.

Thiel, A. (2009), 'Europeanization and the rescaling of water services: agency and state spatial strategies in the Algarve, Portugal', *Water Alternatives*, **2** (2), 225–244.

Verein Rheinischer Braukohlenwerke e.V. (ed.) (1956), 'Braunkohle und Wasser', *Revier und Werk*, Special Edition.

List of Interviews

Arbeitsgemeinschaft der Wasserwirtschaftsverbände in NRW (AGW), 12 May 2009 in Cologne (Interview stakeholder water boards).

Bezirksregierung Düsseldorf, Dezernat 54 Wasserwirtschaft, 6 September 2010 in Düsseldorf (Interview district government Düsseldorf).

Bezirksregierung Köln, Dezernat 54 Wasserwirtschaft, 7 September 2010 in Cologne (Interview district government Cologne).

Bund für Umwelt und Naturschutz, Landesverband NRW, 4 February 2011 in Düsseldorf (Interview stakeholder environment).

Erftverband, Bereich Gewässer, 21 December 2009 and 18 March 2011 in Bergheim (Interviews Erftverband).

Industrie- und Handelskammer NRW, Tätigkeitsfeld 'Umwelt und Energie', 10 May 2011 in Duisburg (Interview stakeholder economy).

Landeswasserverbandstag Brandenburg, 26 September 2011 in Sonneberg (Interview water board association Brandenburg).

Landtagsabgeordneter in NRW der Partei B90/Die Grünen, 13 May 2009 (Interview parliamentarian).

Ministerium für Umwelt, Natur, Landwirtschaft und Verbraucherschutz NRW, Abteilung IV 'Abfallwirtschaft, Bodenschutz, Wasserwirtschaft', 7 September 2010 in Düsseldorf (Environment Ministry, Interview Ministerium für Umwelt, Natur, Landwirtschaft und Verbraucherschutz NRW).

7. Emergence, performance and transformation of Portuguese water institutions in the age of river basin organisations

Andreas Thiel and António Guerreiro de Brito

7.1 INTRODUCTION

What does it take to make reforms of the organisation of water governance effective on the ground and what are the factors that shape the path of development in water governance? This chapter tries to provide insights in regard to these questions based on the study of changes in water management in Portugal throughout the last decade. Recent reforms were part of the process of implementing the European Water Framework Directive (European Community, 2000) which aimed at integrated water resources management. Throughout the last decade, member states of the European Union (EU) changed their national legislative framework to implement these requirements, and day-to-day water planning and management practices started to change accordingly. By 2009 member states were required to present new river basin management plans and programmes of measures are currently being implemented. The application of these prescriptions by European states provides detailed insights in regard to the overarching interests of this book, namely the study of the politics of river basins and their performance. In Portugal, the requirements of the Water Framework Directive with regards to the implementation of river basin management fell on fertile ground because at the time of the transposition of the Directive its introduction had long been considered by Portuguese water experts.

In this context, this chapter reconstructs formal institutional change and its implications for informal institutions of operational Portuguese water management. It tries to give a historic perspective on water governance, introducing the role of path dependence understood as bias towards the role of previously existing institutions because of the transition costs of moving from one institutional set-up to another (North 1994; Challen 2000). First, we want to understand the reasons behind the different ways in which Portugal restructured water management at different moments in time. Second, we address another central question of this book concerning the effect which the formal and informal institutional reconfiguration of water management has had for the attainment of certain criteria such as coordination and cooperation among actors, accountability, transparency of decision making, dialogue in the course of water management, and the kind of legitimacy that water governance has achieved that way. We describe water management reform and the introduction of river basin organisations (RBOs) from the national-level perspective because reform in a unitary state such as Portugal automatically concerns the entire country.

In short, and applying Huitema and Meijerink's framework (Chapter 1, this volume), at the sub-national level, Portuguese water management started out with a very weak type of coordinating RBO. Under this framework, representation at river basin level until 2008 was limited to consultation and approval on proposed river basin planning documents, which were either devised by the national water agency or by coordination commissions that followed territorial boundaries. In 2008, financially and administratively independent River Basin District Authorities (Administrações de Região Hidrográfica), which are autonomous RBOs according to Huitema and Meijerink's typology, were introduced. The competencies of the national Water Institute (Instituto da Água) were adjusted to this new governance scheme. In 2011, the water administration was reformed again and River Basin District Authorities were placed under the direct supervision of the Portuguese Environmental Agency (Agência Portuguesa do Ambiente), approximating them to what Huitema and Meijerink called RBO agencies. As will become clear, five dimensions were of great importance during the institutional reconfiguration of water management in Portugal: (1) the boundaries within which an administration manages water resources; (2) whether the national level determined and coordinated management activities within sub-national management units, or whether sub-national units were financially and administratively independent; (3) the way in which users and stakeholders within basins were allowed to influence water management decisions; (4) the role of pricing of water uses; and (5) the way data about water uses was raised and administered. Overarching challenges that these various transformations of water management in Portugal faced since 1986 were the insufficient human, financial and technological resources to make the regulated formal institutions work operationally. This leads to institutional failures and dysfunctional and problematic performance, although some advances were noticeable. Therefore, while seemingly a flexible trial-and-error approach, as has been advocated by recent work on adaptive management (Pahl-Wostl 2009), in fact neither the time nor the resources were available for any institutional configuration to prove its performance.

In our analysis we will focus on the shift in the Portuguese administrative set-up for water management which took place between 2005 and 2008, and only briefly mention the latest water reform at the end of 2011, because it is too recent to assess. In the following section 7.2 we describe the case study and in section 7.3 the institutional setting upon which water management reform in 2005 built. Also, we introduce the starting point for reform in 2005 before we evaluate it (sections 7.4 and 7.5). In section 7.6 we describe the reform process and refer to the way rule-making was organised and how collective entities were set up. In section 7.7 we describe the new water governance set-up after the 2005–2008 reform and the way RBOs were established in regard to different types of rules and governance functions. Finally, we assess the implications of the reform in section 7.8, and section 7.9 draws the conclusions.

The study is based on in-depth knowledge of Portuguese water management that was acquired through working in and studying the sector for over ten years (Thiel 2004, 2009a; Thiel and Egerton 2011; Brito et al. 2008a). For the preparation of this chapter, ten interviews were conducted in 2009 and 2010 in order to understand the development of the 2005 reform, which was implemented in 2008, and to evaluate its impacts on water management practices. Interviews were conducted in two important water administrations, the transnational rivers Douro and Tagus, out of the five Portuguese river basin administrations altogether. However, as the reform was only two years in place at the time of the interviews, our assessment could obviously not be conclusive concerning all its effects.

7.2 PORTUGAL AND CASE STUDY SETTING

Portugal is located on the south-western end of the Iberian Peninsula and borders Spain to the east and north, with the Atlantic Ocean to the south and the west. In addition to mainland Portugal, it consists of the autonomous regions of the Azores and Madeira, archipelagos located in the Atlantic Ocean. Portugal may be considered a water-rich country with an average annual precipitation of 820 mm (of which 350 mm can be used), but spatial distribution varies to a great extent, from north to south and from the coast to the inland zones. The precipitation regime has an uneven yearly distribution pattern, typical of Mediterranean climate.

Portugal is a downstream country in a transboundary river context: more than 60 per cent of its surface water resources are collected in Spain, and a long tradition of cooperation in the use of water resources exists, dating back to the first bilateral agreement in 1864. In Portugal a major fraction (almost 80 per cent) of water use is to satisfy agricultural demand, while 9 per cent is for industrial use, 7 per cent for urban use and 4 per cent for other uses (Thiel 2009a). These water allocations are highly seasonal with large concentrations in the summer, when agriculture, tourism and urbanised coastal areas have stronger demands (Nunes Correia et al. 1998). This constellation may lead to temporal water services breakdowns in periods of drought. In addition, large floods are a serious matter of concern in Portugal. Transboundary cooperation is necessary to minimise both of these extreme hydrological phenomena, which are expected to occur more often and/or more intensely under future climate change scenarios (Brito and Lacerda 2011).

Groundwater and surface water pollution are still important challenges and ecosystems restoration only recently became a priority issue in Portugal.[1] In contrast, water supply and sanitation services have greatly improved throughout recent decades. Water supply systems reached 80 per cent of the population in 1990 and 97 per cent in 2009. The progress of sanitation infrastructure was even more relevant, with 38 per cent of the population having access to a public sewerage system in 1985 and 83 per cent in 2009. In the same period, bathing water quality improved, from 40 per cent not conforming to bathing water quality in 1994, to 0.7 per cent not conforming in 2010. In 1994, about 50 per cent of drinking water conformed to EU standards; while in 2009, 97 per cent was of good drinking water quality.[2]

With regards to its institutional set-up, Portugal was always a unitary state that was politically very centralised. This tendency was reinforced under the political dictatorship during the years 1926–1974. Following its fall in 1974, after the 'Cloves Revolution', it transitioned to a democratic, unitary state with two autonomous regions (the Azores and Madeira archipelagos). Further, elected local authorities were strengthened, obtaining deconcentrated[3] competencies, for example, in land planning and property taxing. In 1979, in mainland Portugal five Regional Coordination Commissions (Comissão de Coordenação Regional) were established (North, Centre, Lisbon and Tagus Valley, Alentejo and Algarve) with the purpose of coordinating regional planning, territorial management and inter-sectoral cooperation in the mainland. Since the beginning of the 1990s, the central government ministries have decentralised representatives at this level. However, decision making remained rather centralised (Soromenho-Marques et al. 2004).

7.3 THE INSTITUTIONAL SETTING OF WATER MANAGEMENT BEFORE THE REFORM IN 2005

Until the transposition of the EU Water Framework Directive into national legislation, in 2005, Portuguese water law was considered fragmented across a whole set of different legislations since the first Water Law in 1919 (Cunha Serra 2003). Water management competencies were centralised during the dictatorship period. From the 1960s onwards, with increasing efforts to modernise the country, the administration followed what has elsewhere been called the 'hydraulic paradigm' (Sauri and del Moral 2001): utilising water for economic development, specifically irrigation in agriculture and hydropower production, with less consideration for the environmental impacts (see also Jaspers and Gupta, Chapter 2, this volume). From the late 1970s, once the country had started to open itself up after the aftermath of the revolution, significant infrastructure investments funded with external support were launched. This development was accelerated by Portuguese entry into the EU in 1986. EU Directives were transposed into national legislation and funding was directed to industry, fisheries and restructuring of agricultural production, but also to a comprehensive set of infrastructures for water supply and sanitation purposes.

Until 1994, at the sub-national level of administration, water was managed by five Regional Coordination Commissions within their administrative boundaries. They responded to several ministries, principally the Ministry of Public Works. At the end of the 1980s a new Secretary of State for the Environment attempted to undermine the centralised national water authority and to reinforce the river basin approach, trying to introduce river basin administrations in order to abandon highly dependent Regional Coordination Commissions. This was part of advocacy on a more holistic and ecocentric approach to water management (Nunes Correia et al. 1998) which was backed up by academic and expert groups. It aimed to introduce a paradigm shift and strengthen environmental considerations in water management.[4] Nevertheless, these attempts were sidelined in the beginning of the 1990s when water management was allocated to the Regional Coordination Commissions by the Minister of Public Works in order to maintain a strong role for the existing Water Institute (Cunha Serra 2003). Also, by strengthening the Regional Coordination Commissions instead of introducing RBOs, decision makers favoured achieving compliance with European water supply and sanitation standards in regard to urban water uses leaving the improvement of quality of surface and groundwaters to a second phase (Thiel 2009b, 2009a).

In 1993, the newly created Ministry of Environment started operating and environmental issues, including water, were separated from the

Regional Coordination Commissions by the introduction of Regional Directorates for Environment and Natural Resources (Direcção Regional do Ambiente e Recursos Naturais). They had the same geographical boundaries as the Regional Coordination Commissions and competencies for managing water resources were shared between the national Water Institute and these five Regional Directorates for Environment and Natural Resources, but the Water Institute maintained the decisive role.

In 2003, environmental, land use planning and water management competencies of the Regional Directorates for Environment and Natural Resources and the regional development competencies of the Regional Coordination Commissions were again brought together, creating the Regional Commissions for Coordination and Regional Development (Comissão de Coordenação e Desenvolvimento Regional). This indicated the increasing competencies of the Ministry of Environment, Cities and Spatial Planning (2002–2004), and the intention to unite land, water and environmental management, economic development and administrative support to municipalities. On the other hand, it also opened doors for flexibly, shifting priorities and capacities between the various sectors of which the Regional Commissions for Coordination and Regional Development were in charge. This was often considered detrimental to water management because it was argued that these commissions did not pay sufficient attention and resources to water management issues.[5]

7.4 RBOS AND GOVERNANCE FUNCTIONS UNDER THE WATER LEGISLATION BEFORE 2005

In 1994 a set of laws concerning water administration, planning and financing had been adopted that provided the starting point for a complete transformation of Portuguese water management in 2005. Under this regime, coordinating RBOs were established in Portugal, as well as so-called river basin councils. Their territorial delimitation coincided with the responsibility of administrative units (see Table 7.1). With reference to the typology of rules used in this volume for analysing the type of RBO, these basin councils did not have their own authoritative means, nor did they have funds or specific informational resources. Therefore, if at all, river basin entities in that period were weak coordinating RBOs (see also section 7.5).

Instead, under the guidance of the Ministry for Environment, the Water Institute (for transboundary rivers) together with the competent decentralised administrations (for national rivers) held most authority.

Table 7.1 Membership and competencies in consultative bodies at river basin level in Portugal

	1994 water law	2005 water law
Members of consultative body – river basin level	Representatives of: central administration (27%), regional administration (23%), municipalities (19%), water users (25%), non-governmental organisations (5%), universities (1%), and others participants (15).	Representatives of: central administration (22%), regional administration (14%), municipalities (11%), water users (30%), non-governmental organisations (4%), universities (3%), professional associations, water experts, and others participants (16%).
Competencies of consultative body – river basin level	Consultation on river basin plans published in 1999–2001 and, rarely, evaluation of other main studies or projects of interest in the river basin.	Evaluation of annual activity plan and reporting of River Basin District Authorities and consultation on river basin management plans and other water-related plans and strategies, including water taxes.

The most important consultation forum was the National Water Council (Conselho Nacional da Água), the advisory board of the minister. Further, in regard to boundary rules, relevant decision-making bodies were the state agencies in the sector. Information channels involved state agencies and basin councils. In principle, citizens similarly had access to information. Finally, in terms of pay-off rules the system was financed by the central state budget. The Regional Directorates for Environment and Natural Resources, and after 2003 the Regional Commissions for Coordination and Regional Development could augment their income from the penalties and licences they handed out (see also Table 7.2).

In this period, in relation to governance functions, the competent Regional Directorates for Environment and Natural Resources held the authority over interrelated decisions on providing for and coordinating among water uses (including dispute resolution) under the guidance of the Ministry for Environment and the Water Institute. This role was embodied in their planning competencies which were held by the Water Institute for transboundary basins. For national river basins planning competencies were held by the regional level and backed by the Water Institute. Standard licensing competencies were held by the regional level.

Table 7.2 Overview of the institutional set-up of water governance in Portugal since 1994

	1994–2008 (INAG and CCDR as public water actors)	2008–2011 (ARH and INAG as public water actors)
Authority rules	The Water Institute (Instituto da Água, INAG) is the main public actor regarding all administrative water governance functions (planning, licensing, monitoring, inspection and infrastructure management) and is under the direct supervision of the minister. The Comissão de Coordenação e Desenvolvimento Regional – Commission for Coordination and Regional Development (CCDR) acts as regional representatives of all environmental issues (including water), being administratively independent from INAG they depended on the Minister of Environment. CCDR is an agency at the territorial level while at the river basin level coordination is reached through the planning exercise.	The River Basin District Authority (Administracão de Região Hidrográfica, ARH) is the public agency at the levels of the basin with all administrative duties and financial and administrative autonomy. The Water institute (INAG) remained with a strategic role at the national level, encompassing international representation, but without financial autonomy. Both were under the direct supervision of the Minister of Environment. The approach is towards a hydrological model in which ARHs act as autonomous river basin organisations.
Aggregation rules	The National Water Council had a significant role and river basin councils were rather irrelevant.	The National Water Council remained very active and river basin councils were empowered.
Boundary rules	Centralised state decision.	A (slow) process of empowerment of municipalities and users' associations was started.
Information rules	National water information system.	National water information system and river basin water information systems.
Pay-off rules	Water management financed by national budget and to a small extent by penalties raised by the regional-level administration.	Water management mostly financed by user/polluter-pays principle and by external projects. Funding transfers among river basin districts were possible.

Information gathering as well as quality monitoring functions coincided with planning and coordination competencies. Major public waterworks were directly administered by the Water Institute, while most infrastructures for irrigation were administered by the Ministry of Agriculture.

Monitoring of compliance with legal obligations and enforcement rested with the Water Institute, the regional-level administrations and later also with the General Environmental Inspectorate.

Functions of the administration at the national and regional level were to be financed from licence fees paid by users and fines on illegitimate water use. Indeed, the 1994 law prescribed that the water administration received income from water resource charges and from the national budget, but only the last one was used. Furthermore, users were to pay directly for water supply and sanitation services for which local authorities are responsible. Since the mid-1990s, gradually municipalities paid for bulk water from the national water company in public ownership established in 1993, Águas de Portugal, which increasingly also became responsible for municipal wastewater treatment.

Decision making within the water administration was to be in consultation with the national and deconcentrated administration of sectors that were affected by water management (for example, industry and energy, economy, commerce and tourism, agriculture and public health). Projects were also subject to environmental impact assessment and, following the Aarhus Convention, the public had the right to obtain information about on-going projects and plans and to present their suggestions and objections during the discussion periods. The water policy and the national water plan were discussed by the National Water Council. In addition, water planning was to involve all stakeholders and populations without further prescriptions on how this was to be accomplished (Rocha and Vieira 2000).

7.5 EVALUATION OF THE PRE-2005 INSTITUTIONAL SET-UP OF WATER MANAGEMENT

Dialogue between the administration and users and among stakeholders was insufficient. The institutional model described above was assessed as malfunctioning on various grounds: it was considered ineffective with regards to water resources taxation, licensing and inspection, environmental conservation and implementation of the river basin plans that had been released in-between 1998–2001. Horizontal coordination and collaboration between actors and between water, land use and economic development policies within and beyond specific basins were lacking (Nunes Correia 1999; Nunes Correia et al. 1998). Although the regional level was formally in charge of licensing, all strategic decisions were taken at the national level by the Water Institute. One interviewee therefore referred to deconcentrated bodies as 'letter boxes' of the central level (Thiel 2009a).

Concerning accountability, Nunes Correia (1999) and Nunes Correia et al. (1998), who provide the principal assessments of water governance at that time, diagnosed lack of key information and difficulties in access of the public to existing data, as well as unclear competencies. On the whole these authors lament a lack of participatory democratic water culture in Portugal. Occasionally, case studies have shown that data and information were strategically used by the administration in order to ground its priorities (Thiel 2009a). Dialogue between the administration and users and among stakeholders has been said to hardly exist. River Basin Councils only met occasionally, were not attended much by stakeholders and proved largely irrelevant because they were unable to influence the implementation of river basin plans and other sectoral water programmes.

For Nunes Correia et al. (1998), the problems of water management stemmed from limited financial resources, lack of political interest in environmental protection and management, and rigidities of the administration. Specifically, the regional level complained about insufficient funds to carry out their vital implementation functions (Thiel 2010). Intended cost recovery was never achieved because, while users were formally required to pay for water use, in fact they only paid for licences and occasionally they paid fines if they were found to infringe legislation. A main problem was that water use fees were never fixed despite the fact that they were legislated for in the first law on water financing of 1994. Therefore, financial resources were scarce and uneven because the main source of income for water management at the regional and national levels was allocations from the national budget. While management on the ground was not innovating at all, the water elite constantly theorised about ways of organising water management which was largely detached from management problems on the ground, making this a 'wishful thinking' approach (Nunes Correia et al. 1998).

7.6 THE PROCESS OF REFORMING WATER MANAGEMENT IN 2005

Throughout the first half of the last decade, the institutional set-up for water management was up for a significant reform because Portugal needed to transpose the Water Framework Directive. When in late 2004 a new government came into power, a rewriting of the water law had already been discussed for several years. However, a final decision was not reached because the government in that period had changed several times. Still, these extensive discussions led to the fact that a widespread consensus on the new legislation could be achieved relatively quickly among all parties and a new Water Law could be adopted in December 2005. At the heart of the reform were the introduction of a river basin management approach and intensified user participation and consultation. In 2008, economic instruments completed the new water governance architecture. These aspects, and specifically river basin management at a decentralised level, had been advocated by an influential group of Portuguese water experts since the mid-1970s (Veiga da Cunha 1980). Nunes Correia, the minister who came into office in 2004, was part of this group. He strongly believed in these principles, and he had previously failed to introduce basin management when he was director of the Water Authority at the end of the 1980s. In his new position he was able to manage the occasion of implementing the EU Water Framework Directive to introduce River Basin Organisations (RBOs) and a comprehensive set of necessary reforms.

Some initial opposition to the water reform came from the Ministry of Finance because it feared greater costs for water management under a reformed administration. Therefore, in order to overcome its opposition the reform was structured in a way that minimised the burden for the budget. The user-pays principle was introduced in 2008 and the legislation allowed for further sources of income in order to reach self-financing RBOs. In addition, the new administrations were drawn up in such a way as to minimise transition costs for the establishment of these entities because they used previous installations and existing infrastructures. Also, they only relied on technicians from the existing regional-level administrations, further reducing the costs of implementation.

Confrontation to the reform also came from some parts of the Water Institute and the Regional Development and Coordination Commissions, which stood to lose competencies and income after the introduction of autonomous RBOs. However, the minister disposed of a broad set of interrelated decision-making competencies which allowed him to overcome conflicts. In addition, given that Portugal is a unitary state, such a reorganisation of the state administration could be approved of by a simple act of legislation (Benz 2009). Experts, stakeholders and users were fully supportive of the reform and in fact considered it overdue in light of the above-described problems of effective integrated water resource management.[6]

7.7 THE INSTITUTIONAL SETTING OF PORTUGUESE WATER MANAGEMENT AFTER 2005

The Water Law that was adopted in December 2005 with a very large parliamentarian majority significantly restructured the institutional set-up of water management in Portugal. It aimed to harmonise and update previous laws and transpose the Water Framework Directive, but went beyond it. It reorganised water governance through the introduction of River Basin District Authorities (Administracões de Região Hidrográfica) which took over competencies from the regional-level institutions and the national Water Institute. These RBOs were established according to river basin areas and were not territorially congruent with the boundaries of the Regional Commissions for Coordination and Regional Development (see Figure 7.1). The 15 Portuguese river basins were grouped into eight hydrographic areas, which were managed by five specialised and decentralised administrative authorities. In this process, integrated water management of surface waters (coastal, estuarine and freshwaters) and groundwater, but also boundaries of infrastructures concerning water supply and sanitation systems, were considered (Thiel and Egerton 2011). This delineation of RBOs was said to be dominated by technical and ecological considerations while politics took less interest in it (Ministry of Environment, Spatial Planning and Regional Development, Lisbon, 15 May 2009; Aguas de Portugal, Lisbon, 15 May 2009).

The newly founded River Basin District Authorities can be categorised as independent, autonomous, technically oriented RBOs. These organisations concentrated information and expertise on specific water bodies, for which their leadership was considered accountable. Also, they had their own sources of income at their disposal. They adopted governance functions of administering provision and by implication also coordination and dispute resolution. They were to plan and execute water management also in coastal areas and license the corresponding uses. They were to gather information and monitoring activities because they maintained basin-wide and water-related information systems. Also, they were to enforce water regulations. Similarly, they were to execute public water works within their jurisdiction and implement a registry of water uses, and on that basis they were to levy fees for water extraction, pollution and other uses. That way, together with income from penalties, RBOs should be able to raise two-thirds of their income themselves and receive a contribution of the Ministry for the remaining third, if necessary.

The Water Institute was set up as a strategic coordinating body to support River Basin District Authorities and to represent Portugal

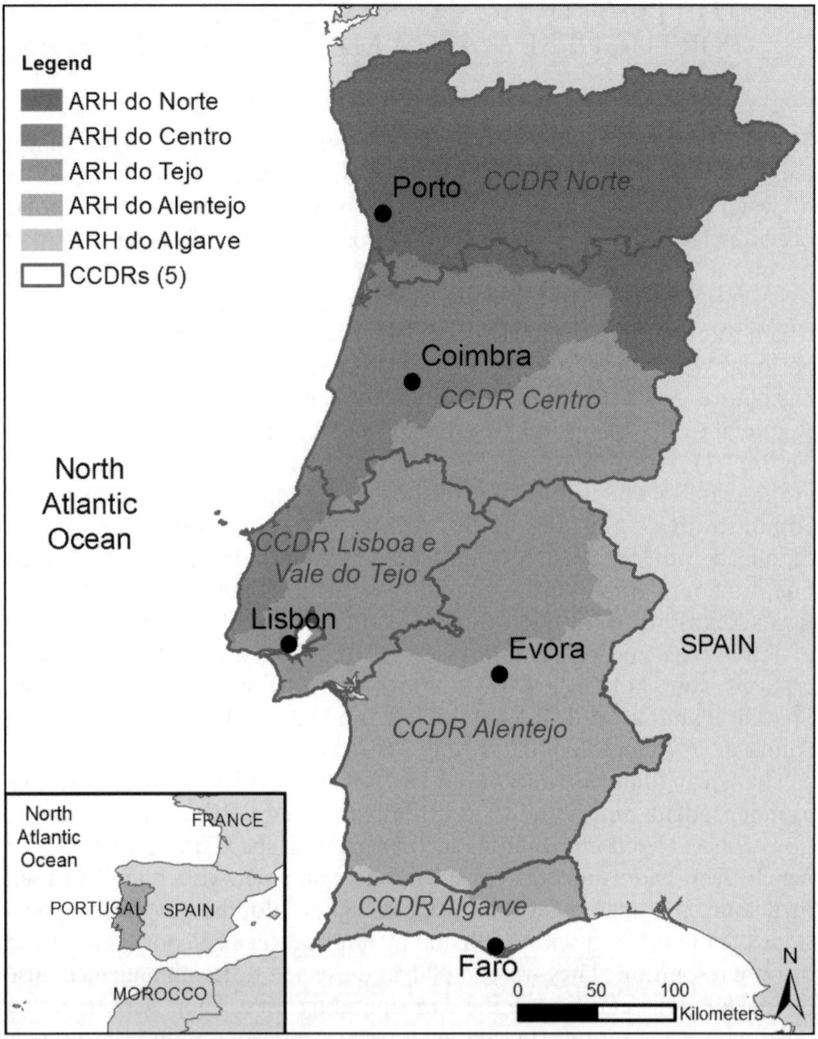

Note: Lines refer to territorially organised water management, grey areas to hydrographically organised water management; the Azores and Madeira regions are not shown.

Source: Adapted from Brito et al. (2008b).

Figure 7.1 Territorially organised water management and hydrographically organised water management in Portugal

internationally. Its mission was to guarantee the implementation of European laws, to coordinate methodologies, to promote infrastructure projects of national interest, to propose water use taxes and to approve river basin management plans prepared by RBOs. The Regional Development and Coordination Commissions lost all competencies in water management.

In the reformed set-up, River Basin Councils (Conselho da Região Hidrográfica) were to provide permanent mechanisms and forums for participation and effective consultation (see Table 7.1) (Assembleia da República 2005). Also the scope of issues presented to them and their rights had been expanded when compared to previous basin councils.

The law regulating financial aspects of water management that was issued in 2008 requested users to pay for all water-related uses.[7] The rate was to be fixed by the RBO in charge and the Water Institute under consultation with the River Basin Councils. Fifty per cent of the amount collected was to support the so-called Fund for Water Resources Protection,[8] 40 per cent the corresponding RBO and 10 per cent to the Water Institute. Table 7.2 shows the changes in rules applied to water governance before and after the introduction of the River Basin District Authorities following the framework exposed by Huitema and Meijerink's introductory chapter (Chapter 1 in this volume).

7.8 EVALUATION OF THE 2005 WATER LAW

In comparison to the previous water governance set-up which relied on centralised management and weak coordination, the 2005 reform allowed for more effective cooperation among water users, the public sector and stakeholders at the level of the river basin. Water management in international basins was facilitated because planning was no longer split between the Water Institute and the river basin level, and a strong interaction with Spanish RBOs counterparts could be initiated. The RBOs had a clear mission and followed a sectoral, basin-wide logic. Previously, several institutions needed to cooperate in order to implement measures, which led to struggles over responsibilities and cost-sharing. The reform was at the expense of coordination at the national level, across hydrographic regions and across sectors.[9] In contrast, the reform enabled management of functionally interconnected water-related issues within one single entity.[10]

The introduction of RBOs added another independent level of competence. As a result, overlaps between authorities increased and competencies were not clear, for example, in regard to coastal zones. This led to initial conflicts and delays in licensing water uses in some areas.[11]

Also, territorial and land use planning activities at regional level were expected to become more difficult in light of, for example, lack of coordination[12] because water management goals were better integrated in previous cross-sectoral territorial coordination entities.[13] In fact, some wondered if regional-level administrations were to be abolished entirely at some point.[14] Also, links to agricultural administrations were not well developed, but specific well-organised farmers' associations cooperated intensely with the river basin administration.[15] Environmental non-governmental organisations supported the creation of RBOs and used the River Basin Councils to express their concerns about water policy. RBOs recognised the importance of municipalities because of their land use planning competence, their local knowledge and administrative presence, and tried to foster an intense cooperation.[16] Indeed, RBOs aimed to progressively delegate competencies of licensing of registering water use and of collecting water use fees for municipalities if well-defined norms and rules were in place. In contrast, the previous relation between regional-level administrations and municipalities was at times characterised by suspicion.[17]

The described specific competencies and better coordination came at the price of less coordination across basins on the national level. First, the Water Institute proved to be rather slow to adapt to the new governance model and in coordinating across RBOs.[18] For instance, it reluctantly provided access to a water-related decision support system, and RBOs decided to design their own geographic water information systems with interoperability requirements. Second, the way in which the national water authority and the ministry were going to coordinate RBOs and exercise the right of approving basin plans and tariffs allegedly did not provide a lever to impose control over RBOs and informal bargaining could be expected to be used.[19] Third, the character of the national water plan was unclear and RBOs aimed at strategic planning not directly conflicting with river basin management plans under preparation at that time.[20] In consequence, the Water Institute concentrated on compliance of River Basin District Authorities with the Water Framework Directive as the lowest common denominator for river basin management.[21] Nevertheless, once investments in inter-basin measures were required by river basin management plans, the Water Institute would gain importance as it provided the required coordination and financial support.[22]

For lack of effective coordination by the Water Institute (or its acceptance), RBOs established an intensive informal consultation process among board members and on the operational level where they exchanged information on management processes and harmonised rules of implementation.[23] Decisions were taken by consensus, making coordination voluntary. However, despite its usefulness true national coordination involving all water administrations did not emerge.[24] Also, coordination across sectors was not addressed that way.

The introduction of basin-oriented administrations had positive implications for coordination on transboundary rivers, although the national water authority maintained its competency in technically representing Portugal at this level. After the Spanish central government had invited its RBOs to the transboundary commission ruling the Albufeira Convention, the Water Institute invited Portuguese RBO counterparts as well. In addition, some of them entered into informal, direct cooperation in the case of emergencies or joint European water protection and management projects.[25]

Three years after basin administrations had been introduced, interviewees recognised that the water administration had become significantly more effective and rational.[26] Water planning consultation and participation with stakeholders had gained significant momentum. RBO administrators themselves viewed river basin councils as important and encouraged the involvement of stakeholders, demanding the approval process of their activity plans and annual reports. Still, council meetings could be ineffective if basin administrations only presented their activities and if they did not involve the river basin council in the decision-making process. Despite these weaknesses, water users attended river basin councils in order to intensify informal bilateral contacts with the RBO.[27] Nevertheless, access to information and capacities of basin administrations to open up to user and stakeholder participation varied significantly. For example, reorganisation also led to the closing of local branches of the water administration inland, making access and information gathering more difficult in some areas.[28] Also, in many regions, users needed to pay for information and data.[29] Still, dialogue and accountability of decision making on the whole seemed to be strengthened. Specifically, large users engaged because basin administrations had become stable and were viewed as regular interlocutors that administered funds whose spending they wanted to influence.[30]

7.9 CONCLUSIONS

In Portugal, water governance has been subject to several significant changes throughout the last ten years. In hindsight, the legal framework that was valid between 1994 and 2005 was never properly implemented because the administration lacked funds to do so, water taxes were never fixed, competencies were overlapping and data about users and status of water resources were insufficient. Public participation and consultation, as well as accountability, were rather ineffective. Until 2005, the Water Institute maintained its predominant role in water management dominating sub-national water administrations that operated within administrative, territorial boundaries with weak, coordinating RBOs. This organisational set-up on water resources was cooperative for the paramount purpose of improving water supply and sanitation systems in Portugal, which were developed independently of hydrographic boundaries. However, basin-wide cooperation and user engagement performed weakly.

In 2005, the need to implement the Water Framework Directive triggered a Portuguese water governance reform in a way that implemented financially and administratively autonomous RBOs (River Basin District Authorities). The aim was to strengthen integrated water resources management at river basin level, stakeholder participation and environmental considerations in water management. The new Water Law in 2005 led to some disorientation due to transition requirements, but also initiated a new dynamic and enthusiasm in the sector as it realised a long-expected reform. At the basin level coordinated management was significantly enhanced. RBOs were very successful in registering water users because their financial sustainability depended mostly on their payments. They also tried to foster a new water culture and participatory culture among stakeholders. Basin-wide consultation became more effective because users had new rights of approval in the River Basin Councils, and because they perceived the RBOs as stable interlocutors. Furthermore, they expected engagement in basin management decisions to pay off because RBOs had administrative and financial autonomy to steer an array of water projects. At the time of the assessment of the reform, in 2010, users and stakeholders seemed to have a relatively clear picture of the new set-up although they still complained about some inertia inherent in implementing the new framework.

The way the reform was implemented minimised the effect on the Portuguese budget by structuring the administration similarly to the previous set-up and by expanding the financial contributions of users. That way, significant path dependence was introduced into the way the sector was reorganised. The Water Institute was left with a national role of oversight and representation of Portugal abroad. Also, it was designated to coordinate RBOs and to aim for economies of scale once RBOs collaborated.

In June 2011 a new government was elected and decided to reorganise the environmental administration. The five River Basin District Authorities and the Water Institute were merged into the Portuguese Environment

Agency. In that way River Basin District Authorities were downgraded to RBO agencies and the Water Institute was extinct. The primary public reason was the need to reduce the number of independent public administrative entities in order to ease the budget. Also, the previous reform had been criticised for maintaining a Water Institute that was considered ineffective in providing strategy and supporting RBOs and interoperability among their common tasks. In order to improve coordination, RBOs were disempowered and lost their financial and administrative autonomy. The Portuguese Environment Agency now holds all water and environment-related competencies, aiming at better coordination and territorial integration of water and all related environmental issues. In consequence, the coordination function of the national water authority in the Portuguese Environment Agency has been much strengthened in this model. Whether this centralised model will reinforce the effectiveness of water governance has yet to be seen. Some fears exist that accountability will be weakened as approval and consultation rules are less evident, namely in River Basin Councils. Another issue is the application of the water resources tax and whether the income distribution will allow for strong RBOs.

With regards to the role of the formal institutional setting for the performance of water governance, the Portuguese case shows that basin management is enhanced by fully fledged RBOs and consultation forums that have a stable role where water stakeholders perceive a gain from participation, financially or through access to actors and information. Therefore, concentration of financing, coordination, monitoring and licensing functions at the level of RBOs is beneficial to river basin management, but potentially detrimental to cross-basin coordination. Furthermore, the case shows that formal institutional reform of water governance in Portugal remains wishful thinking, where neither time nor financial resources allow the establishment of common knowledge and practices among actors for its institutionalisation. Where practices are not yet institutionalised, disorientation and lack of coordination are the consequence. Thus, time and financial resources are seemingly necessary for successful reform and were lacking in the case of Portugal since the start of the worldwide financial and economic crisis in 2009. Further, the case shows that river basin management and the requirements of the Water Framework Directive are extremely demanding with regards to human resources and technical requirements, furthering the need for investment in the institutionalisation of coordination structures and good water governance processes. Recent reforms in Portugal have never been fully completed or institutionalised, due to insufficient strategy, time and resources. As a result, it is hard to conclude about their potential and performance. In the context of a continuing budgetary crisis and limited capacities, we expect the challenges for participatory, sustainable and efficient water resources management in Portugal to persist.

ACKNOWLEDGMENTS

Andreas Thiel's research for this paper benefited from the People Programme (Marie Curie Actions) of the European Union's Seventh Framework Programme (under REA agreement N° 289374: 'ENTITLE') and from grants by the Einstein Foundation Berlin and the Deutsche Forschungsgemeinschaft (DFG).

NOTES

1. Ministry of Environment and Spatial Planning (Ministério do Ambiente, do Ordenamento do Território e. do Desenvolvimento Regional) (2011): Plano de Gestãoo da Região Hidrográfica do Tejo. River Basin District Authority Tejo (Administracão de Região Hidrográfica do Tejo), Lisbon.
2. Available at http://www.ersar.pt/website/ (accessed 22 October 2012) and National Inventory of Water Supply and Sanitation – INSAAR (INAG 2010).
3. Decentralisation is here referred to as the transfer of powers from central government to lower levels in a political-administrative and territorial hierarchy (Agrawal and Ribot 1999). This official power transfer can take the form of administrative decentralisation, also known as deconcentration, that refers to a transfer to lower-level central government authorities, or to other local authorities who are upwardly accountable to the central government (Ribot 2002).
4. In this regard, Portuguese academics and experts usually refer to a specific book that introduced a change in thinking about water management (Veiga da Cunha 1980).
5. Interview: River Basin District Authority North, Porto, 3 September 2010.
6. Interviews: CCDR (Comissão de Coordenação e Desenvolvimento Regional) Lisboa – Vale do Tejo, Portalegre, 17 September 2010; Associacao de Regantes e Beneficiarios Vale do Sorraia, Coruche, 16 September 2010.
7. Decree Law 97/2008, 11 June 2008, establishes the economic and financial framework based on water resources tax, water services tariffs and technical and/or economic protocols. The water resources tax results from the sum of five components, each referring to a specific water use. Under each component, different values were set regarding the different users. In addition, the regulation defines three coefficients of water scarcity and several reductions or exemptions.
8. Its aims are to promote efficiency of water use, decrease in pollution charge, decrease in environmental impact, improvement in ecosystem conditions, to contribute to flood control, improve coastal protection and administrative capacities of the national and basin water authorities.
9. Interviews: Administracões de Região Hidrográfica Norte, Porto, 3 September 2010; Associacao de Regantes e Beneficiarios Vale do Sorraia, Coruche, 16 September 2010; Administracõ de Região Hidrográfica Tejo, Lisbon, 15 September 2010; CCDR Lisboa – Vale do Tejo, Portalegre, 17 September 2010; CCDR Lisboa – Vale do Tejo, Lisbon, 21 September 2010; the National Water Institute (INAG), Lisbon, 6 September 2010.
10. Interview: CCDR Lisboa – Vale do Tejo, Portalegre, 17 September 2010.

11. Interview: Instituto de Conservação da Natureza e Biodiversidade, Alcochete, 8 September 2010.
12. Interview: CCDR Lisboa – Vale do Tejo, Lisbon, 21 September 2010.
13. Interviews: Administracões de Região Hidrográfica Norte, Porto, 3 September 2010; CCDR Lisboa – Vale do Tejo, Lisbon, 21 September 2010.
14. Interview: CCDR Lisboa – Vale do Tejo, Lisbon, 21 September 2010.
15. Interview: Associacao de Regantes e Beneficiarios Vale do Sorraia, Coruche, 16 September 2010.
16. Interviews: River Basin District Authority Norte, Porto, 3 September 2010; River Basin District Authority Tejo, Lisbon, 15 September 2010.
17. Interview: CCDR Lisboa – Vale do Tejo, Lisbon, 21 September 2010.
18. Interviews: River Basin District Authority Tejo, Lisbon, 15 September 2010; National Water Institute, Lisbon, 6 September 2010.
19. Interview: National Water Institute, Lisbon, 6 September 2010.
20. Interviews: National Water Institute, Lisbon, 6 September 2010; River Basin District Authority Norte, Porto, 3 September 2010.
21. Interviews: River Basin District Authority Tejo, Lisbon, 15 September 2010; National Water Institute, Lisbon, 6 September 2010.
22. Interview: National Water Institute, Lisbon, 6 September 2010.
23. Interviews: River Basin District Authority Norte, Porto, 3 September 2010; River Basin District Authority Tejo, Lisbon, 15 September 2010.
24. Interviews: River Basin District Authority Norte, Porto, 3 September 2010; River Basin District Authority Tejo, Lisbon, 15 September 2010.
25. Interviews: River Basin District Authority Norte, Porto, 3 September 2010; National Water Institute, Lisbon, 6 September 2010; River Basin District Authority Tejo, Lisbon, 15 September 2010.
26. Interviews: CCDR Lisboa – Vale do Tejo, Lisbon, 21 September 2010; River Basin District Authority Tejo, Lisbon, 15 September 2010; National Water Institute, Lisbon, 6 September 2010; DG das Pescas e Aquicultura, Lisbon, 22 September 2010; Empresa Portuguesa de Águas Livres, Lisbon, 8 September 2010; Liga para a Proteccao da Natureza, Lisbon, 21 September 2010.
27. Interviews: Associacao de Regantes e Beneficiarios Vale do Sorraia, Coruche, 16 September 2010; National Water Institute, Lisbon, 6 September 2010; Liga para a Proteccao da Natureza, Lisbon, 21 September 2010.
28. Interviews: CCDR Lisboa – Vale do Tejo, Portalegre, 17 September 2010; Associacao de Regantes e Beneficiarios Vale do Sorraia, Coruche, 16 September 2010; CCDR Lisboa – Vale do Tejo, Lisbon, 21 September 2010.
29. Interview: Instituto de Conservação da Natureza e Biodiversidade, Alcochete, 8 September 2010.
30. Interviews: CCDR Lisboa – Vale do Tejo, Lisbon, 21 September 2010; Associacao de Regantes e Beneficiarios Vale do Sorraia, Coruche, 16 September 2010.

REFERENCES

Agrawal, A. and J. Ribot (1999), 'Accountability in decentralization: a framework with South Asian and West African Environmental cases', *Journal of Developing Areas*, **33**, 473–502.

Assambleia da Republica (2005), *Lei da Água no 58/2005*.

Benz, A. (2009), 'Politik in Mehrebenensystemen', Wiesbaden: VS Verlag für Sozialwissenschaften, available at http://www.gbv.de/dms/ilmenau/toc/49787993X.PDF (accessed 1 August 2013).

Brito, A.G., S. Costa, J. Almeida, R. Nogueira and L. Ramos (2008a), 'The

Portuguese institutional reform for water management: the river basin district administration', 6th Iberian Congress on Water Management and Planning, Vitoria Gasteiz, Spain, 4–7 December.

Brito, A.G., S. Costa and R. Nogueira (2008b), *River Basin District Administration: Management Model and Fundamentals*, Lisbon: Ministry of Environment, Spatial Planning Regional Development.

Brito, A.G and M.R. Lacerda (2011), 'Impact of climate change in Portugal: key points regarding transboundary rivers shared with Spain', International Conference on Europe–Asia Transboundary Water Cooperation, UNECE, Geneva, 15 December.

Challen, R. (2000), *Institutions, Transaction Costs and Environmental Policy: Institutional Reform for Water Resources*, Cheltenham, UK and Northampton, MA, USA: Edward Elgar.

Cunha Serra, P. (2003), 'Water services in Portugal', paper presented at European Centre of Enterprises with Public Participation and of Enterprises of General Economic Interest, Coimbra, Portugal, 5 March.

European Communities (2000), 'Directive 2000/60/EC of the European Parliament and of the Council of 23 October 2000 Establishing a Framework for Community Action in the Field of Water Policy', *Official Journal of the European Communities*, L 327/1 of 22 December, pp. 1–72.

INAG (2010), *Plano e Ordenamento do Espaco Maritimo: Volume Sintese*, Lisbon: Instituto da Agua.

North, D.C. (1994), 'Institutional change: a framework of analysis', Economic History, doc. 9412001, available at http://ideas.repec.org/p/wpa/wuwpeh/9412001.html#provider (accessed 1 September 2013).

Nunes Correia, F. (1999), *Water Policy Formulation in Portugal. Water 21 – Eurowater*, Lisbon: Instituto Superior Técnico.

Nunes Correia, F., E. Beja Neves, M. Alzira Santos and J.E. Da Silva (1998), 'Portugal', in F. Nunes Correia (ed.), *Institutions for Water Resources Management in Europe*, Rotterdam and Brookfield, VT: Balkema, pp. 324–348.

Pahl-Wostl, C. (2009), 'A conceptual framework for analysing adaptive capacity and multi-level learning processes in resource governance regimes', Global Environmental Change-Human and Policy Dimensions, **19** (3), 354–365.

Ribot, J. (2002), *Democratic Decentralization of Natural Resources: Institutionalizing Popular Participation*, Washington, DC: World Resources Institute.

Rocha, I. and D.F. Vieira (2000), *Água*, Porto: Porto Editora.

Sauri, D. and L. del Moral (2001), 'Recent developments in Spanish water policy: alternatives and conflicts at the end of the hydraulic age', *Geoforum*, **32** (3), 351–362.

Soromenho-Marques, V., M. Queiros and M. Vale (2004), 'National report on regional sustainable development. EU 5th Framework Programme. REGIONAT – Thematic Network: Strategies for Regional Sustainable Development, an integrated approach beyond best practice', unpublished document.

Thiel, A. (2004), 'Transboundary resource management in the EU: transnational welfare maximization and transboundary water sharing on the Iberian peninsula?', *Journal of Environmental Policy and Planning*, **47** (3), 331–350.

Thiel, A. (2009a), *Environmental Policy Integration and the Development of Water Use in the Algarve, since Portugal's Accession to the EU*, Aachen: Shaker.

Thiel, A. (2009b), 'Europeanization and the rescaling of water services: agency and state spatial strategies in the Algarve, Portugal', *Water Alternatives*, **2** (2), 225–244.

Thiel, A. (2010), 'Institutions shaping coastal ecosystems: the Algarve case', *Coastal Management*, **38** (2), 144–164.

Thiel, A. and C. Egerton (2011), 'Re-scaling of resource governance as institutional change: the case of water governance in Portugal', *Journal of Environmental Planning and Management*, **54** (3), 383–402.

Veiga da Cunha, L. (1980), *A Gestão da Agua. Princípios Fundamentais e sua Aplicação em Portugal*, Lisboa: Fundação Calouste Gulbenkian.

8. The politics of establishing catchment management agencies in South Africa: the case of the Breede–Overberg Catchment Management Agency

Richard Meissner and Nikki Funke

8.1 INTRODUCTION

South Africa is a water-scarce country, a fact aggravated by climate change and international obligations to neighbouring countries with shared watercourses (Claassen 2010). Some of the political, social and economic pressures facing South Africa's decision makers in the water sector include having enough infrastructure to secure water during low rainfall periods and supply areas of high demand, growing enough food to supply the population, and meeting the water demands of energy, industry and mining (Claassen 2010). At present most of the country's water supply has already been allocated. The only remaining 'supply options' available are linked to reallocations between different water use sectors (De Lange 2010). An additional problem that aggravates South Africa's situation of water scarcity is the deteriorating water quality in the country's major river systems, water storage reservoirs and groundwater resources, which results in social, economic and health risks to society (Ashton 2010). Particular problems include acid mine drainage, eutrophication and soil erosion.

In addition to the problems of water scarcity and water quality, the South African government also faces the challenge of redressing the backlog in water supply and sanitation that it has inherited from the apartheid government. The South African government started the process of developing and introducing a number of water reforms when it came to power in 1994, to address this backlog and manage South Africa's situation of water scarcity. In summary, the Water Services Act (Republic of South Africa 1997, Act No. 108) and the National

Water Act (Republic of South Africa 1998, Act No. 36) aim to redress the inequalities of historical racial and gender discrimination, link water management to economic development and poverty eradication, and to ensure the conservation of the ecological resource base for posterity (Schreiner et al. 2002).

The National Water Act is characterised by a number of key principles that set it apart from previous water legislation in South Africa (Funke et al. 2007). One of these principles is the focus on decentralisation, which emphasises public participation in water management-related decision-making processes, and also leans on the subsidiarity principle as enshrined in the South African Constitution (Republic of South Africa 1996, Act No. 108). This principle stipulates that those functions that can be more efficiently and effectively carried out by lower levels of government should be delegated to the lowest appropriate level (Funke et al. 2007).

The principle of decentralisation is particularly relevant for this chapter, as it is directly related to catchment management agencies (CMAs), South Africa's equivalent to river basin organisations (RBOs). In October 1999 the South African government established 19 water management areas (WMAs), whose boundaries are along catchment divides, but do not coincide with administrative boundaries as defined by local government. At the time, the establishment of 19 CMAs was envisaged. The CMAs are meant to perform certain management functions with which they have been tasked, and are also required to cooperate and seek agreement on water-related matters amongst various stakeholders and interested parties. Currently these management functions are still the responsibility of the Department of Water and Sanitation (DWS).[1] A CMA is governed by a governing board to ensure stakeholder representativeness and to prevent powerful parties with vested interests from exercising control over other parties. CMAs also have a mandate to develop catchment management strategies (CMS), which constitute a plan to realise the protection, use, development, conservation, management and control of water resources in their respective WMA (DWAF 2004a; Funke et al. 2007).[2]

Unfortunately, since the National Water Act was promulgated in 1998, the implementation of South Africa's water legislation has been slow and problematic (Funke et al. 2007). These implementation problems also take the form of a serious backlog in setting up CMAs (Hattingh et al. 2004), as to date only two of the originally planned 19 CMAs have been established. These CMAs are the Inkomati CMA and the Breede–Overberg CMA (BOCMA) (DWA 2012).

In 2012, DWA made a decision to reduce the number of 19 planned

CMAs to nine, as stated in its second National Water Resource Strategy (DWA 2012). This reduction of the number of CMAs is the result of a reconsideration of the management model and viability assessments related to water resources management, funding, capacity, skills and expertise in regulation and oversight, and an effort to improve integrated water systems management. The nine CMAs are: Limpopo, Olifants (Mpumalanga Province), Inkomati–Usuthu, Pongola–Umzimkulu, Vaal, Orange, Mzimvubu–Tsitsikamma, Breede–Gouritz and Berg–Olifants (Western Cape Province) (DWA 2012). While we have noted these most recent developments around CMA establishment in South Africa, the focus of this chapter is on BOCMA's establishment process and functioning to date.

The BOCMA case study was selected because it has not yet been extensively studied, particularly when compared to the Inkomati CMA. Therefore there is room for learning more about the dynamics behind the establishment and current functioning of this CMA. We used a mixed-method approach in conducting the research for this chapter. We conducted a desktop study to identify and analyse literature that has been published on CMAs in general and BOCMA in particular. It appears that to date not many studies on BOCMA's establishment and functioning have been completed. To address this information gap, we also conducted semi-structured interviews with individuals who were either involved in the establishment of BOCMA or who are currently conducting research on the CMA. A one-day field trip to the Breede–Overberg area was also undertaken to allow us to get a sense of the circumstances and natural environmental realities that characterise some parts of the WMA.

In the next section we turn to a discussion of the Breede–Overberg WMA, followed by a brief history of the establishment of BOCMA. This is followed by a section on the institutional design of CMAs (8.3), with particular reference to BOCMA, in terms of the various rule types and types of RBOs as outlined in Chapter 1 of this book. In section 8.4, we discuss the politics and strategies involved in the introduction of the CMA concept into the National Water Act and the latest developments around CMAs in South Africa. We then reflect in section 8.5 on what can be surmised about BOCMA's democratic functioning and performance to date before concluding the chapter (section 8.6).

8.2 THE BREEDE–OVERBERG CATCHMENT MANAGEMENT AGENCY

8.2.1 Overview of the Breede–Overberg Water Management Area

The Breede–Overberg WMA stretches over an area of 19 789 square kilometres (km^2); the Breede area covers 12 600 km^2 and the Overberg area covers 7189 km^2. The Breede River Valley is bordered by the Franschhoek and Du Toit's Mountains in the West, the Hex River Mountains in the North, the Langeberg Mountains in the East and the Indian Ocean in the South. Other WMAs bordering the Breede–Overberg WMA are the Berg, Olifants/Doorn and Gouritz WMAs. In the mountain areas the rainfall of the WMA is highest at around 3000 millimetres per annum (mm/a) compared to the central and north-eastern regions where rainfall is as low as 250 mm/a. Occasional snowfalls are not uncommon. The average annual evaporation ranges from 1200 mm in the south to 1700 mm in the north. The mean annual run-off for the Breede–Overberg is

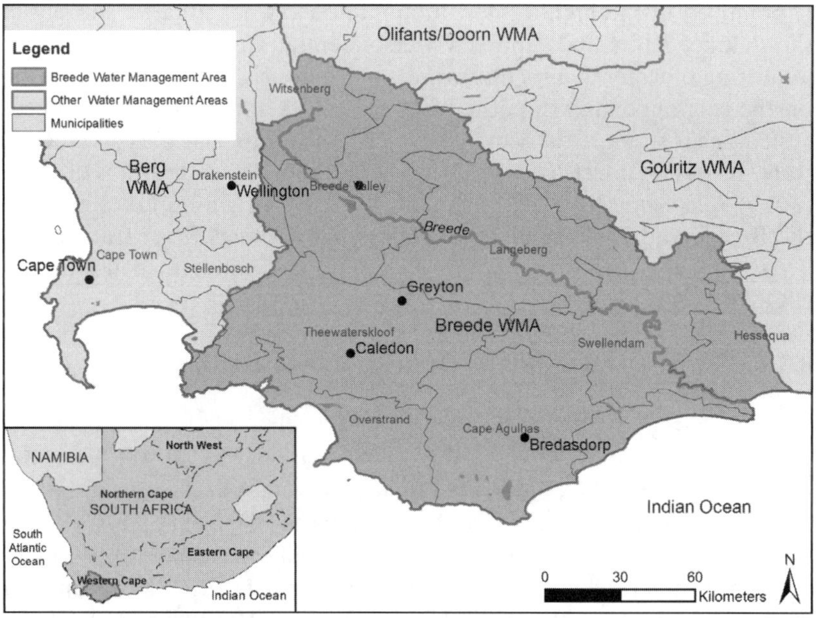

Sources: Based on DWA (2013) and BOCMA (2011).

Figure 8.1 The Breede–Overberg catchment area in relation to South Africa's provinces

2462 million cubic metres per year (mm^3/yr) (DWAF 2004a; BOCMA 2011; Nel et al. 2011).

The estimated size of the population of the WMA is around 500 000 people, with two-thirds residing in towns and villages. In 2004 DWAF noted that demographic projections indicate population growth in the coastal areas only, with a decline in the inland areas. The department therefore envisaged the total population of the WMA to remain constant (DWAF 2004b). Two district municipalities and seven local municipalities are located in the area. The district municipalities are the Cape Winelands and the Overberg District Municipalities. A small portion of the Eden District Municipality also falls within the WMA. The local municipalities are Witzenberg, Breede Valley and Langeberg to the north, Theewaterskloof and Overstrand in the south-west, and Cape Agulhas and Swellendam in the south-east (BOCMA 2011).

The major economic sectors in the WMA are agriculture and coastal tourism, which make the economic base of the area quite water-dependent. Agriculture is dominated by wheat cultivation. Other crops that are cultivated include orchard crops, vineyard crops, citrus and lucerne. Forestry is practiced in the high-rainfall mountainous areas, situated entirely in the Palmiet River and Upper Riviersonderend River Catchments. The allocation of water to agriculture and coastal tourism has a direct bearing on the economic growth paths of these sectors and on the natural ecosystems that support tourism. There are indications that the catchment's water resources are stressed and that aquatic ecosystem health is deteriorating. According to the department, intensive irrigation takes place in the Breede and Riviersonderend Valleys and Palmiet River Catchment. These are also the areas where economic activities are concentrated (BOCMA 2011).

8.2.2 Establishment of Breede–Overberg Catchment Management Agency

BOCMA was established in 2005 under the National Water Act (BOCMA 2011) and the BOCMA Governing Board was established two years later.[3] Prior to the establishment of the CMA and the development of its CMS, the Western Cape Regional Office of DWAF was fully responsible for the management of the catchment's water resources through the WMA's internal strategic perspective (DWAF 2004a). According to McConkey et al. (2005), the establishment of BOCMA was the result of deliberate negotiations between various stakeholders, assisted by the Western Cape Regional Office of DWAF. The main conduit for this stakeholder process was the BOCMA Reference Group, which consisted of representatives from various sectors.[4]

A number of actors were involved in the CMA's establishment process. The most notable actor was DWAF at national and regional level. Others were also involved, specifically as part of the BOCMA Reference Group. Pegasys Strategy and Development, a consultancy, also played a key role in terms of facilitating stakeholder engagement throughout the BOCMA establishment process.[5] BOCMA (2009) notes that the first step in the development of the CMS that was released in February 2011 was the identification of stakeholders and previously disadvantaged groups (for example, emerging farmers[6]). In May 2010, a networking meeting was held with all governmental authorities: provincial and semi-government departments, local and district municipalities and water user associations (WUAs). This was to ascertain how their planning priorities would influence BOCMA's planning processes. The meeting was facilitated by Pegasys Strategy and Development (BOCMA 2010).

8.3 INSTITUTIONAL DESIGN

8.3.1 Introduction

This book makes use of five types of rules in its analysis of different institutional arrangements and, based on these rule types, also discusses different RBO types. This section deals with institutional design by first classifying CMAs and BOCMA in terms of these different rule types, and then proceeding to discuss them according to different RBO types.

8.3.2 Rule Types

The rule types referred to in this book are authority rules, aggregation rules, boundary rules, information rules and pay-off rules (see Huitema and Meijerink, Chapter 1, this volume). In the case of CMAs in general and BOCMA in particular, Table 8.1 summarises the rules that have become evident in CMA establishment and functioning.

Authority rules

CMAs are statutory bodies established in terms of the National Water Act and are able to develop their CMS. Democratic control is also exercised through the governing board, which is representative of all stakeholders and their interests in the WMA for which the CMA is responsible. It is in this regard that BOCMA plays a coordination governance function where it coordinates the actions of relevant stakeholders, including various actions along the governance value chain, such as financially

Table 8.1 Rule types evident during the establishment and functioning of catchment management agencies

Rule-types	Rules in case study
Authority rules	CMAs are quite autonomous and have some internal mechanisms for democratic control. The Minister of Water and Sanitation has a strong influence on the appointment of the governing board. The strategy and functioning of a CMA is contained in its CMS, which must set principles for water allocation, take into account all matters related to water resource protection, use, development, conservation, management and control, be in harmony with the National Water Resources Strategy, and must be reviewed every five years.
Aggregation rules	In the run-up to the development of CMAs, individual views need to be accommodated by means of public participation processes. In the case of BOCMA, two separate stakeholder engagement processes took place, one in the Overberg area and the other in the Breede River Basin. In the Overberg area public participation took the form of a series of public meetings, and in the Breede River Basin public participation was closely linked to the Breede River Basin Study. Decision making at the governing board level takes place through democratic agreement (consensus), cooperation and open discussions in meetings.
Boundary rules	CMAs have a strong focus on stakeholder engagement and are inclusive because provision is made for stakeholders to participate in catchment forums, catchment steering committees and as part of the governing board. A wide variety of actors are involved in the CMA organisational structures to represent as many sectors as possible. These sectors can include environmental interest groups, local and provincial government, business and industry, poor and rural settlements, the agricultural sector and non-governmental organisations (NGOs). However, being represented on an organisational structure is not the same as having power and influence in that structure. In the South African context it is typically those with the most resources and political clout that have the most influence in such structures. BOCMA's area of responsibility is the Breede–Overberg WMA. Two considerations were deemed important at the time

Rule-types	Rules in case study
	of establishing the country's original 19 WMAs: the natural hydrological boundaries and the potential for achieving the CMA's financial viability in the medium to long term.
Information rules	The scientific community played an important role in the development of BOCMA's CMS and the early development of the CMA through the Breede River Basin Study and has therefore influenced the 'language' of the CMA. Essentially the Breede River Basin Study assessed current (early 2000s) and future water use in the basin in an integrated manner by including land use and water demands, surface water resources, groundwater, soil types, climate and crop selection and so on. This study's focus on integration influenced the BOCMA establishment process, which also took place in an integrated manner (through stakeholder participation).
Pay-off rules	A CMA can be funded or recover costs from water use charges levied in its WMA (in terms of the pricing strategy for water use charges set by the Minister of Water and Sanitation), from money from any other lawful sources (e.g. grants, loans), or from money appropriated by Parliament. Some of BOCMA's funding comes from water management charges levied on water users. It also receives seed funding from the Department of Water Affairs. It remains to be seen how benefits and costs will be distributed in the CMA, a process that is not currently regulated. At the moment it seems that powerful actors stand to benefit more than marginalised actors.

sustaining the CMA, distributing benefits and dispute resolution. This coordination takes place on a horizontal, cross-sectoral level. Regarding the influence that other actors have on the functioning of CMAs, reference can be made to the Minister of Water and Sanitation's key role in CMA establishment. The minister also appoints the governing board and can remove board members for good reason, while the CMA oversees its institutional functioning (DWAF n.d.). In this respect the minister plays a rule-making, construction of collective entities, coordination, monitoring and enforcement governance function as well as an initial financing governance function.

While stakeholder engagement is firmly enshrined in the National Water Act and the Breede–Overberg CMS, BOCMA is nonetheless governed by a stringent hierarchy. The highest-ranking decision maker is the Minister of Water and Sanitation, who takes the most challenging and important decisions, such as appointing the CMA board or approving the CMS. After the minister, it is the board of directors that is tasked with making important decisions, which must be in line with the stipulations contained in the National Water Act (for example, equity considerations, stakeholder participation, and so on) and the decisions of the Ministry of Water and Sanitation.

Here it is important to reflect on a particularly important point as far as the autonomous functioning of CMAs in South Africa is concerned. As stated above, the power to appoint the CMA's board rests with the Minister of Water and Sanitation. While one could argue that this extent of power in the minister's hands negates the CMA's autonomous functioning; a counter-argument can be made for the importance of ensuring a reasonable degree of representivity at the CMA board level. This is especially important given the diversity of stakeholders in many WMAs, and the differing degrees of education, capacity, power and so on between, for example, marginalised water users and established commercial farmers.[7] BOCMA's governing board has representation from many corners, and it includes the following groups: emerging farmers, the Western Cape Provincial Government, civil society, industry and business, commercial agriculture, conservation, poor and rural settlements, and emerging farmers (BOCMA 2013).

Brown (2011) substantiates the need for strong government intervention when appointing a CMA board, and argues that given the particularities of the South African context, it may make sense to reassess and in fact strengthen the role of the state in participatory water resources management altogether. In addition Brown (2011: 180) challenges the widely held assumption that 'the higher the level of power devolved to participatory institutions at the local level', 'the greater the outcomes in terms of efficiency and social equity'. The danger exists that strong stakeholder participation may reinforce rather than reduce inequalities in the case of asymmetrical power structures. To tie in with the point above, a *laissez-faire* approach by DWAF could have resulted in a lack of representivity on the board of the two existing CMAs, with the power in the hands of the empowered stakeholder groups at the expense of their marginalised counterparts.

In addition to, and perhaps over and above, representivity, it is increasingly being recognised that it is also vital for a board to have the skills and capacity to function effectively. These considerations are likely to become increasingly prominent in the CMA board establishment process.[8] BOCMA's governing board has a range of competencies including a water resource manager, water resource planner, catchment management coordinator, water allocation reform officer, water use manager, water use specialist, water use officer, licensing clerk, participation and stakeholder engagement manager, water liaison officer, data manager, water data officer and data capturer. While the board has appointed people to fill these positions, and is reportedly doing well to date in terms of operational decision making,[9] it remains to be seen how well and to what effect these positions will be utilised, especially when challenges present themselves (BOCMA 2013). Since these competencies are linked closely with BOCMA's performance as an organisation and since it is still a fledgling organisation that has had a slow start, it is perhaps too early to reflect on this point, and more research into the matter is required.

Aggregation rules

In terms of aggregation rules individual views are accommodated in the public participation phase when catchment forums are formed in the run-up to developing a proposal to establish a CMA. The main purpose of this phase is to ensure that a trusting and constructive relationship is developed between all stakeholders and interest groups, and to find a common vision. While the CMA proposal is being developed, stakeholders may feel that a formal committee, representative of all stakeholders, may be required to guide the CMA establishment process. This would take the form of a catchment steering committee. Once the CMA is established, the minister, responding to the advice of an advisory committee, appoints the governing board, which must represent all stakeholders (including current and potential user groups) and the interests they have in the WMA (DWAF n.d.).

In terms of BOCMA's establishment process two separate stakeholder engagement processes took place: one in the Overberg area and one in the Breede River Basin. These differed slightly from the process foreseen by the government but still fulfilled the purposes of stakeholder consultation and public participation. In the Overberg area information sharing on CMAs and the process of developing a CMA proposal for the WMA formally commenced in 1999. A series of public meetings was held in 1999 and 2000 to inform the public about the CMA establishment process, and to identify individuals, stakeholders and interested and affected parties willing to serve on six catchment steering committees (one for each geographic region in the Overberg area). In June 2000 membership of the Overberg Stakeholder Committee was finalised. The first two meetings of the Overberg Stakeholder Committee were held before the BOCMA Reference Group was formed, which was to take forward the CMA proposal development process. A further round of six public meetings was held in November 2001 before the CMA proposal was finalised. The Overberg Stakeholder Committee was disbanded after the CMA proposal was first submitted in April 2002. Most of the six steering committees are, however, continuing with their meetings and have also changed their names to catchment forums to fit the terminology used in the National Water Act (DWAF 2004a).

Public participation in the Breede River Basin was closely tied to the Breede River Basin Study that started in 1999 and was published in October 2003. The study was conducted by Ninham Shand Consulting Services, MBB Consulting Engineers and Jakoet & Associates, on behalf of DWAF's Directorate of National Water Resource Planning (DWAF 2003). A formal public participation process ran for the duration of the study. The purpose of this public participation process was for the project team to get as much input from stakeholders as possible and to be able to share their findings with the public. The study and the CMA development process were closely linked. This link involved using the stakeholder committee of the Breede River Basin Study, which was formed in 2000, as a key component of the BOCMA Reference Group together with the Overberg Stakeholder Committee. Particular attention was paid to the inclusion of previously disadvantaged groups on the stakeholder committee of the Breede River Basin Study (DWAF 2004a).

The Breede River Basin Study and the Overberg Stakeholder Committees met for the first time on 15 November 2000 as the Breede WMA Stakeholders Forum. This meeting served for the committees to reach final agreement on the CMA proposal development process and to give preliminary inputs on the water resources of the Breede–Overberg WMA and current management arrangements. An agreement was reached that a smaller committee, the BOCMA Reference Group, would take the CMA proposal development process forward, working closely with DWAF and the support team (DWAF 2004a).

After BOCMA's proposal had been submitted to DWAF, a waiting period ensued. During this time the BOCMA Reference Group interacted with the advisory committee regarding the composition of the governing board that was still to be appointed. Meanwhile capacity building of water resource committees and forums continued (McConkey et al. 2005).

Currently decision making within BOCMA takes place by means of democratic agreement, cooperation and open discussions in meetings (Page 2012). Generally the CMA board tries to reach consensus on the decisions it takes.[10] This prevents a situation where strong board representatives are able to overwhelm weaker ones. Egalitarianism is the rationale behind this practice.

Boundary rules

In terms of boundary rules, CMAs are inclusive in that provision is made for stakeholders to participate in catchment forums, catchment steering committees and to be represented on the governing board (DWAF n.d.), as explained in more detail above. During the CMA proposal drafting process all CMAs are required to have a reference group to inform the proposal. This reference group would typically consist of 150 to 200 stakeholders, the aim being to represent as many sectors as possible, from conservation and environmental organisations to ratepayers' associations and forestry and service providers in the WMA. This large group is then split up into smaller groups, each of which has an opportunity to give inputs into the CMA proposal. For BOCMA this process was quite well run and representative.[11] The reference group included representatives from provincial government, local government, the agricultural sector, non-governmental organisations and environmental interest groups. The CMA proposal process was split up into different phases, and the reference group was consulted for each phase. This intensive stakeholder consultation process was very inclusive, but also took a very long time: nine years in total (DWAF 2004b).[12]

In the South African context, however, inclusivity is not only about rules that are stipulated on paper or being included in a particular committee or board, but much more about the agency of different stakeholders to influence decision-making and institutional processes, such as CMA establishment. So, for instance, an established commercial farmer will likely have considerably more power, in the form of resources, experience and clout, to influence decision-making and institutional processes than an emerging farmer, who may have very limited resources and experience at their disposal. Said differently, an emerging farmer can be represented or included in an organisational structure, but will not necessarily have the same power as a commercial farmer.

In addition, the geographical delineation of BOCMA goes back to the development of the National Water Act in the latter half of the 1990s. According to Page (2012), the following factors were important in determining the boundaries of a WMA: watercourse catchment boundaries, social and economic development patterns, efficiency considerations and communal interests within the area in question. Two particularly important considerations were the natural hydrological boundaries and the potential for achieving the CMA's financial viability in the medium to long term (Page 2012). At the time it was Minister Kader Asmal who oversaw the National Water Act drafting process and, together with DWAF, was responsible for the delimitation of the country's WMAs (McConkey et al. 2005). The rationale behind this process was to arrive at a situation where water resources would be developed for the greater good of society through both integrated water resources management and sustainability.[13] With regard to the establishment of the WMAs and the CMAs it is possible that the government at the time was trying to adopt prominent concepts, noticeably integrated water resources management and sustainability, without taking into consideration the potential difficulties of practically implementing these concepts.

Information rules

In terms of information rules, a key document containing important knowledge about the strategy and functioning of a CMA is the CMS that it is required to progressively develop and implement. A CMS must 'set principles for allocating water to existing and prospective users, take into account all matters in terms of the protection, use, development, conservation, management and control of water resources, be in harmony with the National Water Resources Strategy and be reviewed every five years' (DWAF n.d.: 18). In terms of gathering inputs into this strategy, this has to happen in consultation with stakeholders in the WMA and go through a public consultation process (DWAF n.d.). It therefore appears that the experiential knowledge of stakeholders is valued highly in the official language and functioning of CMAs. CMAs, once established, are also expected to communicate relevant information to other actors in the catchment.

An actor that stands out as having had a particularly strong influence on BOCMA's CMS is the scientific community, particularly individuals from engineering, hydrology and soil sciences that are former employees of the department and are now independent consultants. This influence is evident in the fact that the scientific community was involved in the design of CMAs at the national government level in the late 1990s, and has since then also produced a substantial volume of research reports on how CMAs should function and be managed in terms of the National Water Act, equitability (pro-poor governance approaches), integrated water resources management and the notion of sustainability (see Meissner et al. 2013). The scientific community has also identified a host of challenges that CMAs are likely to face in the run-up to their establishment and functioning thereafter. These include human, technical and financial capacity requirements (Pegram et al. 2006). In this regard, the CMA process, including that of BOCMA, is still quite technocratic, with knowledge from the environmental sciences and engineering being highly valued. The majority of knowledge regarding the establishment and running of CMAs is produced by natural scientists (Meissner et al. 2013). This means that the knowledge is likely to take the form of technocratic thinking that will, in turn, influence the establishment and operational processes of CMAs.

In terms of its current functioning and operations, it has taken a while for BOCMA to receive delegated functions and duties from the department and therefore the development of the operational process has been a bit slow. At the moment, there is a focus on getting a better grip on the hydrology within the WMA and to know exactly how much water is being used. There is also a focus on establishing WUAs and on cooperating with local government around water resource protection. There are currently two stakeholder forums, one for the Breede and one for the Overberg, but it is not clear how active they are.[14] It is therefore perhaps a bit too early to comment on exactly which information rules BOCMA has in place.

Pay-off rules

In terms of pay-off rules CMAs have certain inherent powers, that is, 'the powers of a natural person of full capacity', and can therefore open a bank account, enter into contracts and borrow money. A CMA can be funded or recover costs from water use charges made in its WMA in terms of the pricing strategy for water use charges set by the Minister of Water and Sanitation, from any other lawful sources (for example, grants and loans), or from money appropriated by Parliament (DWAF n.d.).

In terms of BOCMA it appears that some of the CMA's funding comes from water management charges levied on water users, and that this is supplemented by funding from DWS as initial 'seed' funding. There has also been some international funding provided by the Dutch Unie van Waterschappen (Page 2012). This funding has, however, been limited and there is a need for the CMA to become more financially independent in order to implement its CMS.

In terms of the question 'Who benefits and for what?' (Strange 1996), or how benefits and costs are to be distributed within the WMA, this is not yet clear in the case of BOCMA. An assumption can be made that strong and well-resourced actors (for example, government, scientists and consultants, agriculture and tourism) are likely to receive most of the benefits. This indicates that the National Water Act has not yet achieved its objectives of socio-economic development and redress, as those who continue to benefit from its implementation are already powerful actors, while poor and marginalised actors continue to be disadvantaged.

8.3.3 River Basin Organisation Types

Here follow some reflections on where CMAs fit into the typology of RBOs that was presented by Huitema and Meijerink in Chapter 1 of this volume, and also, specifically, where BOCMA fits.

CMAs fall somewhere between the different types of RBOs presented in this book. Given the authority rules that characterise them, as discussed above, CMAs are in some respects quite autonomous. However, they also show characteristics of agencies in that they can only be created through the minister's approval to perform water resource management at the regional or catchment level, and the minister controls the appointment of the board. This arrangement could potentially affect the independence of CMA board members. In addition, the initial functions of a CMA mostly centre on coordinating the activities of water users, promoting community participation, advising DWS and coordinating the implementation of its CMS with the water services development plans of water services authorities. The additional powers and duties around water resources management that it may obtain from DWS are likely to be delegated (carried out on behalf of the minister) at first, and only assigned (full transfer of duty to the CMA) once the CMA has demonstrated its ability to carry out the relevant power or duty (DWAF n.d.). BOCMA is progressing towards this institutional arrangement with the Western Cape Regional Office of the department still overseeing management of the WMA.[15]

In addition to having autonomous RBO and agency RBO characteristics, the coordination-focused functions of a CMA tie in with the description of coordinating RBOs. A CMA also has some partnership characteristics in that it can be established on the initiative of the community and stakeholders concerned, and is therefore also accountable to them. Similarly, a CMA board can be seen as a partnership between various water user groups (DWAF n.d.), as is evident in BOCMA's CMS (BOCMA 2011).

In terms of the move from 19 to nine larger consolidated CMAs as per the second edition of the National Water Resource Strategy, it appears that CMAs in South Africa may, at least in the short to medium term, be moving further away from autonomy status and increasingly towards agency status. The strategy states that DWS will assist CMAs in building capacity to manage a range of functions that ultimately are to be delegated to them. These functions include, amongst others, water use authorisation, water resources protection, water quality management and water resources planning. The department will also be considerably more involved in CMA establishment, with 'a dedicated high-level team' to drive the establishment and operationalisation of the nine CMAs by 2016. This high-level team will also communicate to affected water sector institutions and other stakeholders regarding CMA establishment and the devolution of powers and functions (DWA 2012). In addition, our research suggests that private consultants will continue to play an important part in assisting DWS and CMAs with the CMA establishment process and the drafting of CMSs. The department's increased involvement in CMA establishment and functioning may be due to the fact that it is trying to take back control over the process of CMA establishment, which has been lagging behind considerably.

8.4 POLITICS

When CMAs were first introduced into the discussions around the National Water Act there were both propagators and resistors. A key driving force behind the process was the then Minister of Water Affairs and Forestry, Kadar Asmal, who was very strongly in favour of decentralising certain water management responsibilities to the lowest possible level and promoting high levels of public participation. As mentioned earlier this ties in closely with the South African Constitution's subsidiarity principle, which implies that those functions that can be carried out more effectively and efficiently at lower levels of government should be delegated to the lowest appropriate level (Funke et al. 2007). Also, as discussed earlier, concepts such as integrated water resources management and sustainability (compare Jaspers and Gupta, Chapter 2, this volume) were a key influence on the National Water Act and related CMA development process.

At the same time, however, there were resistors to this process. These were individuals in DWAF who were resistant to change in general and to the idea of CMAs in particular. Suddenly these technocrats were faced with new and at the time revolutionary ideas (for example, those of integrated water resources management and subsidiarity), which were very different from the department's day-to-day technical operations which they were familiar with. There were also resistors outside DWAF, especially systems modellers, who up until then had benefited from the use of their models as part of the department's centralised system of water resources management. These modellers resisted demands for the integration of existing models to support the new decentralised system of water resources management. Instead the modellers competed against each other, each promoting their own model for use by future CMAs.[16] These propagator and resistor dynamics are in line with the argument that reform processes are likely to be contested by certain actors, while others try hard to push them through (Kemerink et al. 2011).

The BOCMA establishment process ran relatively smoothly[17] as intensive stakeholder engagement took place in the Overberg area and Breede River Basin, and the National Water Act's guidelines on how the CMA process should be constituted were closely adhered to. The complexities of this process are difficult to establish, and may have found expression in the deliberations of the BOCMA Reference Group. Although one would want to assume that the CMA establishment process was conducted in a fully democratic environment (where every stakeholder's voice was heard and everybody's interest was taken into consideration), it is likely that different groups had different levels of power at various points in time, which is the result of a mixture of hard and soft power. Here it is relevant to refer to Nye (2004: 6), who states that '[s]oft power is a staple of daily democratic politics. The ability to establish preferences tends to be associated with intangible assets such as attractive personality, culture, political values and institutions, and the policies that are seen as legitimate or having moral authority.' To the untrained eye stakeholder engagement might look like a dynamically and democratically inclusive process, but in fact it can still be quite command-and-control driven and/or dominated by certain powerful actors at the expense of more marginalised actors. How this plays out in BOCMA and the WMA in future is an area that needs to be researched further.

The latest nine WMAs are, like their 19 predecessors, designed to take into consideration river catchment and aquifer boundaries, the future financial viability of the CMA (based on the WMAs' economic realities), stakeholder participation and equity. What differs from the original process of identifying WMAs is that the department has had to reconsider the CMA management model as a whole, given the limited success it has had to date. The department has therefore focused particularly on the availability and allocation of funding, capacity, skills and expertise in WMAs. The rationale behind reducing the number of WMAs has been to combine weaker WMAs with those that have the capacity to manage water resources optimally from the department's point of view. In addition, the department argues that combining CMAs will result in improved distribution of scarce technical skills between institutions, stronger revenue streams, shorter CMA establishment processes and more direct cooperation and coordination at regional, provincial and international levels (DWA 2012). In other words geographical considerations and pragmatism have influenced the department's thinking here, based on the very limited success of the CMA process to date.[18] This new development indicates a move away from ideological considerations to more pragmatic considerations in an attempt to address some of the many problems that have been hampering the CMA development process.

8.5 PERFORMANCE

In the case of BOCMA it is important to note that while this CMA became operational in 2007, its draft CMS was only released in February 2011 (Page 2012) and has not been approved by the minister. It may therefore be too early to make definite statements about how democratic its functioning is. Nonetheless some preliminary observations can be made. These observations link closely to the institutional design of the organisation discussed in section 8.3.

BOCMA's governing board seems to be doing well in terms of operational decision making, but has not really had any challenging decisions to make as DWS has not yet delegated many functions to it.[19] As stated earlier the board has a good mix of different representatives and skills, and it remains to be seen how these individuals fare when having to carry out important decisions related to realising the vision and objectives of the CMA.

The vision of BOCMA, as also represented in its CMS, is in line with the democratic ethos of the South African Constitution and focuses on how this CMA can make a positive and meaningful change to the broader social context within which it is situated. Some of the main components of this vision are inclusion and stakeholder participation, as well as mediation between human and environmental priorities in an effort to ensure the availability of good-quality water and to assist in poverty alleviation. This translates into the need to support social redress and economic development while also maintaining the functioning of the vitally important aquatic ecosystems in the Breede–Overberg area. The vision is to be operationalised by means of devolving decision making to the lowest possible level for the benefit of all water users in the catchment. In addition there is a particular focus on the South African context, which centres on addressing water resources reform (aimed at redressing historical inequalities in water access and use) (BOCMA 2011; Page 2012).

The three vision statements that underpin BOCMA's CMS reflect the need to maintain the balance between resource development and protection and are in keeping with the vision statements of the National Water Act itself. These vision statements are:

1. 'protecting our rivers, groundwater, wetlands and estuaries in a healthy and functioning state for nature, people and the economy';
2. 'sharing our available water equitably and efficiently to maintain existing activities, support new development and ensure redress, while adapting to a changing climate and world';
3. 'cooperating to jointly nurture, take responsibility and comply, so that our water resources are well managed, under the leadership of a strong WMA' (BOCMA 2011: iii).

These three vision statements form the strategic areas of focus for the CMA and will be given effect through supporting measures, objectives and actions. Despite the challenges facing the WMA the CMS makes mention of the opportunities that a holistic and proactive approach to water resources planning and management brings with it. These opportunities include: 'a focus on representivity where all demographic groups are involved in the transformation process, participatory decentralisation where decision-making reflects local understanding and needs, cooperative management between sector participants and using water as a catalyst for growth and development' (BOCMA 2011: ii–iii).

From the CMA and its vision statements it becomes clear that the plan is to have stakeholder participation play a key role in the CMA's management of the WMA. The CMS notes that stakeholders need to be engaged in an appropriate manner and that 'overengagement' should be avoided as this could lead to stakeholder fatigue and a less robust process of engagement (BOCMA 2011). This is based on scientific research that warns that there is not yet a comprehensive and functional approach to public engagement in South Africa (Pollard and Du Toit 2005). Implicitly contained within BOCMA's purpose are empathy (through the involvement of stakeholders), prior knowledge engagement (through consultation) and patience (when dealing with diverging views from various stakeholders) (BOCMA 2011). As discussed earlier the process of producing BOCMA's CMS necessitated considerable consultation with a broad and diverse range of stakeholders, in line with BOCMA's vision of playing a central role in the coordination of water resource matters in national, provincial and local government and in consultation with a variety of partners and stakeholders (Page 2012).

Despite the extensive and overall successful efforts to include as many stakeholders as possible in the CMA establishment and CMS processes, stakeholder engagement has nonetheless reportedly been slow and somewhat irregular (Page, 2012). This is likely to have been due to the challenges inherent in involving hundreds of stakeholders. It is assumed that once clear aims have been set and challenges have been identified, it will be possible to decide which other stakeholders should be appointed to the board (in addition to the ones already on it) (Page 2012), and how to better make use of the already existing forums to further involve stakeholders in the functioning of the catchment.

Another challenge with regard to stakeholder consultation is that it is impossible for the CMA to consult every stakeholder on the ground. Stakeholder consultation is also a very time consuming process that can slow down the CMA establishment process. This was the case for both the Inkomati CMA and BOCMA. It appears that future CMA establishment processes (including the proposed merger of existing WMAs into larger WMAs and CMAs) will be conducted by means of more streamlined processes that focus primarily on efficiency and only draw in stakeholders where needed.[20]

Another key element of BOCMA's CMS is strategic adaptive management,[21] a local variant of adaptive management. In terms of strategic adaptive management, the CMS states that 'cooperative and adaptive management will provide the resilience needed to sustain the economy [of the Breede–Overberg WMA], support livelihoods and maintain and improve the WMA's rich environmental heritage' (BOCMA 2011: 96). In addition, integrated water resources management and adaptive management are seen as the underlying theories for the manifestation of a cooperative environment laying the foundation for sustainable development (BOCMA 2011). This narrative indicates that adaptive management is the foundation on which cooperation within the WMA and between BOCMA and other stakeholders will rest. Future research could be conducted to investigate the claim that adaptive management will result in the needed resilience to sustain the WMA's economy.

Apart from BOCMA's focus on stakeholder engagement and strategic adaptive management, a set of values is supposed to underpin the decisions and actions of the board members and employees. These values are: integrity, respect, reliability, accountability, trust, dignity, confidentiality, honesty, approachability, fairness, transparency, equity and passion (BOCMA 2013). A value that is particularly important, given the CMA's focus on stakeholder participation, is accountability. BOCMA is accountable to both its stakeholders and the minister. However, the accountability link to the minister seems to be better developed than the link to the stakeholders. Regular reports are given to the minister, but the forums which the CMA governing board members should use to report to the water users in the WMA are not yet as active and effective as the CMA envisages them to be. In an effort to support the emerging farmers in the CMA, who probably do not have the same skills, knowledge, resources and experience as many commercial farmers do, the chief executive officer of BOCMA has started going on regular 'meet and greet' outings in the catchment where he liaises with these individuals. These 'meet and greet' outings are linked to various programmes in the catchment that are targeted at involving emerging farmers. For instance, there is an initiative to involve some of these individuals in the eradication of alien vegetation, such as Port Jackson and blue gum trees, which is a particular problem in the catchment. Emerging farmers are also receiving information on how to become more productive. For instance, they are learning about how to cultivate different soil types.[22] It seems that the chief executive officer is going above and beyond what is expected of him by taking a proactive stance regarding awareness raising and support to emerging farmers.

As a final reflection on democratic functioning there appears to be a mix of elite and consensus decision making as the Minister of Water and Sanitation is instrumental in establishing the CMA and appointing its governing board, yet board decisions are taken in terms of consensus decision making. As stated previously, such a decision-making mix is essential in the South African context given that in any given catchment stakeholders are very unevenly capacitated. At the same time, however, it is crucial that the board has an even spread of skills and capacity so that it can carry out its functions effectively and manage the catchment in a way that is beneficial to all stakeholders.

While BOCMA has prioritised and continues to prioritise stakeholder engagement, there is criticism against extensive stakeholder involvement, as this can slow down CMA and CMS processes. As mentioned earlier, there is now a move towards reducing the numbers of stakeholders when establishing CMAs and instead only drawing in stakeholders where and when necessary. While this development may speed up the CMA establishment process countrywide, it may have negative consequences for more marginalised stakeholders to participate in the CMA process and to ensure that their interests are represented. There will have to be a particular focus on ensuring that those that are consulted represent the views and needs of their constituency.

8.6 CONCLUSION

As is evident from the discussion above, the establishment of CMAs in general, and BOCMA in particular, is quite a top-down and bureaucratic process, largely under the control of DWS. In addition, the CMA process in South Africa seems to have been and continues to be strongly influenced by scientists from the environmental sciences and engineering. This influence has ensured that the CMA process is still a largely technocratic one (that is, informed by the natural sciences), and perhaps overshadows the potentially valuable role that other disciplines could play in advising the department and others on how CMAs could be run. However, BOCMA reportedly has a well-functioning board and has pulled in stakeholders in all important stakeholder relevant processes to date. This was especially the case when the establishment process started in 1999. Particular windows of opportunity for stakeholder engagement were the Overberg area and Breede River Basin Study stakeholder engagement processes. An important question is how stakeholder engagement will continue to develop in the WMA, particularly given the proposed move to making stakeholder engagement processes more streamlined in future, and only consulting stakeholders if and when necessary. This is likely to become apparent if the establishment of the proposed new nine CMAs goes ahead.

In conclusion it is interesting that some water resource professionals question whether South Africa should have CMAs at all. For instance, Anderson[23] states that while it makes sense to define WMAs according to hydrological boundaries and involve as many stakeholders as possible within a WMA, it has nonetheless been very difficult to establish CMAs in South Africa, especially when considering the lag time between stakeholder engagement and final establishment. Other issues that have made CMA establishment difficult, especially in the case of the Inkomati CMA, have been a water-stressed situation; diversity in stakeholders' resources, knowledge, experience and skills; and extreme socio-economic disparities (Brown 2011). Some of these problems also feature in the Breede–Overberg WMA, though to a lesser extent than in the Inkomati WMA. These difficulties in the CMA establishment process have contributed to only two CMAs having been established in 16 years, and have led to the department reducing the number of CMAs from 19 to nine. With the considerable water governance challenges facing South Africa, and DWS struggling to get many of the fundamentals in place (for example, water use licences), it might be relevant to question whether all the resources that have been spent on the CMA process have made this process worthwhile. This question is particularly pertinent given the many inherent challenges the country faces – such as the inequalities between stakeholders in a WMA – which play out in efforts to establish CMAs and can render the process such a complex one.

Brown (2011) writes about her research focusing on the establishment of the Inkomati CMA and the transformation of existing irrigation boards into WUAs in the Inkomati WMA. She concludes that there may be substantial weaknesses in the participatory governance model, and that such approaches, if not sufficiently controlled by government interventions, may in fact reinforce inequitable outcomes. Based on the findings of her work, Brown (2011) argues that something of a 'cultural revolution' of how different stakeholder groupings in South Africa perceive each other's needs is required before a level of comprehensive and effective stakeholder participation can take place. However, seeing as such a revolution may take many years, and given the very slow progress in CMA establishment to date, Brown (2011: 183) poses a controversial question: 'If the end goal of water reforms was redress, would it have been a more effective solution for a government with such an overwhelming majority to have pursued a direct state-led redress agenda whilst supporting empowerment programmes to expedite the cultural revolution, rather than participatory governance?'

The case of BOCMA is not as controversial as that of the Inkomati CMA, and the process of participatory governance, though a lengthy one, ran more smoothly than in the Inkomati. Brown's (2011) point about the need for a cultural revolution is also pertinent to the BOCMA case, however, as it is imperative to promote equality among stakeholders, not only in terms of representivity, but also in terms of their ability to influence organisational processes.

On a final note, and with reference to CMAs in general, with the move towards an even more centralised approach to CMAs and with DWS being very clear about the need to fast-track the CMA establishment process, it will be interesting to see how CMAs develop in future. To what extent will they embody and realise the principles of integrated water resources management and stakeholder involvement, and to what extent and to what effect will they increasingly be managed by the government?

NOTES

1. Between 1994 and 2009 this department was known as the Department of Water Affairs and Forestry (DWAF) and between 2009 and May 2014 it was referred to as the Department of Water Affairs (DWA).
2. Personal communication, Derek Weston, Pegasys Consulting, 12 September 2012.
3. Personal communication, Derek Weston, Pegasys Consulting, 12 September 2012.
4. Personal communication, Derek Weston, Pegasys Consulting, 12 September 2012.
5. Personal communication, Derek Weston, Pegasys Consulting, 12 September 2012.
6. In the South African context, 'emerging farmer' is a relatively new term commonly used to describe previously underprivileged farmers who are determined to enter the commercial farming space (Jari 2009). These underprivileged farmers typically belong to the 'black' and 'coloured' (South Africans of mixed race) population groups and qualify to lease or own redistributed land, and to access financial and other support from the South African government.
7. Personal Communication, Aileen Anderson, Crossflow Consulting, 7 September 2012.
8. Personal communication, Derek Weston, Pegasys Consulting, 12 September 2012.
9. Personal communication, Derek Weston, Pegasys Consulting, 12 September 2012.

10. Personal communication, Derek Weston, Pegasys Consulting, 12 September 2012.
11. Personal communication, Derek Weston, Pegasys Consulting, 12 September 2012.
12. Personal communication, Derek Weston, Pegasys Consulting, 12 September 2012.
13. Personal communication, Tally Palmer, Director of the Unilever Centre for Environmental Water Quality and the Institute for Water Research, Rhodes University, South Africa, 12 September 2012.
14. Personal communication, Derek Weston, Pegasys Consulting, 12 September 2012.
15. Personal communication, Derek Weston, Pegasys Consulting, 12 September 2012.
16. Personal Communication, Mark Dent, Department of Geography, University of KwaZulu-Natal, 3 September 2012.
17. This process differed from the Inkomati CMA establishment process, which was fraught with a number of difficulties. For instance, Brown and Woodhouse (2004: 34) note that: 'The time that has elapsed and seeming lack of progress in actually establishing the [Inkomati] CMA is of concern to many of the stakeholders, and it is recognised by [the Department of Water Affairs and Forestry's] Institutional Oversight Directorate . . . that initially it was "too optimistic" in believing that the [National Water] Act could be implemented quickly'.
18. Personal communication, Derek Weston, Pegasys Consulting, 12 September 2012.
19. Personal communication, Derek Weston, Pegasys Consulting, 12 September 2012.
20. Personal communication, Derek Weston, Pegasys Consulting, 12 September 2012.
21. Adaptive management is seen as quite appropriate for the management of CMAs as it is a departure from the outmoded command-and-control management style (Denison and Karar 2010). Rogers et al. (2000) and Roux et al. (2009) advocate the use of strategic adaptive management. Strategic adaptive management has two phases. The first is adaptive planning: the establishment of a learning vision in a participative manner and the development of a common understanding of the CMA context as well as operating principles for a number of learning ideals. These ideals include a common future focus, social knowledge sharing, empathy, learning by doing, prior knowledge engagement, patience, experimentation, positive persistence, transdisciplinarity, adaptability and synergism. The adaptive decision-making phase should involve the development of a detailed management plan that realises the specific learning objectives and needs to be monitored regularly, at various levels and through structured reflection (Roux et al. 2009).
22. Personal communication, Phakhamani Buthelezi, chief executive officer of BOCMA, 28 September 2012.
23. Personal Communication, Aileen Anderson, Crossflow Consulting, 7 September 2012.

REFERENCES

Ashton, P.J. (2010), 'The road ahead', *A CSIR Perspective on Water – 2010*, CSIR Report No. CSIR/NRE/PW/IR//2011/0012/A, Pretoria: Council for Scientific and Industrial Research, pp. 64–65.

Breede–Overberg Catchment Management Agency (BOCMA) (2009), 'Stakeholder engagement process', *Newsletter 1*, November.

Breede–Overberg Catchment Management Agency (BOCMA) (2010), 'A vision determined by all', *Newsletter 3*, July.

Breede–Overberg Catchment Management Agency (BOCMA) (2011), *Draft Breede–Overberg Catchment Management Strategy*, Worcester: Breede–Overberg Catchment Management Agency.

Breede–Overberg Catchment Management Agency (BOCMA) (2013), 'Breede–Overberg CMA Website', available at http://www.bocma.co.za (accessed 8 March 2013).

Brown, J. (2011), 'Assuming too much? Participatory water resource governance in South Africa', *Geographical Journal*, **177** (2), 171–185.

Brown, J. and P. Woodhouse (2004), *Pioneering Redistributive Regulatory Reform. A Study of Implementation of a Catchment Management Agency for the Inkomati Water Management Area, South Africa*, Manchester: Centre on Regulation and Competition.

Claassen, M. (2010), 'How much water do we have?', *A CSIR Perspective on Water – 2010*, CSIR Report No. CSIR/NRE/PW/IR//2011/0012/A, Pretoria: Council for Scientific and Industrial Research, pp. 4–6.

De Lange, W. (2010), 'The water situation in South Africa: some inconvenient truths', *A CSIR Perspective on Water – 2010*, CSIR Report No. CSIR/NRE/PW/IR//2011/0012/A, Pretoria: Council for Scientific and Industrial Research, p. 62.

Denison, J. and E. Karar (2010), 'Discussion Note V3 arising from the Water Governance and Institutional Arrangements Think Tank coordinated by the Water Research Commission', Pretoria: Water Research Commission.

Department of Water Affairs (DWA) (2012), *National Water Resources Strategy 2 (NWRS2), June 2013*, Pretoria: Department of Water Affairs.

Department of Water Affairs (DWA) (2013), 'Water management areas', available at http://www4.dwaf.gov.za/wma/ (accessed on 8 March 2013).

Department of Water Affairs and Forestry (DWAF) (2003), *Breede River Basin Study*, Pretoria: Department of Water Affairs and Forestry.

Department of Water Affairs and Forestry (DWAF) (2004a), *National Water Resource Strategy*, Pretoria: Department of Water Affairs and Forestry.

Department of Water Affairs and Forestry (DWAF) (2004b), *Breede Water Management Area: Internal Strategic Perspective*, Pretoria: Department of Water Affairs and Forestry.

Department of Water Affairs and Forestry (DWAF) (n.d.), *Water Management Institutions: An Overview*, Pretoria: Department of Water Affairs and Forestry.

Funke, N., K. Nortje, K. Findlater, M. Burns, A. Turton, A. Weaver and H. Hattingh (2007), 'Redressing inequality: South Africa's new water policy', *Environment*, **49** (3), 12–23.

Hattingh, J., G. Maree, S. Oelofse, A. Turton and E. Van Wyk (2004), 'Environmental governance and equity in a democratic South Africa', conference paper presented at the AWRA/IWLRI International Conference on Water Law Governance in Dundee, Scotland.

Jari, Bridget (2009), 'Institutional and technical factors influencing agricultural marketing channel choices amongst smallholder and emerging farmers in the Kat River Valley', MSc in Agricultural Economics, University of Fort Hare, King William's Town, South Africa.

Kemerink, J.S., R. Ahlers and P. van der Zaag (2011), 'Contested water rights in post-apartheid South Africa: the struggle for water at catchment level', *Water SA*, **37** (4), 585–594.

McConkey, G.E., W.D. Enright, J.A. Roberts and R. Khan (2005), *The Development of a Catchment Management Agency for the Breede River, Western Cape, South Africa*, Cape Town: Department of Water Affairs and Forestry.

Meissner, R., N. Funke, S. Nienaber and C. Ntombela (2013), 'The status quo of research on South Africa's water resources management institutions', submitted to *Water SA*, **39** (5), 721–732.

Nel, J.L., K.M. Murray, A.M. Maherry, C.P. Petersen, D.J. Roux, A. Driver,

L. Hill, H. Van Deventer, N. Funke, E.R. Swartz, L. Smit-Adao, N. Mbona, L. Downsborough and S. Nienaber (2011), *Technical Report for the National Freshwater Ecosystem Priority Areas Project*, WRC Report No. 1801/2/11, Pretoria: Water Research Commission.

Nye, J. (2004), *Soft Power: The Means to Success in World Politics*, New York: Basic Books.

Page, R.R. (2012), 'Description of three environmental co-management systems in the Western Cape', MA in Public and Development Management dissertation, Stellenbosch: University of Stellenbosch.

Pegram, G., G. Mazibuko, B. Hollingworth and E. Anderson (2006), 'Strategic review of current and emerging governance systems related to water in the environment in South Africa', Report No. 1514/1/06, Pretoria: Water Research Commission.

Pollard, S. and D. Du Toit (2005), 'Achieving integrated water resources management: the mismatches in boundaries between water resources management and water supply', paper presented at the International Workshop on African Water Laws: Plural Legislative Frameworks for Rural Water Management in Africa in Johannesburg, South Africa.

Republic of South Africa (1996), *Constitution of the Republic of South Africa (No. 108)*, Pretoria: Government Printer.

Republic of South Africa (1997), *Water Services Act (No. 108)*, Pretoria: Office of the President.

Republic of South Africa (1998), *National Water Act (No. 36)*, Pretoria: Government Printer.

Rogers K., D. Roux and H. Biggs (2000), 'Challenges for catchment management agencies: lessons from bureaucracies, business and resource management', *Water SA*, **26** (4), 505–511.

Roux, D., K. Murray and E. Van Wyk (2009), 'Enabling effective learning in catchment management agencies: a philosophy and strategy', Report No. TT 421/09, Pretoria: Water Research Commission.

Schreiner, B., B. van Koppen and T. Khumbane (2002), 'From bucket to basin: a new paradigm for water management, poverty eradication and gender equity', in A.R. Turton and R. Henwood (eds), *Hydropolitics in the Developing World: A Southern African Perspective*, Pretoria: African Water Issues Research Unit, pp. 127–140.

Strange, S. (1996), *The Retreat of the State: The Diffusion of Power in the World Economy*, Cambridge: Cambridge University Press.

9. Introducing river basin management in a transitional context: a case study about Ukraine

Nina Hagemann and Marco Leidel

9.1 INTRODUCTION

Ukraine introduced river basin management at a time when the country was still in a transitional phase, adapting from a socialist economy to a free market economy. The laws and regulations were often found to be wanting in respect of necessary by-laws to implement new rules. This was also the case for river basin management. In many Ukrainian river basins, international actors enhanced the establishment of river basin organisations (RBOs) by setting up river basin specific regulations. The role of different actors, especially international organisations and their influence in setting up a river basin management concept, are the key aspects of consideration in this chapter.

Water quality of water bodies across Ukraine is severely compromised: all river basins are classified as polluted or heavily polluted (OECD 2006). The highest impacts result from wastewater from old facilities and infrastructure, and depending on the region in question, from industrial water pollution, fertiliser nutrient run-off from agricultural fields and effluents from the mining industry as well. A change in water quality is evident between the period before and after political independence. After becoming an independent country, many industries collapsed due to structural changes and the impact of this sudden change on water quality was significantly reduced. The same can be said for impacts of agricultural land use (FAO 2005; Stålnacke et al. 2003). During Soviet times, sanitary systems had been installed that were state-of-the-art. Yet, with limited maintenance and updating of facilities a continuous deterioration of infrastructure exacerbated the water quality problem. Ukraine also exhibits a discrepancy in water management between rural and urban areas. According to data from the United Nations Children's (Emergency) Fund (UNICEF) and the World Health Organization (WHO) (2012) for 2010, only 89 per cent of the population in rural areas had access to improved sanitation. The resulting negative consequences are also significant for the quality of drinking water that supplies the rural population, since private wells (the major source of drinking water) show contamination by infiltration from inadequate sanitary systems and animal waste.

Even during the times of the Soviet Union legal mechanisms were in place to monitor schemes of water bodies with the main aim of collecting data for the purpose of analysing whether industries, farmers and other water users maintained the necessary standards. During these times tasks in water governance were assigned to specific authorities and were organised centrally. Even though objectives such as the improvement of water quality were set, targets establishing how to archive these objectives were missing. Until recently authorities within river basins were evaluated based on reactive criteria, that is, the amount of fines collected, as opposed to proactive initiatives to reduce water pollution. Driven by a change in international water resources management and increased collaboration with the European Union (EU) and international donor organisations, the Ukrainian government has introduced the concept of river basin management.

River basin management was introduced after independence by the Water Code of Ukraine (213/95-BP) in 1995 (Verkhovna Rada 1995) and has since been added to by several by-laws. The 1995 Water Code of Ukraine requires the introduction of river basin management in compliance with international standards. The 2002 water programme further specified the implementation requirements. For all basins, a River Basin Administration (RBA) and River Basin Council (RBC) had to be designed. However, details of the specific tasks and especially the responsibilities towards and capacities of influencing other authorities have not been outlined further. Additionally the Water Program (which is due to run until 2020) introduced specific calls for an integrated water management system according to the principles of river basin management in 2009. The Law on Strategy of Environmental Policy underlines this need. For the design of the RBAs and RBCs as well as river basin management plans the State Agency for Water Management set up a regulatory body in 2008 called Nakaz No. 56,[1] which defined the principles, methods and necessary content for river basin management plans. At present the scale of implementation and the design of river basin management plans across Ukraine differ.

This chapter takes a closer look at the institutional design and different drivers and challenges regarding river basin management implementation in Ukraine by focusing on one catchment area, the Western Bug River Basin. The key research questions were: (1) What are the driving factors for introducing river basin management in the Western Bug River Basin?

and (2) What kinds of resources did the actors have at hand to enhance or to hinder the process? The analysis is based on a literature review and a set of interviews that were conducted with different stakeholders from political bodies and administrations at the local, regional and national levels, research institutes and non-governmental organisations. Altogether 41 interviews were conducted. These interviews took place between 2008 and 2012 and are complemented by the findings from an RBC meeting as well as a related workshop with key stakeholders that took place in June 2012. The data and information collated provide insight into stakeholder networks as well as institutional obstacles and driving forces for implementation. All interviewees were promised anonymity; therefore, though individual names will not be mentioned, the organisations to which the interviewees belong will be.

9.2 CASE AND CONTEXT

Nine river basins have been defined for Ukraine. Most of them are large in scale and are transboundary. The case study for this chapter – the Western Bug River Basin – is one of the smallest Ukrainian river basins with a catchment area of around 40 000 km^2. It has been selected for study because the requisite data on water quality parameters as well as institutional considerations exists, whereas for other river basins in Ukraine no such data is available (see Figure 9.1).

The Western Bug River originates in the region of Lviv and discharges into the River Vistula in Poland and later into the Baltic Sea. The upstream part of the river basin is characterised by cultivated land; downstream (on Ukrainian territory) the river flows through forest areas as well as swamps. The basin is characterised by low water quality in some sections due to the mixing of wastewater and agricultural effluents, but it also has a high self-purification capacity (Ertel et al. 2012) and a low to moderate hydromorphological degradation state (Scheifhacken et al. 2012). Wastewater, especially from the city of Lviv, is the major source of pollution for the Western Bug River and a reduction would require a high level of investment in the sewage system covering urban and rural areas (World Bank 2006). Though local and regional authorities are legally in charge, this amount of investment would require national and international support.

The river is used for different wastewater discharges – domestic industrial wastewater and mining effluents – but not for drinking water abstraction or irrigation. There is no competition over water use, and the downstream users are not directly affected by low water quality, as the

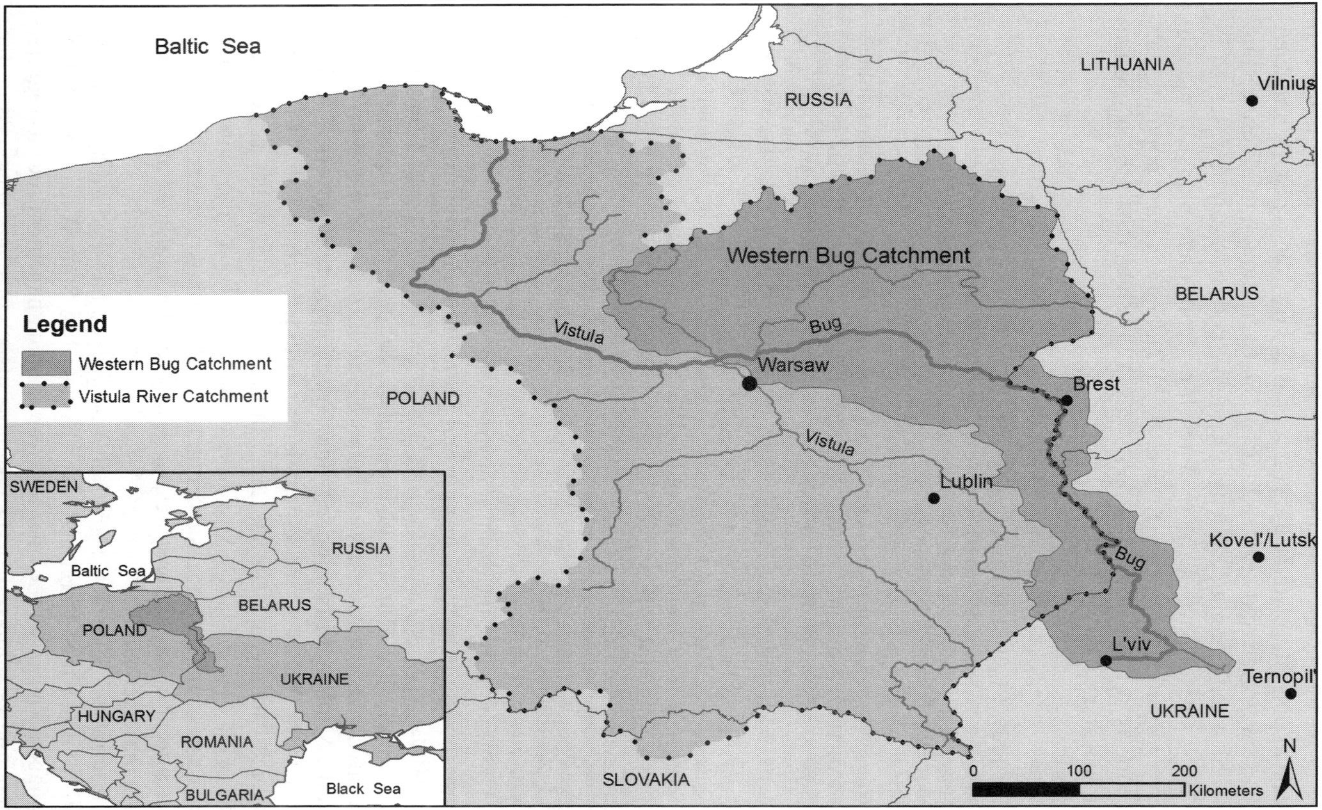

Figure 9.1 The Western Bug river catchment

river flows through forests and swamp land and human access is limited. The local population in the upper river basin, however, is heavily impacted by wastewater discharges (UNECE 2011). Even though the Western Bug has a high self-cleaning capacity, secondary pollution can be observed in the downstream part of the river resulting from phytoplankton blooms of the Dobrotvir Reservoir (Ertel et al. 2012), and the rural population in Ukraine as well as communities on adjacent Polish land have been affected by low water quality.

9.3 INSTITUTIONAL DESIGN

9.3.1 Introduction

In this section the different rule types that apply for river basin management in the case study of the Western Bug are presented to provide a picture of the overall organisational structure. The different rule types that were presented by Huitema and Meijerink (Chapter 1, this volume) will be applied and subsequently presented. These rules are predominantly derived from the 1995 Water Code of Ukraine, the National State Program for the Development of the Water Industry, the Law on Strategy of Environmental Policy and the Nakaz No. 56. Table 9.1 provides an overview of the rules in the case study.

9.3.2 Authority Rules

In Ukraine many different ministries are in charge of water-related issues: the Ministry of Environmental Protection, the Ministry of Health, the Ministry of Agriculture, the Ministry of Housing and Communal Services and the Ministry of Emergency. Indirectly other ministries are also involved: the Ministry of Finance, the Ministry of Foreign Affairs and the Ministry of Economic Affairs are all important players. The State Agency for Water Management is the authority responsible for water resources management in Ukraine. In the past, water policy issues were dealt with by a specifically dedicated ministry. Today the State Agency for Water Management is subordinated to the Ministry of Environmental Protection but has exclusive responsibility for the implementation of river basin management.

In terms of governance functions the agency has a coordination function covering the established river basins. The State Agency for Water Management can set the rules for river basin management such as requirements and deadlines for river basin management plans. They assign the

Table 9.1 Overview of rule types present in river basin management in Ukraine

Rule-types	Rules in case study
Authority rules	• The State Agency for Water Management is the authority in charge at the national level. • The river basin administration is responsible at the catchment scale for the organisation of river basin council meetings and presides over it. They also have to design river basin management plans and collect and prepare data for publication.
Aggregation rules	• The river basin administration prepares the river basin management plan and the river basin council officially takes the decision on it by majority rule. The State Agency for Water Management has the power of veto and either agrees or asks for amendments. • Decision-making rules have not been outlined further
Boundary rules	• The boundaries of the river basins were defined in the 1960s based on the hydrological boundaries. • The river basin council of the Western Bug River Basin consists of members from two different territorial units in Ukraine within the catchment boundaries such as the regional and local authorities, non-governmental organisations (NGOs), industrial representatives and scientists. Poland is not part of the river basin council.
Information rules	• Regional authorities from the catchment area have to collect data on water quality and provide these to the river basin administration. • The river basin administration has to publish data on the water quality of the river basin. • Citizens' interests are represented by NGOs in river basin councils.
Pay-off rules	• The State Agency for Water Management is vested with a budget to finance river basin administration staff. • The river basin management plans have to be financed by state and local budgets as well as water use and pollution fees, and other funds such as international support.

head of the RBA as well as the budget for the river basin authorities. Furthermore, potential river basin management plans have to be confirmed by the State Agency for Water Management and the State Agency can ask for amendments.

Key actors at the river basin level are the RBA and the RBC. The RBA

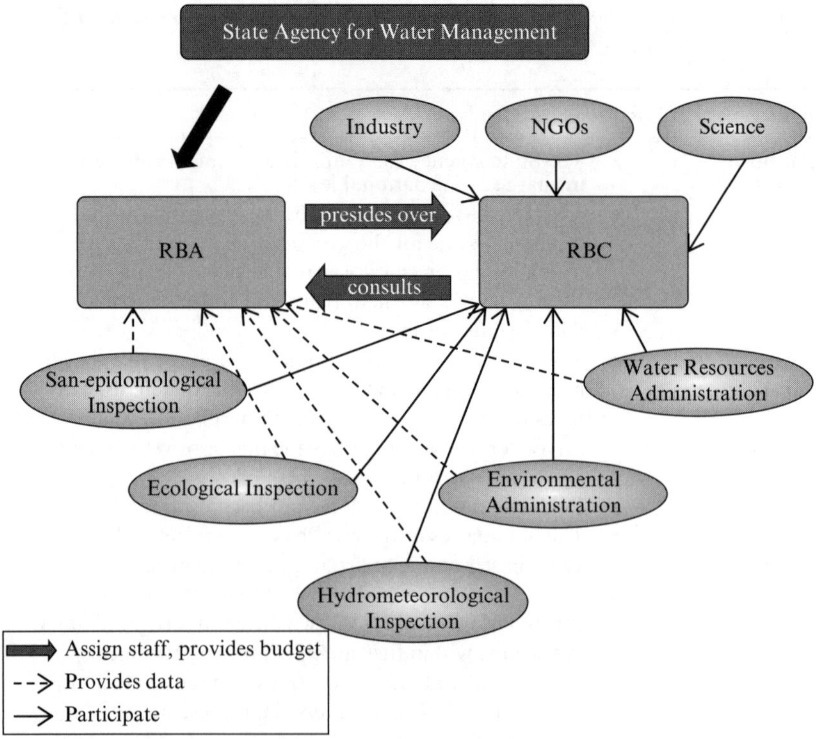

Figure 9.2 Organisation of river basin management in Ukraine

is the key actor at river basin level (see also Figure 9.2). The main task of the RBA is to coordinate all the activities concerning river basin management and they are listed thus: (1) to collect and prepare data from respective authorities for a coherent analysis of rivers water assessment to be able to take decisions on measures to fight water pollution; (2) to analyse the data as a basis for potential river basin management plans and to provide certain data on water quality on the Internet to be freely accessible for everyone who is interested; (3) to initiate and preside over the RBC, which means they prepare and chair the sessions; and (4) to develop a river basin management plan (together with the RBC). Both the RBA and the RBC have no specific functions in terms of water quality and quantity, except for developing plans that increase water quality and quantity, but for implementing actions the regional and local authorities within the basin are responsible.

According to the Nakaz No.56, RBCs have three main functions: (1) to make decisions with regard to water resources management; (2) to develop river basin management plans including specific measures; and (3) to arbitrate water use conflicts.

However, the Nakaz regulation does not provide specific guidelines for the implementation and setting up of these functions, nor does it set schedules for plan approval and the achievement of goals. These decisions are left to the RBA and RBC but they often lack the power and experience to carry out their tasks (Leidel et al. 2012). In the end the RBC has a theoretical rule-making function but it is too dependent on other authorities such as the RBA and the State Agency for Water Management. Members of the RBC are representatives from all the authorities that monitor and analyse water quality in the river basin, sectors affected by or affecting water quality, non-governmental organisations (NGOs) and officially scientists too. Financially the RBC is dependent on the assignment of a budget from the RBA.

9.3.3 Aggregation Rules

In terms of water management at the national level, individual ministries claim their rights in terms of water use and management defending their own interests. No authority integrates the interests of all ministries at the national level as is the case with the Water Committee in Mongolia (see Dombrowsky et al., Chapter 11 in this volume). The State Agency for Water Management does not directly communicate and coordinate activities with other ministries. Very little is known about decision making and internal processes within the State Agency for Water Management.

At the river basin level the RBA has the greatest power in terms of setting the agenda for RBC meetings and drafting the river basin management plan. What is missing is a monitoring and enforcement function; the RBA has no instrument at hand to require the data from the authorities. Instead it is reliant upon their good-will. In terms of power the director of the RBA has great influence on the processes regarding the RBA and RBC, as he initiates and chairs the RBC meetings which means he also sets the schedule for the meetings. Based on the Nakaz No. 56, the RBA would also be responsible for arranging financing, such as collecting water use and pollution fees which are to be reinvested in water management measures. However, for as long as the budget collected at local and regional level goes directly to Kiev and is redistributed from there, this function is only a theoretical one (see also section 9.3.6).

At the basin level the final decision-making power lies with the RBC. However, it is unclear if the majority rule or unanimity rule has to be applied for decision making. Furthermore the individual authorities represented in the RBC often have a limited view and focus mainly on the interests of their authority and their territorial boundaries. The dispute settlement role of the RBC has not been practised so far and it is unclear how such a system could be initiated and enforced. The RBA takes the lead over decision-making processes with a focus on the entire catchment area. In the end it is the State Agency for Water Management which has the final say on the river basin management plan. It examines the proposals and either agrees with the plan or asks for amendments.

9.3.4 Boundary Rules

Water management in Ukraine has a long tradition and goes back to Soviet times when in the 1960s the first river basins were defined according to their hydrological boundaries. Research within the International Water Research Alliance Saxony (IWAS) project shows that the boundaries of the Western Bug River Basin are delineated according to surface waters and in some cases do not fit with the groundwater bodies because of the karstic system. This displays an important issue for water balancing within the catchment, yet from today's institutional point of view this plays only a minor role for the overall river basin management. The territorial structure has not been amended since that time and the river basin boundaries of the Western Bug continue to cross two administrative units (regions). However, except for the Tisza River Basin, all other Ukrainian river basins stretch across two or more regional units. A region is the most important subordinated administrative level in Ukraine. According to these boundaries two territorial units, the Oblast Lviv and Volyn with all their water-relevant authorities, are party to the RBC. They send representatives to RBC meetings; in most cases the directors represent their authority at such meetings. Further members of the RBC such as the above-mentioned representatives also represent either regions or the whole catchment scale. Even though Ukraine has a frequent exchange with Polish experts on water management issues for the Western Bug River Basin, Poland is not represented in this RBO. The competencies for the design and implementation of the river basin management plan lie with the RBC, it has limited power as many decisions are taken outside the sphere of the RBC. These decisions include tariff setting for water use and licensing; that is, for water enterprises national authorities are responsible. The same holds true for the budget competencies that lie with national-level authorities.

9.3.5 Information Rules

The Western Bug RBC consists of representatives from administrations that gather data on water. They include the san-epidemiological inspections, hydro meteorological inspections and ecological inspections, the State Environmental Administrations and the regional branches of the State Agency for Water Management and the Water Resource Administrations (see Figure 9.2). All these authorities are supposed to transfer water quality data to the RBA to be collected and prepared for official public provision and as a basis for the set-up of river basin management plans. An important information-gathering function also lies in the publication of data. One major aim is to inform the public about water quality issues in the Western Bug River catchment area. Prior to the establishment of the RBA and RBC in the Western Bug, the above-mentioned authorities transferred the data to their respective ministries in Kiev. During the implementation process in 2006 a contract was signed between all authorities, which stated that this data be transferred to the RBA and also collected by the State Environmental Administration. However, the exchange of sample data at the river basin level provides obstacles because, historically, communication is organised on a vertical axis. This means that data travels up from lower-level to higher-level authorities, but not horizontally between different sectors. The same holds true at the river basin level. Even though the different authorities should theoretically work together in the RBC and jointly provide plans for the whole basin, the focus is often on the territorial unit and therefore individual contribution and position is strongly dependent on the superior authority.

The Western Bug is a transboundary river and as such contact with and exchanges between the Polish and Belarusian authorities are important. Ukrainian authorities within the basin, including representatives of the RBA, describe the cooperation and communication with Polish and Belarusian authorities as fruitful and flourishing; a Ukrainian–Polish Commission for transboundary waters with different working groups exists where experts from all three countries exchange information on water quality data.[2] However, in the end it is an exchange without any binding character. There are no legal obligations for data provision or collaboration, even if an agreement between the governments of Poland and Ukraine has existed since 1996 in the field of water industry on transboundary waters. An agreement between Belarus and Ukraine on joint use and protection of transboundary water came into effect in 2002. Although several trilateral collaboration projects have already been carried out, there is still a need for a transboundary water commission with all neighbouring countries for the Western Bug River Basin based on internationally established best-practice examples (UNECE 2011; Leidel et al. 2012).

9.3.6 Pay-Off Rules

Budgetary provisions for water management problems of all kinds represent a crucial issue in Ukraine. At present the state is under an austerity budget because Ukraine is dependent upon the International Monetary Fund's support. To finance river basin management, Nakaz No. 56 stipulates that the measures defined in the river basin management plan have to be financed by state and local budgets, by water use and pollution fees as well as other sources such as international support. Only the RBAs are directly funded by the State Agency for Water Management. It is vested with its own budget to finance river basin management tasks. We learned, from interviews, that the RBA is financed by the State Agency for Water Management to cover payroll costs, but that no money has been provided for financing river basin management plans because of a lack of sufficient money transfer from the central state to the river basin authorities, despite the fact that financial contributions by the state have been promised in order to help establish the respective organisations. Local authorities are not able to cover these costs because of the budget law: revenues from most of the taxes and other income in the river basin are collected by the local and regional administrations and sent to Kiev. From there they are redistributed to the regions and municipalities, but the majority is allocated to certain non-RBO tasks. Only 30 per cent of the fees from water users are made available for the State Agency for Water Management to be invested.[3] The problem of insufficient redistribution of money from state level to local level is a general problem within the Ukrainian fiscal system (Khmelko 2012). Some RBOs in Ukraine earn additional money by provision of services such as water for irrigation purposes, but these are exceptions. Often the fees are low and 1 cubic metre costs only €0.01. An example where money was provided is the Tisza River Basin: in 2007 after a severe flood in Western Ukraine, the Ukrainian government provided €20 million for flood protection measures. This was an exception, however, because according to a Ukrainian scientist the general perception is that there is enough water and the quality is of secondary importance.[4] Another example shows that the availability of money alone does not solve the problem. According to the personal communication of a consultant, financial support is provided by the EU to implement, among other things, river basin management plans under the condition that the plans are set up and have a clear approach on what should be done, but the money is not claimed and invested. The interview partner assumed that the money is not transferred because the authorisation is subject to provisions and some actors feel that these requirements are too laborious and that there are easier ways to receive funding.

Water treatment plants currently make use of technologically inferior equipment which in themselves are the major reason for pollution of the Western Bug, and the amount of investment needed to improve the infrastructure and facilities is high (Ertel et al. 2012; World Bank 2006). As a result, the RBA is restricted in its actions because in terms of the major polluter it is unable to react with sanctions or fines. Incentives for investment in new technologies are low[5] and new investment, for instance, for water treatment technology, has been postponed due to the serious financial situation in Ukraine. Furthermore, investment in facilities is unpopular considering the limited returns because of low tariffs (see section 9.4). New incentive instruments for investment in new technologies such as the introduction of tax credits failed because the budget was never provided. In addition, programmes with a focus on improvement of water management from 2001 to 2004 received only 10 per cent of the funds that were necessary for implementation, and in the same period 'none of the planned centralized water supply systems for rural communities materialized' (UNECE 2007).

9.4 POLITICS

9.4.1 Introduction

For the analysis of the politics and strategies in designing river basin management, two different levels have to be looked at. Both levels involve different actors and actor groups with distinct interests: (1) the political design level; and (2) the implementation level. Both will individually be described in this sub-section even though both are of course to some extent interdependent. For both levels hardly any opposing strategies have been identified that directly concern river basin management but more general and fundamental political and legal conflicts have had a spillover effect on river basin management design and implementation strategies. Therefore the broader institutional and governance context will be taken into account.

9.4.2 Policymaking and Strategies at the National Level

Environmental policy in general does not yet feature prominently on the political agenda in Ukraine.[6] Even though society is concerned about the state of the environment, people are occupied with more urgent issues than taking actions for the environment (Nazarov et al. 2004). The introduction of implementing river basin management in Ukraine in the mid-1990s was an attempt to move closer to the EU in terms of amending laws and regulations to better resemble the EU's *acquis communautaire*. During this time, the EU tried to establish links with all countries in Eastern Europe. Furthermore Article 13 of the Water Code of Ukraine requires that water management in Ukraine should be established according to European and international guidelines. The law has been amended several times but even today the requirements and obligations are not explicit in the respective laws and by-laws. From today's perspective, it is hard to trace back the detailed negotiation processes of the founding period of river basin management in Ukraine. Many of the actors involved have changed office since that time and documentation of the process is rare.

The lack of by-laws for the implementation of river basin management in Ukraine is a result of the as yet unconsolidated political situation in the country. The current legal framework in Ukraine, the result of the rocky transition process, plays an important role. The transition period did not follow the consequent path of Poland and other new EU member states. One result is that Ukraine entered a 'third way' of transformation which is between capitalism and communism, by on the one hand establishing 'social democracy' but on the other hand establishing secure 'national protectionism of the state producers' (Stadnytskyi and Nobis 2008: 33). The results are still visible today, considering the allocation of responsibilities between several different administrative levels as well as different bodies at the national level which lack transparency. The confusing legal basis makes it hard to successfully implement legislation. Territorial restructuring and decentralisation were envisaged at the beginning of the transformation process and even initiated, but never finalised. Since the 1990s, the country has been stuck in a legal and political quagmire where old Soviet regulations are still in place – such as monitoring requirements for surface waters – together with new legislation that has not yet been fully implemented. Water is an example: the Water Code was introduced but remained unimplemented for several years because the by-laws were missing. Every now and then, political initiatives are started at national level with new laws and regulations to enhance the restructuring process. Also, these policies often do not reach fruition due to a high rate of exchange of ministers and administrative staff, to name but a few. Every new minister or head of department has his own political goals, and because of the frequent changes, proposals often do not reach the stage of being discussed in Parliament. A prominent example is the reform of the water tariff system. As mentioned in the introduction, river pollution by domestic and industrial wastewaters is the major source of pollution for the Western Bug River as well as other rivers in Ukraine. One crucial aspect here is that low tariffs, which are a legacy from the Soviet times, do not cover the operation and maintenance costs. As a result, investment in the facilities is too low to maintain the infrastructure. Subsequently the cleaning capacity of the facilities decreases rapidly and polluted wastewater is discharged into rivers. An initiative by the former Minister for Environmental Protection, Ms Zapatrina, was designed and prepared for official discussion but as she had to leave office this proposal was never taken up again because the new minister had different preferences.

The above mentioned loopholes in the current legislation can be regarded as one of the reasons why the river basin management approach has not been detailed further. Unfortunately this haphazard approach seems to be the rule rather than the exception. Many legal acts also lack the necessary specifics to make them viable. To put it more succinctly, this means that RBAs are required to design river basin management plans, but what the preparation deadlines are and what they should look like are moot points. The State Agency for Water Management attempted to amend legislation in this respect several times,[7] but decisions were frequently postponed due to overall political changes which also resulted in staff changes at the State Agency. At present it is aiming at a strategy to empower RBAs to become more active in the design and finalisation of river basin management plans but no real action has been taken so far.

9.4.3 Implementing River Basin Management on the Ground

Due to the weaknesses in legislation the implementation process of river basin management in Ukraine did not start until external players took action. External pressure can be regarded as the most important driving factor for enhancing the implementation of the river basin management idea. The necessity stems from the fact that none of the legal requirements provide a sufficient basis for RBOs to actively start designing plans and measures for river basin management. In all of Ukraine's river basins the driving forces have been international actors such as neighbouring countries (for example, Tisza and Dniper) or international support (for example, Western Bug and Dniester). These international actors often foster the implementation of river basin organisational structures because they have a vital interest to promote integrated water resources management by encouraging the empowerment of local and regional actors in water management. Two main actor groups on the ground play a key role and influence the development of river basin management: the local and regional administrations, and the RBA. The local and regional administrations are the key actors at the basin level. As mentioned above they are supposed to contribute data to the RBA on a regular basis. However, the

RBA is a cross-regional unit without any real power to demand data from local and regional authorities.

There is no clear opposition to river basin management implementation, but hindering factors can be identified and one is the generally negative performance of water-related sectors such as domestic and industrial water treatment: the high pollution loads that are discharged into the river are the dominant factors of river pollution. Yet as these problems are so overwhelming, local and regional actors are powerless before them and as such they make small initiatives at the basin level almost obsolete.

River basin management in the Western Bug River Basin

In the specific case of the Western Bug, the RBA and RBC were established in the context of the European Technical Assistance to the Commonwealth of Independent States (TACIS) project (UNECE 2002). The project served as a driving force for setting up the RBA and initiated the first meeting of the RBC in 2006. The TACIS project could at that stage rely on the support of all actors from state to regional and local level. As the river basin stretches across two regions – Lviv and Volyn – it required the cooperation between these two administrative units. The TACIS project aimed at enhancing the cooperation between the administrations from both regions and succeeded because a contract of data exchange was designed and enforced. Furthermore, with the support of the project, RBA staff were trained and an RBC meeting was initiated. The diversity of actors involved in the RBC comes close to the inclusive stakeholder participation as promoted by the Water Framework Directive (Zingstra et al. 2009).

Shortly after the TACIS project phase ended, some arrangements were changed. The contract on data exchange was terminated and the river basin management plan was not finalised. Until June 2012 no further RBC meeting had taken place. A variety of reasons may explain this situation. The Bug RBC does not have the data and monitoring results from which they could derive recommendations, mainly because of the little cooperation between the different administrations. Several interviewees stated that there is no real data exchange within the region and that different authorities apply different measurement techniques. In addition, data is not gathered or stored with one authority but with many different administrative authorities and units (Zingstra et al. 2009).

The process has almost stagnated over the last few years and that is how things stand at the moment. One interview partner who was part of the process in 2006 said that many of the directors have changed since the agreement was signed and the new directors are not aware of the old agreement.[8] They keep working on their assigned tasks and follow the old institutional structures from Soviet times. This means they are strictly subordinated to their superiors, particularly the ministries. Due to the highly centralised state system in Ukraine, certain authorities also lack proactivity in becoming engaged in the RBC, and decide on programmes or other initiatives for several reasons that range from inexperience in participatory approaches to historically accustomed mind-sets of hierarchical thinking. Yet even if municipalities would take over more responsibilities regarding water management, they would still have the problem that they lack the fundamental personal capacities and expertise. An OECD (2009: 72) report further defines the shortcomings local actors face when it comes to:

> legal interpretation, contractual arrangements, interactions between utility providers, tariff procedures, regulatory impact assessment as well as raising external financing for infrastructure development. At the same time, the number of staff in the local state administrative bodies responsible for coordination and control of the water supply sector is insufficient, leading to a lack of coordination and support from the rayon (oblast, republic) state administrations.

Some actors are more proactive and try to foster the implementation of river basin management and the coordination of authorities, but these are often scientists and NGOs that have limited power in terms of river basin management.

Examples and background for enhanced river basin management in other river basins

Two other examples show that the driving forces for the implementation at the catchment scale differ widely. In the Tisza River Basin, for example, a window of opportunity was opened with two major flood events in 1998 and 2001. These floods resulted in actions between Ukraine and the downstream neighbouring countries. Ukraine signed a Memorandum of Understanding with other Danube River Basin countries entitled: '[the] strengthening of Tisza River Basin cooperation. Towards the implementation of the Integrated Tisza River Basin Management Plan supporting the sustainable development of the region.' The most prominent innovation is an automatic water-quality monitoring station on Ukrainian territory at the Tisza River. With the financial support of the International Commission for the Protection of the Danube River, a modern monitoring system along the river was set up that is also part of a flood warning system (UNECE 2007). Data is transmitted not only to Ukrainian authorities but also to other EU countries to be prepared for flood events should they occur.

The other example is the Dnister River Basin. Unlike the Western Bug River, in the Dnister Basin, surface water is also used for drinking water purposes not only in the upper part but also for large cities such as Odessa. Furthermore, the upper basin is also prone to flooding (Kovalchuk et al. 2008). The river basin management approach in the Dnister River Basin is significantly better developed than in the Western Bug River Basin. The RBC meets at least once a year to discuss basin-relevant issues and the authorities provide an overview of their activities within the previous month. Interviews revealed that many directors of authorities regard their presence at such meetings as a priority. The reason for frequent meetings is that the Dnister River is of high importance for the country because it stretches across eight regions and flows from Ukrainian territory into the Black Sea. There are incentives for the affected regions to become actively involved in river basin management because water pollution directly affects the population in Ukraine.[9] From the Ukrainian side the river basin is regarded as being of high importance for economic reasons, as it serves as a drinking water reservoir for one of the largest cities in Ukraine and it attracts tourists. Yet despite these incentives, a river basin management plan has not yet been designed for the Dnister River Basin either.

Even though the origin of the Dnister River basin management is similar to that of the Western Bug and was developed in the framework of an international project, the difference to the Western Bug is that the efforts continued once the project ended. The financial situation is also problematic for the Dnister basin as again investment is necessary to enhance water quality. Technically defunct equipment impacts on the water quality but also dykes have to be repaired to protect against flood hazards (Kovalchuk et al. 2008).

9.5 PERFORMANCE

It is necessary to look at how the state functions in general to best understand the current state of implementing river basin management in Ukraine, because the role of democracy and the aligned features such as transparency and participation play a major role in this regard. Today's territorial structure goes back to Soviet times and has not essentially been changed since independence. As a result different forms of administrative units with different responsibilities coexist today, and in addition, these administrative units are not in line with today's economic and demographic situation (Morgner 2010). Furthermore, 'not all features of the decentralization process have been adequately implemented' (OECD 2009: 72). No government has yet expressed the will to finalise decentralisation. In the case of Ukraine, there is no tradition of local self-governance and the institutional setting at local level is not stable. The Ukrainian state is highly centralised despite the fact that the Law on Local Self-Governance provides for a more decentralised approach. The administrative structure consists of four different levels: national, regional (*Oblasti*), districts (*Raions*) and municipalities. All levels have their own legislative branches. However, the regional and district levels do not have their own administrations with executive powers, instead the state administration at the respective level fulfils administrative obligations.

The municipalities have self-governance authority only in a few minor areas and, in addition, they often lack the resources, capacity, knowledge and money to execute power. The legal basis for municipalities is weak because laws and regulations are sometimes contradictory and, as a result, enforcement is almost impossible. Several responsibilities have not been clearly distributed between different levels and this makes it easy for central government actors to interfere in local and regional affairs. As a consequence they are often not held accountable for their actions because it is common knowledge that many authorities lack the resources to fulfil their tasks. In terms of accountability the State Agency for Water Management as a subordinated body of the Ministry of Environmental Protection has the final say on the river basin management plan, but it is only indirectly legitimised.

The shortcomings that result from the insufficient decentralisation process have implications for local government budgets. Importantly, as fiscal powers have not been decentralised, municipalities lack the possibility to raise money from taxes in order to more efficiently provide the local population with their respective public services (OECD 2009). The different administrative levels in Ukraine have only limited budgets because they are to a large extent 'essentially "nested" within the budgets of their corresponding higher-level governments' (D'Anieri et al. 1999: 102). 'If the degree of freedom depends on the length of one's leash, the leash in this instance refers to the financial resources left at the council's disposal' (Navruzov 2002: 124). Furthermore, municipalities are not allowed to borrow money (OECD 2009). However, it is not fiscal federalism alone that would be able to reduce the problems, but in addition, administrative and territorial reforms have to be initiated (ibid.).

Consequently, from a legal perspective and also in practice the RBOs are highly dependent on the good-will of the regional administrations competing with a list of other urgent problems; the Western Bug is not a high priority, mainly because it is not used for water abstraction and it represents no severe risk to Ukrainian territory. To increase the awareness of the RBO for the Western Bug River an interview partner (State Environmental Administration) said that the heads of the administrations within both regions have to actively promote its positive role and set incentives for people to become engaged again. At present the RBO is unable to fulfil its coordination function because it is not accepted as a superior authority to the local and regional administrations, and as such much information is not transferred from these administrations to the RBA. From a democratic point of view the general set-up of decision-making power concerning river basin management is questionable. The decisions on river basin management plans are taken by the RBC which is, in the case of the Western Bug, dominated by the authorities. These authorities are state authorities and are directly controlled by the President of Ukraine. Only the representatives of the cities are directly elected and therefore accountable to the people in the region. Also NGO participation as representatives of society is low.

Today, the old governance structures are still part of the overall political and legal system and are challenged by new ones, and therefore democracy, democratic institutions and procedures are not yet fully accepted and integrated (Bos 2010). Informal structures as well as personal relationships often have more power than legal structures and the required resources are concentrated with the executive (ibid.). Informal rules prevail in cases where transaction costs for actors are low, which means that as long as informal rules serve the actors who are in charge and a change to formal rules reduces their benefits they will not change the system. Especially when control mechanisms such as jurisdictions and civil society are either not interested in interfering or do not have the respective power. While even for well-established institutional structures in industrialised countries implementing river basin management is a challenge, it is even more so for transition countries such as Ukraine. The legal framework lacks aspects of accountability and the legal system is subject to revision. The sometimes unclear assignment of competencies and communication between authorities opens the door to corruption, which is also a serious issue in Ukraine (Khmelko 2012). Powerful leaders play an important role, they are dominant at all levels and their interpretation of rules is often decisive. Especially powerful actors perceive legal guidelines as not necessarily binding and the reputation of the jurisdiction has been tarnished by several scandals and corruption (Bredies 2009). The Water Code of Ukraine, for example, requires buffer zones for housing and agricultural activities, but often these regulations are ignored due to a sustained level of corruption. Depending on personal contacts with local and regional administrations, house owners may get permission to establish their houses within the buffer zones. Similar activities have been observed in agricultural cases.[10] The buffer zones are sometimes just ignored and fields are cultivated close to the river.[11]

While the law requires that data has to be transparent for RBAs, as well as NGOs and other civil society groups, in practice data is hard to access. Up-to-date information is often only accessible when it is paid for – only old data is accessible for free.[12] Even though a list of requirements exists within the Nakaz No. 56 on what kinds of elements belong to a river basin management plan implementation mechanism, such as timelines, the definition of timed milestones is missing and it does not have a budget to fund such a programme. The lack of a budget for funding activities was one of the reasons why the Western Bug RBC did not sufficiently fulfil its task and as a result lost its credibility and importance.[13]

The environmental performance of the river basin is characterised by high pollution loads from domestic and industrial wastewater as outlined above, but also by the high self-purification capacity of the downstream part of the river (Ertel et al. 2012). However, as the RBO has not yet become effective the environmental performance of the RBO itself cannot yet be characterised. When the envisaged instruments of river basin management are implemented it will have great potential to contribute to the improvement of the environmental performance of the river basin, because a comprehensive assessment of collected data will be able to identify as yet undetected sources of pollution and further more jointly decided plans to enhance the pressure to implement the necessary actions.

9.6 CONCLUSIONS

River basin management is envisaged in Ukraine but the process is still in its infancy. New laws and programmes have been designed, new authorities established at national as well as at catchment scale, but regulations often remain imprecise and are therefore often neglected. The legal basis is in many respects too weak to serve as a guideline for the implementation of river basin management in Ukraine. The outdated administrative structure as a result of the stagnated reform process is a serious political challenge, not limited to, but especially for, water management. Thus, one of the key legal drawbacks is that the set-up of river basin management and RBOs was not accompanied by a rearrangement of power and responsibilities. The design of the organisational structure for river basin management with the set-up of the RBA and the RBC is reasonable even though the legitimacy of members in the RBC is questionable. The official procedure is that the river basin management plan is set up by the RBC together with the RBA and sent to the State Agency for Water Management for approval. The initiative for river basin management in Ukraine started in 1995 but except for the Tisza River Basin no complete river basin management plan has yet been set up (Zakorchevna 2011), a fact that raises serious questions about the hindering factors. One answer to the question is that many governance functions remain with national authorities or are not precisely defined for key actors at the basin level such as the coordination or dispute resolution function. In short, the RBOs have been installed within an existing system without having their interrelationship to the other authorities clearly defined.

International actors were identified as the most important driving factor for implementing river basin management on the ground. Yet this can only be regarded as the starting point. The case studies from the Tisza and Dnister river basins show that – even though here international assistance was also provided – the motivation has to come from society to initiate a sustainable process. Unfortunately this is often limited to reactions to accidents or crucial environmental needs, when in such cases the support is broader and public awareness higher. For the Western Bug River Basin there are no such driving forces. Some background problems have to be solved to increase the incentives and motivation of actors to become engaged. These include investment in old sanitary facilities to reduce the high pollution loads from wastewater treatment plants. Financial support from external actors or national authorities is a key requirement for investment.

While participation and transparency of actions are major aims of river basin management and the RBC acts as representative of the region, it is not accountable for its cooperation in the river basin. This presents a major obstacle to implementation as no crucial feedback processes are embedded within the system. Despite the fact that a contract exists for the collaboration of regional authorities with the RBA, the RBA is often not powerful enough to request data from these authorities and enforcement mechanisms are non-existent. Consequently, only limited data is made publicly accessible by the RBA. Participation in council meetings is also difficult; they are officially open to the public but the announcement of the meetings is kept within limited boundaries.

Knowledge and available capacities are another crucial aspect when a country wants to implement river basin management, but in Ukraine the knowledge of river basin management, and even more the perception about what river basin management is and what the requirements are, are often misleading and not in line with the official definition and practise. A consultant described the situation in an interview as follows: 'in Ukraine everyone says we want the river basin management approach, but there is no definition [of] what it means and what river basins are. The question is what you want to manage. It can't be only water quality'.

ACKNOWLEDGEMENTS

This work was supported by funding from the German Federal Ministry for Education and Research in the framework of the project IWAS – International Water Research Alliance Saxony (grant 02WM1166 (TUDD) and 02WM1165 (UFZ)). We thank Aletta Bonn, Dave Huitema and Sander Meijerink and all authors within this book for their helpful comments and suggestions to improve the chapter. We are also very grateful to the people who agreed to be interviewed.

NOTES

1. Based on an unpublished legal analysis by Moynihan (2013).
2. Personal communication, hydro meteorological inspection.
3. Personal communication, consultant.
4. Personal communication, scientist.
5. Personal communication, Ivan Franko University, Lviv.
6. Personal communication, think tank, Lviv.
7. Personal communication, State Agency for Water Management.
8. Personal communication, State Environmental Administration.
9. Personal communication, State Environmental Administration.
10. Personal communication, State Agency for Water Management.
11. Personal communication, National University of Life and Environmental Sciences of Ukraine, Kiev.
12. Personal communication, National University of Life and Environmental Sciences of Ukraine, Kiev.
13. Personal communication, State Environmental Administration.

REFERENCES

Bos, E. (2010), 'Stabile Instabilität, dynamische Blockade, Das politische System der Ukraine', *Osteuropa*, **2–4**, 77–88.

Bredies, I. (2009), 'Zur Verfassungsordnung ohne Verfassungs- und Rechtstaatlichkeit in der Ukraine und ihre Auswirkungen', *Ukraine-Analysen*, **66**, 6–7.

D'Anieri, P.J., R. Kravchuk and T. Kuzio (1999), *Politics and Society in Ukraine*, Boulder, CO: Westview Press.

Ertel, A.M., A. Lupo, N. Scheifhacken, T. Bodnarchuk, O. Manturova, T.U. Berendonk and T. Petzoldt (2012), 'Heavy load and high potential: anthropogenic pressures and their impacts on the water quality along a lowland river (Western Bug, Ukraine)', *Environmental Earth Sciences*, **65**, 1459–1473.

Food and Agriculture Organization of the United Nations (FAO) (2005), *Fertilizer Use by Crop in Ukraine*, FAO: Rome.

Khmelko, L. (2012), 'Administrative decentralization in post communist countries: the case of water management in Ukraine', *Journal of Political Science, Governments and Politics*, **1** (1), available at http://www.scientificjournals.org/journals2012/articles/1512.pdf (accessed 9 September 2013).

Kovalchuk, I., J. Quast, A. Mykhnovych, J. Steidl, V. Ehlert and Y. Andreychuk (2008), 'Concepts of the sustainable water use and flood protection in the Upper Dnister Floodplain', in M. Roth, R. Nobis, V. Stetsiuk and I. Kruhlov (eds), *Transformation Processes in the Western Ukraine*, Berlin: Weißensee Verlag, pp. 431–440.

Leidel, M., S. Niemann and N. Hagemann (2012), 'Capacity development as a key factor for integrated water resources management (IWRM): improving water management in the Western Bug River Basin, Ukraine', *Environmental Earth Sciences*, **65**, 1415–1426.

Morgner, M. (2010), 'Reformbedarf. Kommunale Selbstverwaltung in der Ukraine', *Osteuropa*, **2–4**, 163–174.

Moynihan, R. (2013), 'Potential contribution of principles and approaches of international and European water law to Ukraine's management of the Western Bug River', unpublished report.

Navruzov, Y. (2002), 'Local government in Ukraine', in I. Munteanu and V. Popa (eds), *Developing New Rules in the Old Environment*, Budapest: Local Government and Public Reform Initiative, pp. 109–159.

Nazarov, N., H. Cook and G. Woodgate (2004), 'Water pollution in Ukraine: the search for possible solutions', *International Journal of Water Resources Development*, **20** (2), 205–218.

Organisation for Economic Co-operation and Development (OECD) (2006), 'Financing water supply and sanitation in Eastern Europe, Caucasus and Central Asia', available at http://www.oecd.org/dataoecd/29/46/36388760.pdf (accessed 31 May 2013).

Organisation for Economic Co-operation and Development (OECD) (2009), 'Dealing with post-decentralisation implications in the sector. Based on country experience cases', available at http://www.oecd.org/env/outreach/44096445.pdf (accessed 31 May 2013).

Scheifhacken, N., U. Haase, L. Gram-Radu, R. Kozovyi and T.U. Berendonk (2012), 'How to assess hydromorphology? A comparison of Ukrainian and German approaches', *Environmental Earth Sciences*, **65**, 1483–1499.

Stadnytskyi, Y. and R. Nobis (2008), 'Special problems of transformation in the Ukraine', in M. Roth, R. Nobis, V. Stetsiuk and I. Kruhlov (eds), *Transformation Processes in the Western Ukraine*, Berlin: Weißensee Verlag, pp. 31–41.

Stålnacke, P., A. Grimvall, C. Libiseller, M. Laznik and I. Kokorite (2003), 'Trends in nutrient concentrations in Latvian rivers and the response to the dramatic change in agriculture', *Journal of Hydrology*, **283**, 184–205.

United Nations Children's (Emergency) Fund (UNICEF) and World Health Organization (WHO) (2012), 'Progress on drinking water and sanitation: 2012 update', available at http://www.unicef.org/media/files/JMPreport2012.pdf (accessed 31 May 2013).

United Nations Economic Commission for Europe (UNECE) (2002), 'Bug Report No 2: Identification and review of water management issues', UNECE Water Convention, Pilot Project Programme on Transboundary Rivers.

United Nations Economic Commission for Europe (UNECE) (2007), *Environmental Performance Reviews. Ukraine. Second Review*, New York and Geneva, Switzerland: United Nations.

United Nations Economic Commission for Europe (UNECE) (2011), *Second Assessment of Transboundary Rivers, Lakes and Groundwaters*, New York and Geneva, Switzerland: United Nations.

Verkhovna Rada (1995), 'Water Code of Ukraine of 06.06.1995 No 213/95-BP', available at http://cis-legislation.com/document.fwx?rgn=8684 (accessed 16 August 2012).

World Bank (2006), 'Ukraine: addressing challenges in provision of heat, water and sanitation', available at http://siteresources.worldbank.org/INTUKRAINE/147271–1089983407712/20931047/HeatandwatersectorEng.pdf (accessed 31 May 2013).

Zakorchevna, N. (2011), 'Ukraine: existing situation', unpublished report.

Zingstra, H., V. Simeonova and K. Kitnaes (2009), 'The Bug River. Corridor in the Pan European ecological network. A feasibility study', BBI-Matra Project 1006/015.

10. River basin organisations in Northern Afghanistan: the holy trinity of contemporary water management in practice

Jeroen Warner and Vincent Thomas

10.1 INTRODUCTION

While, at the time of writing in December 2013, their country prepared for an uncertain economic and political transition as a consequence of the withdrawal of most international troops by 2014, farmers in Afghanistan continue irrigating their crops to make a living and feeding the population. Afghanistan is not water-scarce, it has many water resources and its geography provides significant opportunities for their exploitation. War has, however, dwarfed the area under irrigation. Water resource development and irrigation are key to Afghanistan's National Development Strategy 2008–2013 which claims that, 'throughout, this transition can only take place with the full participation of all stakeholders' (IROA 2008: 151).

The Afghan Water Law of 2009 proposes water sector reform leading to the delegation of the national Ministry of Energy and Water's management responsibilities to participatory river basin and sub-basin committees to promote principles of integrated water resources management, multi-stakeholder consultation and basin-level decision making – which Warner (2007) has called the 'holy trinity' of water management. River basin and sub-basin committees are a novelty for Afghanistan, and it is unclear when the committees will gain their mandate.

This chapter reports on research into the functioning of river basin participation processes in two sub-basins of the Panj-Amu River basin in Northeast Afghanistan (see Table 10.1 for some 'fast facts'), carried out by the Afghanistan Research and Evaluation Unit in 2011–2012, to see how the process of devolution to the basin level and multi-stakeholder deliberation functioned.[1] Predicated on progressive liberal principles of equality, accountability and participation, basin platforms can trip over

Table 10.1 The Panj-Amu River Basin, fast facts

	Panj-Amu River Basin
Annual discharge	22 000 billion m^3
Water availability per capita per year	7412 m^3/capita/year
Proportion of water use	24%
Estimated irrigated area	402 100 ha
Estimated settled population	297 million
Percentage of surface water (of national)	39%
Percentage of irrigated land (of national)	13%
Percentage of rain-fed land (of national)	29%
Percentage of settled population (of national)	14%

Sources: Based on Beekma and Fiddes (2011) and Favre and Kamal (2004).

non-egalitarian power structures in the communities they are proposed for, as, for example, Bangladeshi experiences show (in Warner 2007). This means that the level playing field is not guaranteed. While Edmunds and Wollenberg (2001) have suggested levelling processes, these do not extend to wider power relations. However, they incorporate elements that translate better to non-Western settings, such as consociationalism in a highly divided society (Warner and Simpungwe 2003). Seats are not allocated according to relative power relations, but to a diversity of interests, so that multiple perspectives may be incorporated in the final decision.

The implementation in Afghanistan of the river basin organisation (RBO) model, however, has been patchy: the 'pure', imported version was largely ignored, but a sporadically emerging ad hoc variety performed reasonably well. Illustrated by the developments in the dry years of 2008 and 2011, this chapter analyses the way RBOs and ad hoc committees for the Panj-Amu conform to taken-for-granted 'Western' principles of participatory integrated water management at basin scale. As a drought is the ultimate test for effective water allocation to provide security, we will briefly touch on their resulting impact on water penury in times of drought – seven years after the 'good water governance principles' were first being piloted.

The analysis reported here is based on 118 interviews with all major stakeholders[2] in both sub-catchments to assess the impact of the allocation arrangements on the ground in an arena where even the most fundamental data is missing for lack of measuring equipment and experience. Most RBOs are a type of multi-stakeholder platform (Warner 2006).

Afghanistan has five river basins and 34 sub-basins (see Figure 10.1). The two sub-basins selected for closer scrutiny (the Taloqan sub-basin and

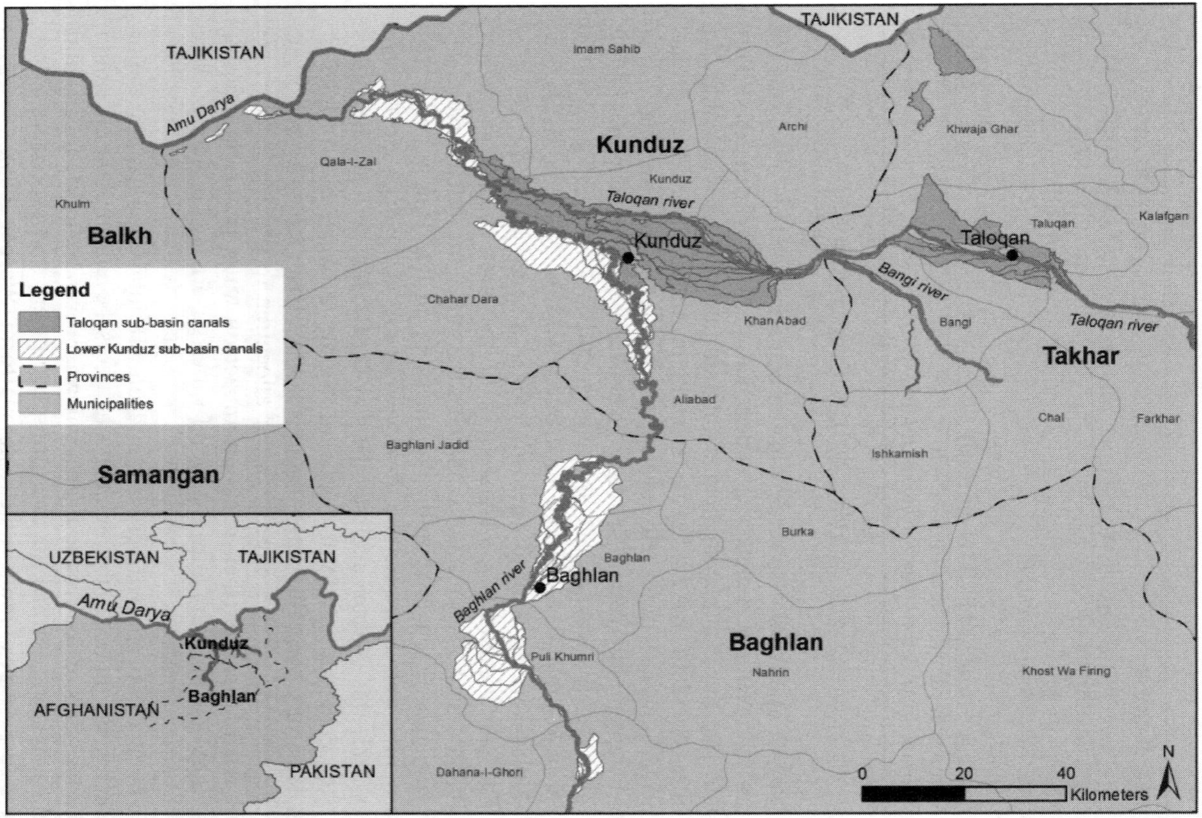

Figure 10.1 *The Lower Kunduz and Taloqan sub-basins*

the Lower Kunduz sub-basin), comprise three provinces (Takhar, Kunduz and Baghlan), collectively known as the 'grain basket' or 'bread basket' of Afghanistan.

The two sub-basins were the first two to be selected by the Afghan government and its donor (the European Union) to pilot the introduction of the 'good' water governance principles that are now enshrined in the 2009 Water Law. Both sub-basins have the longest exposure to the introduction of the system. Thus, it was expected that the study of these two sub-basins would provide more lessons which could inform development in other sub-basins as the water reform unfolds in other parts of the country.

Both case study areas were settled in the 1960s. Ethnic communities do not easily mix, and family and tribal ties prevail. After a period of growth in the 1970s, the Afghan wars (1980–2001), conflict between factions and the emergence of warlords eroded social ties, encumbering collective action (Thomas and Ahmad 2009).

As the provincial economies are so dominated by agriculture, water shortage directly affects the majority of the population. Both rivers under scrutiny are unregulated with high discharge variability within and between seasons. Parts of Kunduz only have access to rain-fed irrigation. The remainder depends on irrigated agriculture. Irrigation for the first crop of the season (mainly wheat) is complementary to rainfall in April and May, but the second crop almost entirely depends on it as there is almost no rainfall in summer. The river does not have physical regulation structure (dams) to regulate its flow. There is also no flow measurement system. Water availability is characterised by high levels of uncertainty.

The layout of the irrigation system shows upstream–downstream dynamics in the different provinces comprising the two sub-basins: upstream Takhar (upstream in the Taloqan sub-basin), Baghlan (upstream in the Lower Kunduz sub-basin) and Kunduz downstream in both sub-basins. A key role is set aside at canal level for the traditional water master (*mirab*), and upstream–downstream relations are organised by discretionary water grants (*abandâz*)[3] rather than fixed water rights.

As we shall see, this does not go smoothly in dry years, and according to interviewees,[4] water allocation in 2008 and 2011 gave more trouble than in earlier (dry) years. The allocation issue moreover attracts the intervention of political outsiders both within and outside the basin. Switching water management from administrative boundaries to hydrological boundaries has therefore not been as 'natural' as policymakers assumed. Instead, incidental multi-stakeholder processes emerged to address acute water allocation crises in dry years.

Key questions addressed in this chapter are the following:

1. How were RBOs founded in Afghanistan designed in terms of their institutional set-up and how were their geographical boundaries chosen?
2. Who propagated and who resisted the founding of RBOs, and through which strategies did they try to accomplish their goals?
3. How well do the RBOs function, in a democratic sense (accountability, transparency, and so on), and to which degree can this be attributed to the institutional design of the organisations and their geographical delineation?

We shall first sketch the background of the Water Law and the pilot project on the Lower Kunduz, in section 10.2. This history addresses the first two of the above questions. Section 10.3 assesses this development in light of the 'holy trinity' of basin scale, the multi-stakeholder platform and integrated water resource management. Sections 10.4 and 10.5 then look at the actual operation of the RBO and its sub-levels. Responding to the third question, section 10.5 again feeds the analysis of this case back into the 'holy trinity'. A conclusion (section 10.6) asks if a 'non-fit' to the policy model is necessarily a bad thing. The implementation of the RBO model in Afghanistan has been patchy: the 'pure', imported version was largely ignored, but a sporadically emerging ad hoc variety mixing formal and informal institutions performed reasonably well. Illustrated by the developments in the dry years of 2008 and 2011, the chapter analyses the way RBOs and ad hoc committees for the Panj-Amu conform to taken-for-granted 'Western' principles of participatory integrated water management at basin scale. Perhaps unsurprisingly (Mosse 2004) this shows a 'policy gap' between the theory and practice of RBOs. But the crucial question underlying this analysis is whether this gap is a problem or an opportunity. This is briefly discussed at the very end of this chapter.

10.2 HISTORY OF WATER LAW REFORM IN AFGHANISTAN

Since 2004, policymakers and international donors, such as the European Union (EU) and GIZ (former GTZ, the German technical agency for development cooperation) have introduced and promoted 'good' water governance concepts in the course of attempts to reform Afghanistan's water sector. As a consequence, the Water Sector Strategy of February 2008 and the Afghan Water Law of April 2009 formally adopted the 'holy trinity' (Warner 2007) of integrated water resource management, river basin management and participation via multi-stakeholder platforms, which can be said to be the cornerstones of RBOs. These formed part of a push toward greater devolution of decision-making power, both from the central government to sub-basin platforms, and from the government to water users.

In the early years following the fall of the Taliban, there were virtually no comprehensive studies on existing water management practices and governance at the river basin level (an exception is Lee 2006). Consequently, there was little grounded understanding of what problems the introduction of integrated water resource management, river basin management and multi-stakeholder platforms would address, and what added value the model would bring to the institutional arrangements already in place. Consequently, attempts to justify the introduction of this new governance model relied mainly on broad and often vague generalisations. Afghan officials in charge of piloting water sector reform argued that the new model was necessary because three decades of war and years of droughts had left 'a shortage of efficient institutions, organisational capabilities of staff and effective rules and regulations in regards to water use' (Mahmoodi 2008).

Beyond this, however, few efforts were made to justify why integrated water resource management, river basin management and multi-stakeholder platforms – as compared to other possible approaches – were best suited to tackle the specific problems of post-Taliban Afghanistan. In fact, when reforms were being considered during the early 2000s, foreign advisors and donors tended to view water governance in Afghanistan as a blank slate. Consequently, they assumed that – given enough capacity building and support – new, 'proper' models would inevitably improve on existing, 'defective' institutions. This would in turn lead local and national actors to accept them as a logical improvement. In a country where almost the entire national budget comes from donors,[5] external influence on the development norms and models to be followed is huge. This international overlay enabled 'global entrepreneurs' to gain an easy entryway to promote ideas such as the 'holy trinity' in Afghanistan in the early 2000s.

Both Afghans and expatriate advisors considered RBOs in Afghanistan as a novelty. The driver for the Water Law Reform leading to the creation of RBOs was emphatically external. In the years 2003–2004, representatives from the European Commission 'sold' what we have termed the 'holy trinity' of water management in the early twenty-first century (Warner 2007, Warner et al. 2008) to Afghanistan: river basin management and multi-stakeholder participation in RBO to the Afghan Ministry of Energy and Water as part of a package that would include large-scale infrastructure rehabilitation (intakes, headworks, and so on, but not dams). The RBO organogram came under the German development agency's Water Sector Reform Project. The team leader reported to the Ministry of Energy and Water. The pilot Panj-Amu River Basin programme, also funded by the EU, was awarded to the Landell Mills consultancy. From the European Commission side, the *chargé d'operations* was very involved, as the Panj-Amu River Basin programme was the European Commission's flagship in Afghanistan. In practice, the team leader of the Panj-Amu River Basin programme and the European Commission were deeply involved in GIZ work, and were very influential in writing all documents related to Afghanistan's water policy reform. The Panj-Amu River Basin programme team leader and GIZ leaders were influential members of the working group designing the Water Law, basically writing drafts that were more or less accepted as such.[6]

The intervention met with equanimity rather than resistance. Almost no change is to be found between the law's early drafts of 2004 and the final version of 2009. Once the Water Law had been passed, other projects similar to the Panj-Amu River Basin programme started, such as the Western Basin project funded by the Asian Development Bank. They were implementing the RBO model as designed by the donor rather than founding or reforming it.

The pilot, the Panj-Amu River Basin programme, failed to establish itself as a legitimate problem-solving actor. Its main focus concerned infrastructure and regulatory structure. The Panj-Amu River Basin programme was also involved in technical advice. An example is predicting water availability from snow cover (Beekma and Fiddes 2011). Variations in snow coverage of the mountains lead to discharge variations by a factor of five. This made it logical to develop a monitoring and early-warning system predicting such variations, especially low flow. This is especially relevant for the second crop (rice, melons, mung beans, corn and cotton). This crop is grown in summer when there is hardly any rainfall, no springs and high evapo-transpiration. Warnings of limited water availability, however, were not needed. These early warnings also did not lead to convening Water Allocation Committees in the provinces before the start of the irrigation season.

The Landell Mills consultancy hired by the European Commission in 2004 established a platform consisting of actors fitting the profile of future basin authority and council members. Despite no intersectoral conflicts being reported, the Panj-Amu River Basin programme unhelpfully insisted on multi-user mapping.[7] In the absence of water control structures and measuring equipment that would have provided quantification for planning, and with no new regulatory infrastructure or other concrete projects to address flood issues (the absence of regulation infrastructure makes for high levels of uncertainty for farmers) the Kunduz River Basin programme[8] seemed both impractical and irrelevant to both the Ministry of Energy and Water and water users. The project's information rules were thus externally decided and did not resonate with the engineering experts in the ministry, which moreover proved largely uninterested in its governance aspects.

The resulting lack of ownership makes a farce of the 'authority rule' attributing the set-up's mandate from the Ministry. Nobody at local or national level commented when the sub-basin working groups stopped meeting for three years from April 2008, and the registration of water user associations in Kabul was delayed indefinitely. Only when construction of the new building for the river basin agency (line ministry staff, from departments directly under Ministry authority) ran into delays, in 2009 to 2010, did the Ministry of Energy and Water complain to the European Commission in Kabul.

Between 2008 and 2011 informal bodies and platforms were consulted and avenues explored, and the project was mainly busy making regulations; its working groups failed to meet. In 2009 and 2010 water was abundant enough to forestall major conflict, but 2008 and 2011 were very dry years, in which one would expect stakeholder coordination to be most crucial. Even when sub-basin working groups were convened again in January 2011, the lead in taking allocation decisions (mainly upstream–downstream issues) and dispute resolution steps continued to be assumed by informal water user committees in the crucial summer months. The governor and key actors (including the Water Management Department and water user representatives) at the time clearly did not consider them relevant when they formed the alternative Water Allocation Committees.

The above-mentioned lack of buy-in makes the RBO conform to the 'agency' model of RBO identified in the book's introductory chapter (see Huitema and Meijerink, Chapter 1, this volume), totally dependent on external funding and impetus, and it is highly unclear how the RBO is to be funded in future (pay-off rules). The monitoring of progress inherent in this model moreover is questionable: since 2005 the Panj-Amu River Basin programme has not been monitored, and only evaluated in internal documents reviewing short periods, while only one independent review was carried out (reported on in Varzi and Wegerich 2008). A progressive phasing-out of European support for the Panj-Amu River Basin programme is planned for the end of 2014.

10.3 THE PANJ-AMU RIVER BASIN PROGRAMME AND THE 'HOLY TRINITY' (INTEGRATION, PARTICIPATION, RIVER BASIN SCALE)

10.3.1 Introduction

GIZ and Landell Mills, an international consultancy, both funded by the European Commission, can be said to have expressly based their design of the RBOs for Afghanistan on the 'holy trinity' of integrated water resource management, river basin management, and multi-stakeholder representation at basin scale. We will discuss each of these in turn.

10.3.2 River Basin Scale? Hydrological, Not Administrative Boundaries

In conformity to the boundary rules befitting the 'autonomous' RBO type identified in this book, Afghanistan's Water Law brought a shift from administrative to hydrological boundaries, to arrive at a river basin approach.[9] The river basin approach assumes that the river basin is the natural unit for water management (Warner et al. 2008). River basin agencies and river basin councils were created to run the basins. Provincial water management boards would fall under a river basin agency. The river basin agency and river basin council are both RBOs, and supposed to work together. The river basin council has a decision-making power and is more representative of water users. River basin agencies are more technical support and advisory actors, and are more representatives of line ministries (that is, departments directly under the Ministry's authority).

For the basin level, GIZ and Landell Mills believed that one single organisation (a combination of river basin agencies and river basin councils) would not be effective. They therefore decided to split the basin into sub-basins and that in each sub-basin a working group would be established (Varzi and Wegerich 2008). Some sub-basins, however, still proved either too big or were considered too heterogeneous in terms of land and water management. As a result they were split again, although this would be considered nonsensical from a purely 'hydrological boundary' point of view. This is the case for the Lower Kunduz and Dushi-Bamyan sub-basins, which form one 'natural hydrological unit' but have been split into two sub-basins. While the Dushi-Bamyan sub-basin is an upper catchment with steep slopes and agricultural land concentrated in narrow valleys, the downstream Baghlan sub-basin has flatter land.

The assumption that different water management activities would be required under different physical characteristics was a central rationale to justify the split into two sub-basins. The split also responded to logistical issues for the programme's implementation. The Panj-Amu River Basin programme infrastructure rehabilitation did not include Bamyan in the first phase. So leaving Bamyan off the project map conveniently sidestepped potential awkward questions from the Ministry as to why no infrastructure rehabilitation work was planned for Bamyan.

By dividing up the sub-basins, the sponsors also sought to reduce logistical costs. These include the costs of transportation for participants and the costs of organising meetings. Considering that sub-RBO management plans need to be approved by RBOs, logistical issues – that is, bringing people from Bamyan to Kunduz – are still going to be huge. The overall layout of the sub-basins, with two large pockets of irrigated land clearly separated by provincial boundaries, was already a sign that the sub-basin was unlikely to be the 'natural' scale at which water allocation is considered in practice.

What is 'natural' from a hydrological point of view may not be so natural in farmers' minds, particularly when it comes to tackling a practical problem in countries that have poor communication (that is, roads) and low financial resources to travel long distances for meetings, monitoring agreements, and so on. Natural hydrological boundaries did not pass the test of qualitative cost–benefit assessments by farmers. When Kunduz faced water shortages in the past, farmers always complained about Baghlan Province, but never about Bamyan, which is upstream of Baghlan. Although farmers are well aware where the river comes from, they never considered Bamyan as being part of the 'problemshed' – the area that matters when it comes to realistically addressing the problems of water sharing at sub-basin level. Bamyan simply seems too far away (costly to monitor) and also not so significant in terms of water use, therefore little benefit is to be gained from complaining anyway – and there are no flow measurements to assess whether or not this perception is justified.

10.3.3 Devolved Multi-Stakeholder Platform Decision Making?

While the Water Law promotes devolution, this is not practiced much in Afghanistan. The concept of RBOs was received in Afghanistan with relative indifference: neither applauded nor resisted. It appears from interviews[10] that Afghan decision makers (including the Ministry of Energy and Water) generally seem to have a very positive view of the traditional *mirab* system,[11] believing that water user associations (or irrigation associations) should just formalise it. The interviews reveal that to the mind of some expatriate consultants the *mirab* system was dysfunctional. The two sides ended up agreeing that water user associations were to build on the venerable traditions of the *mirab* system, but left its operationalisation for later. One of our concerns is whether such a balance between the traditional and the modern is indeed feasible.

The same logic applied to the Panj-Amu basin studied here appears to have been used at other sites in Afghanistan. Once the Water Law had been passed, other projects similar to the Panj-Amu River Basin programme started, such as the Western Basin project funded by the Asian Development Bank. In each of these cases, the organogram of river basin agencies, river basin councils and water user associations does not immediately make sense. For the people in charge of implementing these programmes it was not clear at first what form of organisation existed on the ground (from canal to river basin level). Therefore, the logic of the organogram – that is, what issue this organogram was supposed to resolve and whether it was structured in a way that actually addressed shortcomings identified and acknowledged by local users and local government – was not evident to practitioners, local government or water users.

This lack of clarity has had a strong impact on the actual working of the multi-stakeholder set-up. Given its largely voluntary character, participation appears to follow a 'partnership' logic. A multi-stakeholder platform is a 'decision-making body (voluntary or statutory) comprising different stakeholders who perceive the same resource management problem, realize their interdependence for solving it, and come together to agree on action strategies for solving the problem' (Steins and Edwards 1998). Multi-stakeholder platforms allocate seats to various interests rather than on a numerical basis, to prevent minorities being outvoted. The assumption made is that the agenda discussed is relevant to all stakeholders and positively affects their situation by solving or mitigating a commonly experienced challenge. This precondition is not necessarily met here.

In light of overt, if belated, political interference, the RBO clearly does not have an 'autonomous' orientation. While the Ministry of Energy and Water was leading the process through its expatriate consultants, the Ministry of Irrigation and Livestock's interest and contribution was extremely limited, despite being encouraged to participate. It emerged only just before the Water Law was to be sent to the Parliament, and argued that it should have its own associations at canal level, organisations which have a representation in the RBO. This led to a coexistence of water user associations and irrigation associations. The boundaries of responsibility (boundary rules) between these water user associations and irrigation associations remain highly unclear.

10.3.4 Integrated Water Resource Management?[12]

The creation of the sub-basin committees discussed in this chapter follows on from the Panj-Amu River Basin programme and the Western Basin project. Both sought: (1) major infrastructure rehabilitation and development; (2) water institution building at the canal level (water user associations, irrigation associations) and at the middle level (river basin agencies and river basin councils); (3) on-farm water management; and (4) upper-catchment rehabilitation and conservation.

These elements taken together would seem to make for an integrated package. Yet the Afghan government's interest in committing to devolution appears mainly related to access to construction projects for companies associated with its ministries; as one spokesperson noted, all they talk about in the Ministry of Energy and Water is 'dams, dams, dams', as they are interested in resuming their 'hydraulic mission' (a concept coined by Reisner 1986), started in the 1970s but which ground to a halt due to the different wars. Given Afghanistan's low storage capacity, this striving for dam projects has merit. But background issues of equity in allocation and conflict resolution remained unaddressed.[13] At a recent national water conference in Kabul in January 2013, President Karzai gave an introductory speech, restating that infrastructure development (meaning large dams) was the number one priority in the water sector. The governance issue did not feature in the conference presentations.

10.4 THE LOWER KUNDUZ SUB-BASIN: QUESTIONING DEVOLUTION OF DECISION-MAKING POWER

In semi-arid zones, a dry year puts water allocation arrangements under extra stress. It therefore makes sense to briefly elaborate what happened in both sub-basins in 2008 and especially in 2011, two very dry years highlighting mechanisms of coordination.

Water allocation in the Lower Kunduz sub-basin illustrates contrasting differences between the actual organisational arrangements on the ground and the formal water management policy model (see Table 10.2).

In the Lower Kunduz, water flows from Baghlan down to Kunduz. Tensions and conflicts between upstreamers and downstreamers have often emerged during dry years in the recent history of the sub-basin, but they ran particularly high in 2008. Downstream Kunduz felt the upstreamers were lacking in support. A high-ranking member of the Ministry of

Table 10.2 Model and realities for the Lower Kunduz sub-basin

Sub-river basin agency/sub-river basin council model	Actual practices observed in 2011
Sub-river basin council as formal single decision-making platform for water allocation at sub-basin level.	No sub-basin-level platform. Various arrangements with demarcations along provincial boundaries (Water Allocation Committee in Kunduz). No Water Allocation Committee in Baghlan but ad hoc meetings. Evolutive composition of the Water Allocation Committee. Borders of participation in decision making shaped by the geographical and political borders of the problemshed.
Line ministry staff (sub-river basin agency) as technical advisors. Undefined role of governors. Water users (sub-river basin council) as key decision makers.	Central importance of governors and the Water Management Department (WMD) in shaping if not imposing decisions. No or limited direct discussions between water user representatives from both provinces. In Baghlan, composition of ad hoc platforms and level of water users' participation shaped by local political interests and power dynamics. Participation of water users as option by default for WMD.
Separation between advisory role and decision making role.	Within the Kunduz Water Allocation Committee, preference for consensual agreements among both water users and government.
Decentralising decision making, from national agencies/actors to sub-basin institutions/actors.	Pressure from central ministry level (Ministry of Energy and Water) on local WMD to shape water allocation decisions.
Representation of multiple water sectors on the sub-river basin council.	Irrigation as the only represented water sector.

Energy and Water came from Kabul to Baghlan during the 2008 irrigation season to personally convey his dissatisfaction, and Kunduz voiced its complaints up to the central government in Kabul. In 2009, a letter from the vice-president urging the resolution of water allocation issues between the two provinces provided an additional incentive for the Baghlan Water

Management Department to demonstrate improved results. In 2011 this led to more pressure from Kabul senior officials in the Ministry of Energy and Water to put some pressure on the local water department.

In Baghlan, only ad-hoc meetings were organised at the initiative of the local government. In Kunduz, however, a formal Water Allocation Committee was first instated by the provincial governor early in the 2011 irrigation season to look after water allocation issues on canals within the province exclusively. In practice, the role of the governor was critical in legitimising the decisions on *abandâz*. The governor, interviewees[14] noted, is considered the father of the province, and nothing happens without his say-so. Apart from representatives from the governor's office and ministerial departments, water user representatives of irrigation canals sat on the Water Allocation Committee. Not all irrigation canals were represented, but interviewed[15] canal-level stakeholders said they felt represented by those that were.

When the water level fell really low in the upstream part of Kunduz, the members of the Kunduz Water Allocation Committee decided to file a request for *abandâz* to Baghlan. At first, the discussion was mainly between governors and Water Management Department (WMD) directors of each province. Baghlan water users were not involved in this first meeting. Thus, there was no direct discussion among water users' representatives from both provinces as the sub-basin river basin council model suggests. In this meeting, the Baghlan governor promised the Kunduz Water Allocation Committee members that an *abandâz* would be given. He handed the responsibility to the WMD director to make good on this promise. The WMD director assured Kunduz Water Allocation Committee members that an *abandâz* of at least seven days would be starting within a few days.

From this formal engagement followed a series of ad-hoc intra-provincial meetings and platforms among Baghlan stakeholders at the initiative of the local government. The meetings were for dealing with conflict on the execution of the *abandâz*. For some canals, the WMD imposed the decision on the duration of *abandâz*. Representatives from these canals were not asked to sign any agreement, no Kunduz representatives were invited and a seven-day *abandâz* was settled as a *fait accompli*. For other canals, however, water user representatives were able to oppose and change the initial decision of the WMD director through a series of discussions. In some of these meetings, Kunduz water users were involved, if mainly as observers. The outcome was a three-day *abandâz* for most canals, and only two days for the largest canals of the province.[16]

The difference in both process and outcome can be well explained by differences in local power relations. While the policy model emphasises the leading role of water users (in the sub-river basin council) in decision making, the practice in Baghlan shows that the composition of ad-hoc platforms and the level of water users' participation was shaped by local political interests and power dynamics – the decision-making process was far from being a level playing field. During the whole process of decision making, new representatives (as compared to the original Water Allocation Committee composition) from different canals were brought in depending on the issue at stake – basically depending on which canal and who was involved in defaulting agreements. Thus composition of the Water Allocation Committee remained flexible throughout the process.

In sum, in contrast with the RBO set-up proposed by the Water Law, water allocation was not decided through one single sub-basin platform, but mediated through various platforms with demarcations along provincial boundaries. While in the policy model the governor's role and responsibilities are not mentioned at all, he proved crucial to conflict resolution and allocation decisions.

10.5 THE TALOQAN SUB-BASIN: CHALLENGING DECENTRALISATION

The case of the Taloqan sub-Basin, in which water flows from Takhar down to Kunduz, provides particularly useful insights into the limits of the decentralisation paradigm regarding water allocation at sub-basin level. In 2011, for the first time in the sub-basin history, a presidential decree was issued stipulating fixed rights for each province. These water rights were expressed as a percentage of the river flow. President Karzai signed the decree a month after the official publication of the Water Law. The decree overruled the principle of decision making by a decentralised platform.[17] The origin of the decree dates back to the 2008 crisis, when 60 per cent of the downstream section, in Kunduz Province, could not be irrigated for the second crop while only around 20 per cent of the upstream province of Takhar suffered from shortage – and mainly due to technical constraints rather than resource scarcity. The 2008 *abandâz* that Kunduz obtained from the upstream Takhar Province was ill-defined and poorly implemented. Allegedly, the root of the problem was the governor of Kunduz, who had Takhar origins, as well as land and connections in the upstream province. Thus, in 2008, his interest in supporting Kunduz was limited. During the course of the irrigation season, Kunduz elders attempted to mobilise their provincial council representatives as well as members of parliament in Kabul in order to change a water-sharing situation perceived as inequitable. It was also a sign that the *abandâz* system was no longer considered as the panacea given the increase in water demand over the past decades. Although MPs actively tried to raise the issue at the level of the Ministry of Energy and Water, as well as with the presidential office, their action could not lead to tangible results as it was already late in the course of the irrigation season when their voice started to be considered. For 2008, part of the slow reaction at local government and ministry level was because most people in charge had not been exposed to such drought, most of them being away during the Taliban years – the time of the previous similarly severe drought.

However, MPs continued to push the concerns of Kunduz constituencies until they found a way to access the president, who ordered in April 2009 the setting-up of a commission to resolve the matter. This commission comprised mainly ministry representatives (Ministry of Energy and Water) and security commanders for the north. Their mandate was to define *haqabah* (water rights) in the Taloqan sub-basin, which has always practiced traditional *abandâz*.

Interviewees[18] are unanimous that the making of the decree was not a participatory process. The commission sought to extract data, not to consult water users about their perspective and suggestions. Not even governors were asked to comment. In Takhar, suspicions of hidden political motives were high when the Kabul delegation arrived. Thus the neutrality of the commission was put into question from the start. The head of the WMD openly expressed misgivings about manipulation along ethnic lines.

The decree contained a number of technical flaws that heightened the perception of a rushed, non-concerted decision, but also possibly bias favourable to Kunduz. The decree defined a percentage of almost two-thirds of the Taloqan River's surface flow to Kunduz. Rice was given three times as much irrigation rights as non-rice crops; questionable reasoning when increased traditional rice cultivation over the past decades is recurrently singled out, particularly by downstreamers and local government, as a cause of upstream–downstream inequity in water access. Several high-level actors claim the Takhar Ministry of Energy and Water director was under pressure to provide largely outdated cropping data, and therefore underestimated rice coverage in the province. For Takhar community leaders and elected representatives, the Ministry of Energy and Water probably tried to avoid trouble given that a Kunduz member of parliament (MP) close to the president had a strong influence over the presidential committee. The head of the commission did not hide that the content of the decree had been designed to correct 'an injustice to Kunduz'.[19] At ministry level, the endorsement of the decree by the Ministry of Agriculture, Irrigation and Livestock is unclear: the deputy minister refused to comment on persistent claims from high-ranking officials that he was not given much choice other than signing the decree despite his initial disapproval.

In the course of 2009, a Takhar protest letter citing technical flaws in the process and content of the decree went unanswered. A delegation from Takhar went to Kabul to see the president, whom they managed to reach through a high-level contact. In response to the complaint, Karzai ordered the formation of a new committee for better fact-finding. But as there was enough water in the river in 2009 and 2010, the initial sense of urgency faded. In 2011, however, the issue returned with another drought.

Early in the irrigation season, a Kunduz Water Allocation Committee – similar to that described in the Lower Kunduz Sub-basin case but composed of entirely different individuals – went to Takhar to demand water allocation based on the 2009 decree. A meeting was convened comprising the formal Kunduz Water Allocation Committee and an informal group of Takhar water users, line ministries and provincial councillors, and the governor.

While the Kunduz Water Allocation Committee pushed for applying an official decree, Takhar provincial councillors and water user representatives strongly opposed its implementation, arguing about the biased and misinformed decisions of the presidential commission. In the end a consensual agreement on a follow-up procedure was found. Both parties agreed to leave water allocation decisions and implementation to a joint inter-provincial Water Allocation Committee. Officially, the joint Water Allocation Committee was mandated to apply the decree. In practice, the committee was informally requested to find an acceptable compromise between *haqabah* (fixed rights) and *abandâz* (traditional rights). Strictly applying the decree would have meant endangering the relationship between the provinces. But at the same time, ignoring it to the extent that it would lead to a similar crisis (in terms of water access for the downstream province) as in 2008, would have been unacceptable to the authorities in Kabul; a perspective that the Takhar governor[20] and WMD could not afford, considering that their hierarchy had designed and signed the decree. At the end of the meeting, four water user representatives from both provinces were chosen to form a joint committee. In Takhar, upstream canals were over-represented: three members out of four. As it was anticipated that most defaulting on the water allocation compromise would come from those large canals, it was felt that more community leaders from this area would be needed to prevent such actions. The platform was formally recognised by the governors of the two provinces. Although not formally members, the governors' office and WMD were requested to assist water user representatives of the joint Water Allocation Committee in case it failed to enforce the agreement.

The joint Water Allocation Committee defined operation rules for opening and closing gates at the canal headworks. The few modern headworks available were only a few years old, and delegates had no technical training in their operation and on downstream effects on Kunduz. For traditional canals, negotiations focused on how many sandbags to remove from traditional endogenous intakes. The smallest canals and those located in the upper valleys were ignored, as the transaction cost of monitoring them would be too high. Thus, the joint Water Allocation Committee actions focused only on the large irrigated plains of Takhar and Kunduz.

During the implementation, several irregularities were reported. For instance, Takhar delegates permitted adjusting headwork gates at times when Kunduz delegates could not monitor them. Takhar joint Water Allocation Committee members conveniently depicted a difficult security environment, warning about some headworks locations being no-go areas at night, particularly for people from outside Takhar and government officials. The adjustable gates gave upstreamers better control of the river than before.

In sensitive cases, such as when a Takhar gatekeeper was beaten by frustrated water users, the Kunduz joint Water Allocation Committee members would arrange for police support under the injunction of the Takhar governor. On other occasions the WMD director made personal visits. Kunduz members of parliament, who have no formal authority in these water affairs, would also bring their influence to bear on the governors, reminding them through phone calls of the presidentially sanctioned agreement.

Overall, a balancing act between local and national interests shaped the way the joint Water Allocation Committee functioned informally, with the monitoring and support of the Takhar authorities and the interventions of external actors (that is, MPs). The introduction of the presidential decree through the involvement of MPs, the President's Office and the Ministry of Energy and Water in Kabul pushed the governor and WMD of Takhar to align their interests more closely with those of Kunduz water users; at least, more so than in 2008. In other words, the decree created room for cooperation and support to the downstream province. However, the governor and WMD director also had to take into consideration the rejection of the decree by Takhar actors in order to avoid or limit unrest in the province, which is directly a responsibility of the governor. The matter was complicated by the clan ties between the WMD director and the most influential community leaders of upstream Takhar canals, who at first were particularly virulent in their opposition to the decree. Rather than satisfying the terms of official *haqabah*, the governor used short-term 'pacifying tactics'. Interviewees[21] in Takhar predicted bigger conflicts for 2012 if a drought was to strike again.

Within the downstream province of Kunduz, different forms of water user organisation have been created. Prior to the initial engagement with Takhar stakeholders, a first Water Allocation Committee has been created to allocate water within the three zones (that is, groups of canals) supplied by the Khanabad River barrage – rehabilitated by the EU between 2008 and 2011. The formation was formally led by the WMD director under the supervision and approval of the governor. The Water Allocation Committee was chaired by the representative of upstream canals, as his influence in the upstream area of Khanabad would help in persuading upstreamers to respect the application of agreements with downstreamers. Government representatives from the Ministry of Energy and Water and the National Directorate of Security as well as members of the provincial council were represented, but left decision making to water users and the WMD. Time-based rotation between the three different zones supplied at the barrage were decided on a consensus basis and formally recorded. No complaints were recorded regarding the decision-making process or the content of agreements.

Soon after the formation of the first Water Allocation Committee, a second platform was created in order to deal with 'water turns' (distribution of water to users, in succession) inside the largest zone supplied by the Khanabad barrage. This time the district WMD director was selected as chairman. This selection of a local government officer – as opposed to a water user as in most other cases – was not contested, considering his acknowledged technical understanding of the barrage structure operation and his good relationship with all local leaders among water users. This 'internal Water Allocation Committee' was much more informal in its composition and decision-making process. Although nine members (mainly canal *mirabs*) were first selected, most meetings included additional water users and *mirabs* depending on the group of canals, mainly concerned by the decisions on water turns. Ad-hoc changes in the composition were also triggered by the occurrence of cases of defaulting and conflicts which required additional meetings to correct plans of action. In such meetings, agreements were made verbally.

Implementation difficulties did arise, however, and the police and governor were not always willing to enforce the agreement in an area with armed people. Defaulters could therefore proceed with impunity, at the cost of downstream water access. Improved water control of the barrage made it possible to route the flow exclusively to the left- or right-bank canals, excluding downstream canals. When it became clear that they would not get water from the barrage as per the agreements, it became a case of everyone for themselves. Users resorted to coping mechanisms such as diversion of resurgence flows. Downstream *mirabs* with good contacts upstream brokered deals with left-bank canals leaders to let some water drain back to the river, so that it could be channelled towards downstream canals.

Tables 10.2 and 10.3 summarise the analysis comparing the original Panj-Amu River Basin programme set-up and the ad-hoc platforms

Table 10.3 River basin organisation model and realities for the Taloqan sub-basin

Sub-river basin agency/ sub-river basin council model	Actual practices observed in 2011
Sub-river basin council as formal single decision-making platform for water allocation at sub-basin level.	Water allocation mediated through various platforms ('Water Allocation Committee'), with a clear demarcation along provincial boundaries.
Water users (sub-river basin council) as key decision makers.	Compromised plan balancing national authorities plans and upstream water users' interests in the status quo.
Line ministry staff (sub-river basin agency) as technical advisors.	Central importance of governors and the Water Management Department (WMD) in balancing national and local interests.
Undefined role of governors.	
Decentralising decision making, from national agencies/actors to sub-basin institutions/actors.	Significant influence of members of parliament, Kabul senior ministry officials (Ministry of Energy and Water) and the President's Office. Decree defining water rights of provinces issued by the President's Office in Kabul after a non-participatory process.
Representation of multiple water sectors on the sub-river basin council.	Various arrangements with demarcations along provincial boundaries (different Water Allocation Committees in Kunduz; no Water Allocation Committee in Takhar but joint Water Allocation Committee at interprovincial level).
Composition of sub-river basin council should cover the whole watershed (or sub-basin) geographical area.	Evolving composition of the Water Allocation Committees in Kunduz. Borders of participation in decision making shaped by the geographical and political borders of the problemshed. Borders of decision making stretching to the capital city.

Table 10.4 Elinor Ostrom's rules for common-pool governance applied to the Panj-Amu River Basin programme model and the ad hoc Water Allocation Committees for the Lower Kunduz and Taloqan sub-basins

Rule type	Panj-Amu River Basin programme model	Water Allocation Committees for the Lower Kunduz and Taloqan sub-basins, 2011
Authority rules	Devolution of decision-making power to sub-basin level; authorities relegated to technical advisor role. Allocation and infrastructure domain.	Overlay of external actors making final decisions. Allocation only.
Boundary rules	River basin scale. Wide stakeholder representation.	Provincial scale. Selected stakeholders gathered on ad hoc basis, with apparent consent of the excluded.
Aggregation rules	Democratic decision making.	Authoritarian decision making.
Information rules	Externally constituted technical knowledge, joint fact-finding but little internal resonance.	Informal information exchange, ignoring knowledge and policy there was (danger of, or actual) violence if the policy were implemented.
Pay-off rules	Entirely dependent on external funding, not related to output.	Reasonable upstream–downstream equity.

on the Lower Kunduz and Taloqan, then, in light of their apparent similarity. Findings were grouped according to the 'Ostrom rules' for identifying institutional design identified in Chapter 1 in this volume (Table 10.4).

In terms of the criteria outlined in the present book's introductory chapter (see Huitema and Meijerink, Chapter 1, this volume), clearly Water Allocation Committees enjoyed greater legitimacy and support despite, or perhaps thanks to, the limited scope for democratic debate and the dominance of figures of authority, than the Panj-Amu River Basin programme model. The willingness to rely on patronage and political capital ensured external support in times of conflict. When the outcomes of that process seemed impractical or to be fuelling conflict, as in the case of the 2009 presidential decree for the Taloqan, they were to a degree ignored.

As for environmental performance, environmental considerations were not part of the Water Allocation Commission's concerns, nor was it part of water users' preoccupation. There are no issues of pollution to speak of in those rivers. Nobody (that is, upstream or downstream water users, local government officials, governors, facilitators from the Panj-Amu River Basin programme, central government) ever raised the question of the need to maintain a minimum environmental flow in the river. During the irrigation season, however, the Kunduz River (fed by both provincial rivers in Takhar and Baghlan) did not reach the Amu Darya any more. Only small resurgences appeared in the most downstream section of the river but these were captured by downstream water users to make small pools from which they (or at least those who could afford it) could pump water to their fields.

10.6 HAS THE MODEL TAKEN ROOT?

How well do the RBOs function, in a democratic sense (accountability, transparency, and so on), and to what degree can this be attributed to the institutional design of the organisations and their geographical delineation? And has the RBO model taken root in Afghanistan? Perhaps it is too early to say. While the RBO model in theory exemplifies 'good governance' in terms of accountability and transparency, it has failed to interest local and national actors to the extent that alternative water management organisations (that is, water allocation commissions) were formed to solve practical issues, which we may term 'ad-hoc RBOs'. In a democratic sense, however, the ad-hoc RBO discussed in this chapter can be said to malfunction: that is, top-down imposition by decree, and local governors and WMD directors forcing decisions (when possible) to fulfil their own interests. A high degree of extra-provincial meddling was possible because the region's strategic importance for Afghanistan's food security mobilises political interests projecting far beyond the basins themselves.

The lure of funding for infrastructure appears to have been the main reason why the European Commission could successfully promote the RBO model in Afghanistan. As noted, the Panj-Amu River Basin programme was dropped on North Afghanistan from the outside and its governance component did not address farmers' interests. The lead actor, the Ministry of Energy and Water, appeared uninterested in governance aspects, at least far less than the infrastructure rehabilitation component of the programme. It stands out in this case how water control infrastructure plays an important role in the functioning of RBOs, both as a priority for ministries to get a piece of the cake and pursue their hydraulic mission, and at lower levels as an instrument of diversion towards or away from others, as conduits of power (Turton 2000). Let us take a second look at the 'trinity' in light of the case study:

Integrated water resource management: Warner (2007) has summarised integrated water resource management as 'decompartmentalising water management' in response to the limitations of sectoral approaches. On the other hand, the issues were largely monosectoral. Integrated water resource management makes most sense in countries where supply-oriented approaches have shown their limits, and environmental considerations are high on the agenda of many civil society organisations. Afghanistan by contrast is resuming its 'hydraulic mission' initiated more than four decades ago, and still has scant environmental awareness. RBOs in Afghanistan so far have tended to be monosectoral, thus not conducive to integrated, balanced processes. Confirming the findings of earlier studies in different contexts (Warner et al. 2008; Cohen and Davidson 2012), the attempted shift from administrative to hydrological boundaries would also appear to provide no added value, if not being counterproductive in the current environment.

Basin scale: As noted, administrative rather than hydrological levels dominate, so that there can hardly be a comprehensive ecosystems approach. Creating water rights, moreover, another liberal paradigm, would seem to exacerbate rather than reduce tensions in this context. In Northeast Afghanistan the *abandâz* tradition, which defines temporary rather than fixed rights granted from the upstream to the downstream province, further confirmed the consideration of administrative borders in water allocation. Both sub-basins present cases that underline the significant importance of governors in shaping – or least supporting and legitimising – decisions over water allocation. Provincial actors (that is, governors) are still seen as a legitimate and relevant actor in conflict resolution, contradicting the shift from administrative to hydrological boundaries.

In this light, the level that actually appears to dominate (boundary rule) is the 'problemshed' (Allan 1998). While issues are largely defined locally as province against province, not all irrigated areas within these provinces are defined as being part of the problem or the solution. Only the main large pockets of irrigated land are concerned (that is, those that make a real difference in terms of water use, and that are more easily accessible if canals have to be controlled). The maps that matter in practice are delineated by the 'problem boundaries', rather than the provincial boundaries or the sub-basin boundaries.

Multi-stakeholder platform: The RBOs and sub-basin organisations studied in Afghanistan are hardly models of devolution and democratic accountability. The envisaged composition seems overambitious anyway, trying to respond to all types of water use (irrigation, livestock, environmental flows, municipal use, industries, and so on), all types of actors (water users, government, elected bodies) except the governors, in all geographical areas of the sub-basin (from upper catchment valleys, to irrigated plains). Thus, it actually tries to represent all interests and tries to prepare for all kind of problems. But in doing so, it fails to address the core problem: water allocation between the two pockets of irrigated plains in dry years. This is because: (1) the actual problems do not concern all sectors and do not concern all provinces and geographical areas of the sub-basins; and (2) nobody who is affected or concerned by the problem will accept that people who are not directly concerned by that problem from the area will take a decision on their behalf.

For instance, in the case of the 2011 dry year, the 'hypothetical' river basin council sub-basin council would have had only four out of 19 relevant people that water users have considered legitimate in taking decisions on how to address the problem of water allocation.[22] The sub-basin council (one level lower) would have been more relevant but still not in line with the preferred choice of representation.

It could be argued that any composition fixed in advance is likely to be problematic because it would never be tailored to the existing problem at hand. Contrary to the proposed decentralisation, especially in the Taloqan sub-basin parliamentarians, central ministries and even the President's Office were a driving force behind water allocation during the dry year of 2011, suggesting an attempt at re-statisation instead of a de-statisation: in this case it is not so much about the national government refusing to relinquish the power it had, but about regaining it. The government cannot afford to stand idly by when faced with complaints from farmers, as it did in 2008 and 2011, and so it will not hesitate to intervene if need be. The intervention of the government is contrary to the idea of having water users as decision makers. In the context of Afghanistan, where the government has not been a powerful player in the management of water resources during the decades of war, such intervention cannot be interpreted as a sign of refusal to relinquish power. It could, however, be interpreted as an attempt by the central government to regain power which it had (including in the sub-basins under study) in the years prior to the decades of war (pre-1979).

Moreover, as Fischer (2000) notes, the mandate does not match the responsibilities. In relation to the research question, the reason for the presidential decree to have emerged is because farmers think their problems are more likely to be solved through a 'poor governance' process than through the principles of the RBO model.

In light of its hydrological definition, the RBO model in Afghanistan generally neglects the role of national actors and provincial governors. So far, the Water Law and subsequent regulations have neglected to define the role of the governor in the sub-river basin agency and sub-river basin council model, and to represent all key stakeholders – beyond project site elites. Considering his mandate, it is unlikely that the governor's position in the sub-basin governance set-up would be limited to that of a technical advisor in times of water scarcity crisis.

It may be useful to rethink the composition and structure of decision-making platforms at the sub-basin level, and explore the possibility of multiple and flexible multi-stakeholder platforms at different scales rather than a single, fixed sub-river basin council.

Finally, content matters. Ignored by much institutional and participatory analysis, the local situation is not only about social-institutional, but also about material realities. Participation literature is often process-oriented (if due process is observed, it is successful), ignoring what the participatory process is about. A proper understanding of the political and physical context is key to a useful multi-stakeholder platform outcome (Verhallen and van Densen n.d.). One pitfall of the Panj-Amu River Basin programme was that the agenda set for the development of river basin organisations was often at odds with the situation on the ground.

10.7 CONCLUSIONS

The above paints a muddy, mixed picture in terms of the RBO typology advanced in this book. In fairness it was to be expected that integrated water resource management as 'good policy' would not fully translate on the ground. Integrated water resource management can be considered an overambitious, unattainable 'nirvana' concept (Molle 2008), especially in a conflict-ridden country with an incomplete state-building process. Mosse (2004) has cogently argued that 'policy gaps' between prescriptive models and realities to the ground are to be expected and may even be functional. In this light, the observation that ground realities do not match the model is no immediate reason for concern. A more realistic approach, integrated water resource 'management-lite' (Butterworth et al. 2010), has been proposed, aiming at partial rather than full integration. By and large, we can say the same thing about RBOs: expectations may be excessive.

However, the whole package does not make much sense without considering the interconnection with the local situation, leading to seemingly unanticipated (dis)empowerment effects for specific actors and institutional *bricolage*. The 2011 case study shows that the involvement of the governor and the presidential decree in forcing decisions to make things work contributed to relatively positive results in terms of water access for downstream water users, and limiting inter-provincial conflicts.

In light of our third research question, scrutinising the RBO model's 'democratic functioning' we may see the political involvement of governors in the decision making as proper to a democratically legitimate administrative process. After all, one of the critical mandates of governors is to limit unrest and preserve security in a province. Considering the sensitivity of irrigation water allocation in dry years, their involvement is considered legitimate by water users.

But the way in which the governors and also WMD directors were involved did not always follow the democratic high road. Interviewees[23] recognised this 'non-democratic' process as a better alternative than the RBO model alone, which expects water users to decide among themselves. The 'poor democratic performance', which saw the governors, WMD directors and president forcing decisions, could be seen as a sign of rejection of the institutional design of the 'pure' RBO model, and with a better outcome. The governors' mandate moreover is strictly based on administrative rather than hydrological boundaries, which is unlikely to change despite the water sector reform.

The gaps between policy models and on-the-ground practices observed in the Lower Kunduz sub-basin do not prejudice the satisfactory performance of the water allocation commissions when it comes to limiting inequity in water access between provinces (based on the analysis of satellite imagery, and confirmed by interviews) and in terms of limiting conflict escalation. Results have been evaluated positively in 2011, especially in comparison to the crisis of 2008 (Thomas et al. 2012).

Practical implementation means that the envisaged composition of river basin councils is hardly representative of the spectrum of interests: it would include two upstream and two downstream representatives and 19 representatives from outside the areas concerned. The representatives would therefore have been unlikely to have made a difference. Existing sub-basin meetings have failed to solve any of the problems.

The RBO model being largely ignored, and the formation of an ad hoc alternative, suggest bad news for the transfer of RBOs as an institutional 'technology'. This is not necessarily a problem when it comes to output and performance. But in terms of process it may be faulted as the non-democratic aspect of the process; a lack that appears to have been key to the very success of the alternative structure. The relatively good performance (that is, limiting conflicts and limiting inequity) of the ad hoc RBO, however, should not be attributed only to the structure of the institutional arrangements but also to opportunistic behaviours.

The RBO model may not have worked so well in 2011, perhaps because the system was not democratic and because it eluded the institutional design of the EU and the Panj-Amu River Basin programme. But nothing is stable here. Another drought, another socio-political context, may bring an entirely different process.

In the context of Afghanistan, there is no obvious choice for design. Including local government and governors in decision making cannot be completely ignored because they can be part of the solution (as observed in 2011) when it comes to achieving results. Yet they may also be part of the problem if their interests do not align with those of downstream water users. Whatever design choices are made, their relevance will vary as the local and/or national political context changes (Box 10.1).

BOX 10.1 SUMMARY OF THE MAIN FINDINGS

1. The Panj-Amu River Basin programme was dropped on North Afghanistan from the outside without even creating awareness among key actors of the multi-level RBO system and its envisaged functioning, despite requests on the Panj-Amu River Basin programme for awareness-raising. This confirms the finding in Varzi and Wegerich (2008): outside agendas dominate. Not all stakeholders are represented – only project site elites.
2. Practical implementation means that the envisaged composition of river basin councils is hardly representative of the spectrum of interests; it would include two upstream and two downstream representatives and 19 representatives from outside the areas concerned. The representatives would therefore have been unlikely to have made a difference. Existing sub-basin meetings have failed to solve any of the problems.
3. The region's strategic importance for Afghanistan's food security mobilises political interests projecting far beyond the basins themselves.
4. Integrated water resource management does not make much sense without considering the interconnection with the local situation, leading to seemingly unanticipated (dis)empowerment effects for specific actors and institutional *bricolage*.

5. This local situation is not only social-institutional, but also material. Participation literature is often process-oriented, ignoring the substance and physical context of the participatory process.
6. The higher the relative power of the Water Management Department (WMD), the less room for decision making among users, but the more advantageous the decisions for Kunduz in terms of the *abandâz* it received.

NOTES

1. While the Panj-Amu is a transboundary river, there are almost no fora for transboundary water cooperation. Hence this chapter focuses on river basin management on the river section within Afghanistan (King and Sturtewagen 2010).
2. The research project also includes the Normative Difference Vegetation Index to access the consequence of the allocation agreement in terms of productivity and upstream–downstream equity, but this will not be elaborated in this chapter.
3. *Abandâz* is a temporary voluntary restraint on the part of upstreamers to permit access to downstreamers, a humanitarian gesture. *Abandâz* does not grant water rights, but is a temporary agreement by upstreamers not to fully acquire water by closing their intakes and using all available water, for the benefit of downstream canals – between provinces, down to the tertiary irrigation level. Actors at the community level, the meso-level comprising governors and, occasionally, even national-level ministers, are called on informally for *abandâz* and enforce its implementation. They are therefore crucially involved despite formal devolution. Despite a context of recurrent violent non-water conflict, the traditional water allocation system, *abandâz*, has proved enduring. While the influence of authority figures is naturally unstable in unstable times, *mujahedin* did not go against *abandâz* except in Kunduz due to infighting between communities from different canal systems, and under Taliban rule this tradition was restored. It had never really gone away, but meso-level actors just failed to support its implementation.
4. Based on interviews with *mirabs* and government officials from both sub-basins. Names available at request.
5. A recent World Bank report (World Bank 2012) claims that '97 per cent of Afghanistan's roughly $15.7 billion gross domestic product comes from international military and development aid and spending in the country by foreign troops'.
6. The only level at which there has been strong resistance was below the canal level. The Ministry of Agriculture, Irrigation and Livestock was strongly opposed to the Ministry of Energy and Water having its own water user associations, while the Ministry of Agriculture, Irrigation and Livestock would have no farmers' organisation under its jurisdiction.
7. The lead consultant of Landell Mills provided a number of lectures and exercises which constituted a 'blueprint' curriculum for RBO development, based on the principles of integrated water resource management. As the assumption of the facilitator was that RBO have to deal with multisectoral water allocation according to quantifiable water volumes, the facilitators considered it important to conduct a mapping exercise of water uses and users, and to conduct exercises on volumetric (that is, quantified) water allocation by sector. In the absence of intersectoral conflicts and lack of measurement structures, such exercises seemed irrelevant to the participants. They also felt that there

were more urgent issues to look at such as quality issues with constructions of canal infrastructures, emergency responses to flood damage, and so on.

8. Early name of the Panj-Amu River Basin programme.
9. Interviews with the Director General of Water Affairs Management at the Ministry of Energy and Water (MEW) (6 October 2011) and a National Senior advisor for MEW (11 September 2011).
10. At RBO level, the Water Law's early drafts of 2004 and the final version of 2009 are very similar. The only level at which there was dissent was at canal level for water user associations.
11. Based on interviews with 11 *mirabs* and Water User association (WUA) chairmen and one governors' representative (Kunduz Province). Interviews were conducted between the 14 August 2011 and 22 August 2011. Names available at request.
12. Based on interviews with three *mirabs* from most downstream canals in Kunduz Province. Interviews were conducted between 15 August 2011 and 21 August 2011.
13. Defining the so-called '*mirab* system' as 'traditional' and 'community-based' is conceptually questionable. A historical perspective of one relatively big canal located in the Lower Kunduz sub-basin (Thomas and Ahmad 2009) showed that the '*mirab* system' has meant very different things at different times. The question then becomes, which '*mirab* system' do you want to restore? The one from the 1970s when the government footprint was quite important? Or the one from the 1990s where the influence of warlords was a key driver in the way water management was functioning?
14. Integrated water resource management is about 'the coordinated development and management of water, land, and related resources, in order to maximise the resultant economic and social welfare in an equitable manner without compromising the sustainability of vital ecosystems' (see Global Water Partnership, www.gwp.org).
15. For example, when a river barrage was finally built in Kunduz, allowing the exclusive diversion of water to the left- or right-bank canals, downstream canals were excluded, despite earlier agreement that they would benefit from the flow.
16. While formally open to a second *abandâz* to help Kunduz water users faced with water penury, in practice Baghlan authorities resorted to delaying tactics. They made no promises to Kunduz, as they knew that Baghlan water users would show much stronger resistance this time; a potentially conflictive situation.
17. The decree was also endorsed by the Ministry of Energy and Water, the Ministry of Irrigation and the Supreme Court. There are doubts regarding the endorsement of Ministry of Agriculture, Irrigation and Livestock.
18. Based on interviews which include the water management directors of Takhar (upstream province) (22 September 2011) and Kunduz (downstream province) (24 August 2011), Water Allocation Committee members of Takhar (20 September 2011), provincial council members of Takhar (19 September 2011) and other key informants (22 September 2011). Names available at request.
19. Anonymous interview (17 October 2011)
20. Provincial governors are directly appointed by the president. They are mainly responsible for coordinating security, and the overall administration in their provinces.
21. Based on interviews with the water management director (22 September 2011), the deputy governor (25 September 2011) and four water allocation commission members (20 September 2011).
22. These criteria belong to the province concerned by the issues and being someone directly involved in the issue (that is, irrigation in that case). According to the composition, 15 people would have been from other provinces or representing sectors which are not part of the problem.
23. Based on interviews with five Water Allocation Committee members of Kunduz (downstream province) (interviews conducted between 14 August 2011 and 21 August 2011). Names available at request.

REFERENCES

Allan, T. (1998), 'Watersheds and problemsheds: explaining the absence of armed conflict over water in the Middle East', *Middle East Review of International Affairs Journal*, **2** (March), available at http://biu.ac.il/SOC/besa/meria/journal/1998/issue1/jv2n1a7.html (accessed 1 October 2013).

Beekma, J. and J. Fiddes (2011), 'Floods and droughts: the Afghan water paradox', Background Paper Afghanistan Human Development Report 2011, Center for Policy and Human Development, available at http://www.cphd.af/Downloads/Background%20Papers/Flood%20and%20drought%20-%20The%20Afghan%20water%20paradox.pdf (accessed 1 September 2013).

Butterworth, J., J. Warner, P. Moriarty, S. Smits and C. Batchelor (2010), 'Local approaches to integrated water resources management', *Water Alternatives*, **3** (1), 82–101.

Cohen, A. and S. Davidson (2012), 'The watershed approach: challenges, antecedents, and the transition from technical tool to governance unit', *Water Alternatives*, **4** (1), 521–534.

Edmunds, D. and E. Wollenberg (2001), 'A strategic approach to multistakeholder negotiations', *Development and Change*, **32** (2), 231–253.

Favre, R. and G.M. Kamal (2004), *Watershed Atlas of Afghanistan*, Kabul: Ministry of Irrigation, Water Resources and Environment.

Fischer, F. (2000), *Citizens, Experts, and the Environment*, Durham, NC: Duke University Press.

Islamic Republic of Afghanistan (IROA) (2008), 'Water sector strategy', Kabul: IROA.

King, M. and B. Sturtewagen (2010), 'Making the most of Afghanistan's river basins: opportunities for regional cooperation', EWI-East West Institute, available at http://www.ewi.info/system/files/Afghanistan_Water.pdf (accessed 1 October 2013).

Lee, K. (2006), *Water Management, Livestock and the Opium Economy: Social Water Management*, Kabul: Afghanistan Research and Evaluation Unit.

Mahmoodi, S.M.M. (2008), 'Integrated water resources management for rural development and environmental protection in Afghanistan', *Journal of Developments in Sustainable Agriculture*, **3** (1), 9–19.

Molle, F. (2008), 'Nirvana concepts, narratives and policy models: insight from the water sector', *Water Alternatives*, **1** (1), 131–156.

Mosse, D. (2004), 'Is good policy unimplementable? Reflections on the ethnography of aid policy and practice', *Development and Change*, **35** (4), 639–671.

Reisner, M. (1986), *Cadillac Desert: The American West and Its Disappearing Water*, New York: Viking.

Steins, N.A. and V.M. Edwards (1998), 'Platforms for collective action in multiple-use CPRs', paper presented at Crossing Boundaries: The Seventh Annual Conference of the International Association for the Study of Common Property, Vancouver, 10–14 June.

Thomas, V. and M. Ahmad (2009), *A Historical Perspective on the Mirab System: A Case Study of the Jangharoq Canal, Baghlan*, Kabul: AREU.

Thomas, V., W. Mumtaz and M.A. Azizi (2012), 'Mind the gap: local practices and institutional reforms for water allocation in Afghanistan's Panj-Amu River

Basin', Afghanistan Research and Evaluation Unit Case Study Series, Kabul: AREU.

Turton, A.R. (2000), 'Precipitation, people, pipelines and power: towards a political ecology discourse of water in Southern Africa', in P. Stott and S. Sullivan (eds), *Political Ecology: Science, Myth and Power*, London: Edward Arnold, pp. 132–153.

Varzi, M.M. and K. Wegerich (2008), 'Much ado about nothing – sub-basin working groups in Kunduz River Basin, Afghanistan', *Central Asian Waters*, 47–61, available at http://water.tkk.fi/English/wr/research/global/material/CA_chapters/05-CA_Waters-Varzi&Wegerich.pdf.

Verhallen, A. and W. van Densen (n.d.), 'Leerzame complexiteit. Debatteren en informeren over de toekomst van het Schelde-estuarium', unpublished manuscript.

Warner, J. (2006), 'More sustainable participation? Multi-stakeholder platforms for integrated resource management', *International Journal for Water Resources Development*, **22** (1), 15–35.

Warner, J. (2007), *Multi-Stakeholder Platforms for Integrated Water Management*, Aldershot: Ashgate.

Warner, J. and E. Simpungwe (2003), 'Stakeholder participation in South Africa: power to the people?', paper presented at the 2nd International Symposium on Integrated Water Resources Management: Towards Sustainable Water Utilization in the 21st Century, ICWRS/IAHS, Stellenbosch, South Africa, 22–24 January.

Warner, J., P. Wester and A. Bolding (2008), 'Going with the flow: river basins as the natural units for water management?', *Water Policy*, **10** (S2), 121–138.

World Bank (2012), *Afghanistan in Transition: Looking Beyond 2014*, Washington, DC: World Bank.

11. Evolving river basin management in Mongolia?

Ines Dombrowsky, Annabelle Houdret and Lena Horlemann

11.1 INTRODUCTION

As a decentralised, participatory and cross-sectoral approach to water management, river basin management has recently attracted greater attention in Mongolia. In socialist times, the Mongolian People's Republic was strongly influenced by the Soviet system, centralised planning and top-down implementation prevailing in the water sector as elsewhere. A largely peaceful democratic revolution in 1990 then marked the beginning of the transition to a multi-party system, a market economy and, from 1992, growing decentralisation. Water use has since risen dramatically; not only the boom in the mining industry, but also the expansion of irrigated agriculture and livestock production, urbanisation and changing lifestyles have all contributed to resource depletion and growing pollution. In response to the mounting difficulties in water governance, the Mongolian government adopted the concept of integrated water resources management to improve the sustainability of water use. While the first statements on this date back to the late 1990s, actual decisions to implement integrated water resources management, including river basin management, and to create appropriate institutions were not taken until the 2004 Water Law, which provided for the establishment of river basin councils, and in 2012, when a new Water Law providing for the creation of river basin administrations was passed.

Generally speaking, the implementation of integrated water resources management and river basin management in Mongolia is still in its infancy (Horlemann and Dombrowsky 2012). This chapter therefore explicitly investigates the institutionalisation of river basin management in Mongolia with a focus on the institutional design of river basin organisations and the views on different design options. On the basis of the analytical approach developed by Huitema and Meijerink (Chapter 1, this volume), the chapter explores five different rules at play in this process of institutional design. It also considers the positions taken by the various actors at the local and national levels during this process with a view to throwing more light on its politics and dynamics, and the implications of institutional design for the balance of power and vice versa. However, as Mongolia is a country which, in many respects, is still in political transition, this analysis of river basin management and river basin organisations (RBOs) also takes the broader governance environment into account, which is characterised *inter alia* by an incomplete decentralisation process and limited scope for participatory approaches (Houdret and Dombrowsky 2013).

Mongolia is particularly interesting for a case study of the design and introduction of RBOs for two reasons. Firstly, the (formal) implementation is has been faster than in many other countries, which may be triggered by increasing environmental degradation caused by the rapid expansion of the mining sector and changing political circumstances. Secondly, the implementation is happening in midst of several other far-reaching transformation processes from a predominantly rural and nomadic to a largely (peri-)urban society and from a socialist to a liberal economic system. The influence of the larger political environment on the implementation of river basin management is therefore particularly relevant.

The chapter focuses on the design and implementation of river basin management in this highly dynamic environment. It is based on an extensive literature review and empirical research carried out between 2009 and 2013 in two projects: the International Water Research Alliance Saxony project on water management in Mongolia, and the Integrated Water Resources Management in Central Asia – Model Region Mongolia project, both financed by the German Federal Ministry of Education and Research. Over 100 qualitative, semi-structured in-depth interviews were conducted with stakeholders from relevant ministries, administrations and non-governmental organisations at national and local levels, with representatives of international organisations, scientists and water users (nomads, farmers and drinking water users). Additionally representatives from river basin organisations in the Khar-Lake-Khovd basin in Western Mongolia, the Kharaa basin in the north of the country and the Tuul basin, including the capital city Ulaanbaatar, were interviewed.

The chapter is structured as follows: section 11.2 begins with a short review of the institutionalisation of integrated water resources management in Mongolia. Section 11.3 focuses on the institutional design of river basin management and the choice of geographical boundaries for the basins and analyses their institutional design rules. Section 11.4 takes a closer look at the various stakeholders involved in the reform process, at their respective views and at the implications of the design rules for them. Section 11.5 evaluates how well river basin management performs in terms of mobilisation of funding, enhanced coordination, accountability, legitimacy and support. Finally, section 11.6 draws conclusions.

11.2 INTEGRATED WATER RESOURCES MANAGEMENT IN MONGOLIA AT A TIME OF TRANSITION

The first document to mention the concept of integrated water resources management in Mongolia was the national action plan for the implementation of Agenda 21, Sustainable Development of Mongolia, in 1995. In connection with this, the government requested support from the United Nantions (UN) Capacity 21 Programme, which was launched to help developing countries build their capacities for integrating the principles of Agenda 21 into national development.[1] In 2000 the National Water Committee was created under the auspices of the Ministry of Nature, Environment and Tourism.[2] In line with the cross-sectoral approach of integrated water resources management, this institution is responsible for the coordination of water-related activities of relevant ministries and for the preparation of a National Water Programme. Other relevant actors in the water sector at national level include several ministries such as the Ministry of Construction and Urban Development, the Ministry of Mining, the Ministry of Industry and Agriculture, and others (see Figure 11.1, which gives an overview of the key actors in the Mongolian water sector).

In 2004, a new Water Law was adopted and formed the basis for the establishment of the Water Authority in 2005 under the auspices of Ministry of Nature, Environment and Tourism. One of its tasks was to introduce integrated water resources management in Mongolia. In addition, the 2004 Water Law reassigned a number of water management responsibilities from central government to local government, which is now responsible for financing and implementing national legislation and policies at local level. Provincial and district governors have since been responsible for the implementation of measures related to the use, protection and restoration of water resources. Municipal governors have the task of enforcing the legislation on the protection of water resources, while provincial parliaments approve water management plans drawn up by the river basin institutions and governors and adopt the budget for their implementation. The provincial and district departments of the Ministry of Environment and Green Development are responsible for implementing environmental policies, which include measures to prevent and detect

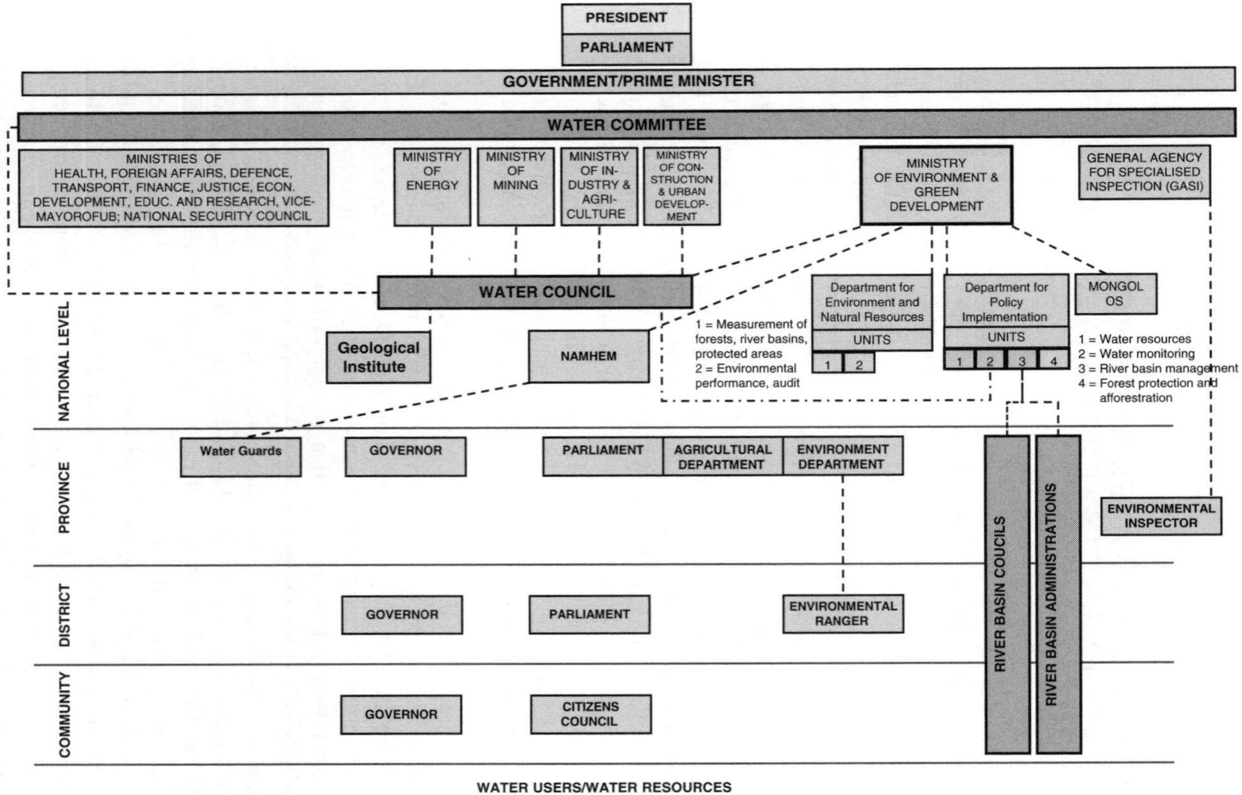

Note: Only water-related entities are shown.

Source: Authors' own compilation.

Figure 11.1 *Key actors in Mongolia's water sector*

breaches of relevant legislation. However, like Mongolian environmental governance in general, local water governance is very weak. The lack of financial and human resources, unclear mandates, deficient regulation and weak enforcement impose serious limits on the scope for action at local level (Tortell et al. 2008; UNDP 2010).

Furthermore, the Water Law of 2004 opted for the introduction of river basin management, including the establishment of river basin councils. The water sector thus underwent two separate decentralisation processes: water governance tasks were redistributed in the process of political decentralisation, and river basins were created as new hydrological and political units under the 2004 Water Law. These two processes happened at very different levels and were not necessarily coordinated: administrative entities were assigned new tasks in water governance, but at the same time and at another level new institutions for water governance – the river basin agencies – were implemented.

In 2010, the National Water Programme, coordinated by the National Water Committee, was approved by the national parliament. It too refers to the establishment of river basin councils and a nationwide integrated water resources management strategy and calls for 'common methodologies for developing river basin management' (Parliament of Mongolia 2010: Article 3.5).

In May 2012, a new Water Law was passed by the national parliament and signed by the President. The 2012 law opts for the creation of river basin administrations in addition to the river basin councils, which exist at least formally, and shifts some responsibilities from the latter to the former (see section 11.3). Other important changes concern the introduction of regulations on water prices, water pollution and water monitoring, and the creation of Mongol Us, a public company engaged in the construction and management of water infrastructure. Moreover, the Water Authority was dissolved, its main tasks being integrated directly into the new Ministry of Environment and Green Development. A specific sub-unit of the Ministry of Environment and Green Development's Department for Policy Implementation has now been created for river basin management. The National Water Committee was enlarged and granted legal status and now operates under the Prime Minister's auspices.

Additionally, a Water Council was created at the ministerial level which is responsible for granting licences for water use and for research on water resources. The State Secretary for the Environment presides over the Water Council, other members are the head of the Ministry of Environment and Green Development's department for water resources, deputies of the ministries for Energy, Mining, Industry and Agriculture, Construction and Urban Development, the head of the National Water

Committee, the Ministry of Environment and Green Development's Climate Change Coordination Office and the Institute of Geology and Mineral Resources.

Given the rapid changes that Mongolia is undergoing and the growing competition for water, the recently created river basin organisations potentially have an important role to play in improving resource governance. Besides making for more efficient and sustainable water management, they may also facilitate the democratisation of water governance by strengthening decentralised and participatory approaches. However, as Horlemann and Dombrowsky (2012) show, the introduction of integrated water resources management in Mongolia is also a highly political process. The vested interests of the various local, national and sometimes even international stakeholders, and the necessary renegotiation of the balance of power that this rescaling of water governance entails, partly explain the slow introduction of river basin management in the country. The following section (11.3) describes the institutional design of RBOs in the country and analyses the different rules which guided their implementation and operation. On the basis of this understanding of the 'rules of the game' of river basin management in Mongolia, section 11.4 will then analyse contested issues in this process, illustrating the positions of the different actors involved.

11.3 THE INSTITUTIONAL DESIGN OF RIVER BASIN ORGANISATIONS IN MONGOLIA AND THE CHOICE OF GEOGRAPHICAL BOUNDARIES

11.3.1 Introduction

This section analyses the process of identifying river basins in Mongolia and the institutional design of the RBOs in terms of authority, aggregation, boundary, information and pay-off rules. Each rule is presented as it is under the present Water Law (2012), and where meaningful changes have occurred, as it was under the previous Water Law (2004).

11.3.2 Boundary Rules: The Definition of River Basins and Membership in River Basin Organisations

In our case study, we distinguish between two types of boundary rules: those referring to the geographical boundaries of the river basins, and those for the membership in RBOs. On the first aspect, there had previously been no generally accepted definition of the borders, scope and ultimate number of basins in Mongolia. Following the call for river basin councils in the 2004 Water Law, these therefore had to be identified. The Water Authority, the Ministry of Environment and Green Development and the consultants involved agreed on a combination of hydrological, political and economic criteria to be taken into account. This led to the designation of 29 river basins. Economically relevant rivers were granted the status of discrete river basins even when, from a hydrological perspective, they actually belonged to larger catchments. As the Figures 11.2 and 11.3 reveal, a definition of river basins based on hydrological criteria alone (16 basins) would have produced a very different arrangement from that actually chosen (29 basins). In the southern part of the country, most of which forms part of the Gobi Desert, there is no surface water. However, in view of the rapidly increasing exploitation of groundwater for the mining industry, river basins were also defined for this region. As not enough was known about groundwater aquifers for the catchment areas to be determined, the river basins were defined by reference to provincial boundaries.[3]

Our interviews suggest that political and economic criteria combined with the physical and environmental characteristics of the basins formed the basis for decisions on river basin boundaries in Mongolia. They also revealed that the environmental and political urgency to address water problems was a major factor in this process, such as in the case of high pollution and increasing water scarcity in and around the capital city.

The process of defining river basins was thus highly political, which confirms two statements by other authors in this regard: Huitema and Meijerink (2012: 8) observe that this 'means that defining the boundaries of a basin requires choice, and this implies a role for politics'. In a similar vein, Schlager and Blomquist (2008: 15f.) state: 'Drawing boundaries is the first step in determining who decides and how and with what effects. Different boundaries imply different decision makers and different effects.' The second aspect of boundary rules referring to membership in the RBOs illustrates these political implications. In Mongolia, two types of RBOs currently co-exist: river basin councils (introduced in 2009) and river basin administrations (introduced in 2013). As the following shows, both forms of organisation imply different decision makers, and political control ranges from a localised but weak approach to river basin management in the river basin councils to a stronger and more centralised approach in the river basin administrations.

River basin councils have been created since 2009 as participatory stakeholder fora in which all relevant stakeholders and districts crossed by a river are represented and citizens are involved in water resources

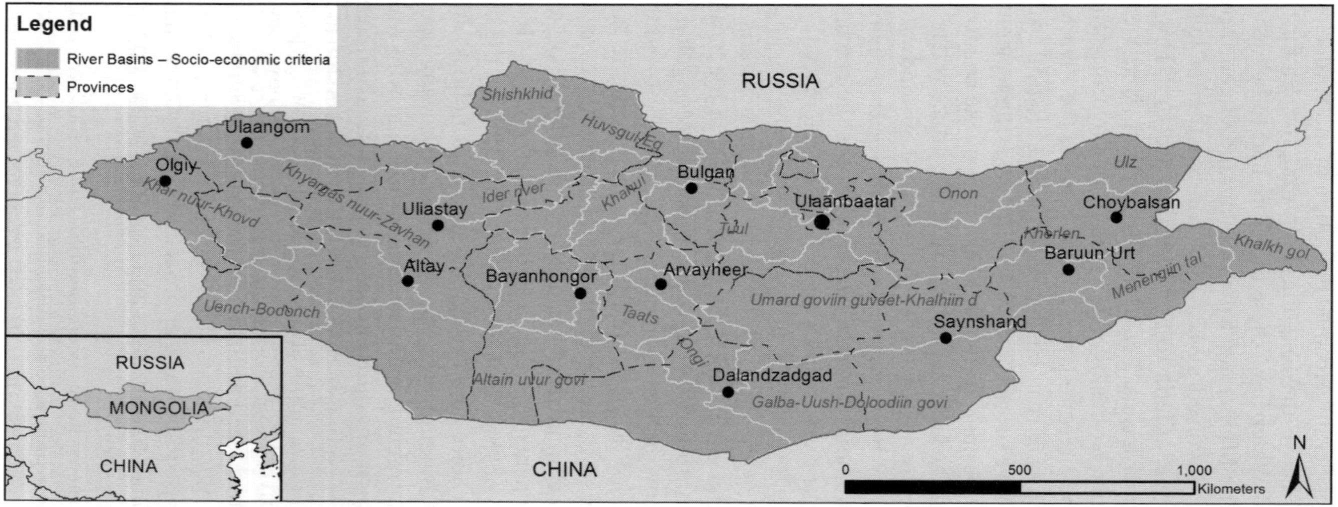

Source: Ron Wunderink, based on the original kindly provided by Mr Batbayar, Water Authority of Mongolia.

Figure 11.2 River basins in Mongolia as defined by politico-economic criteria

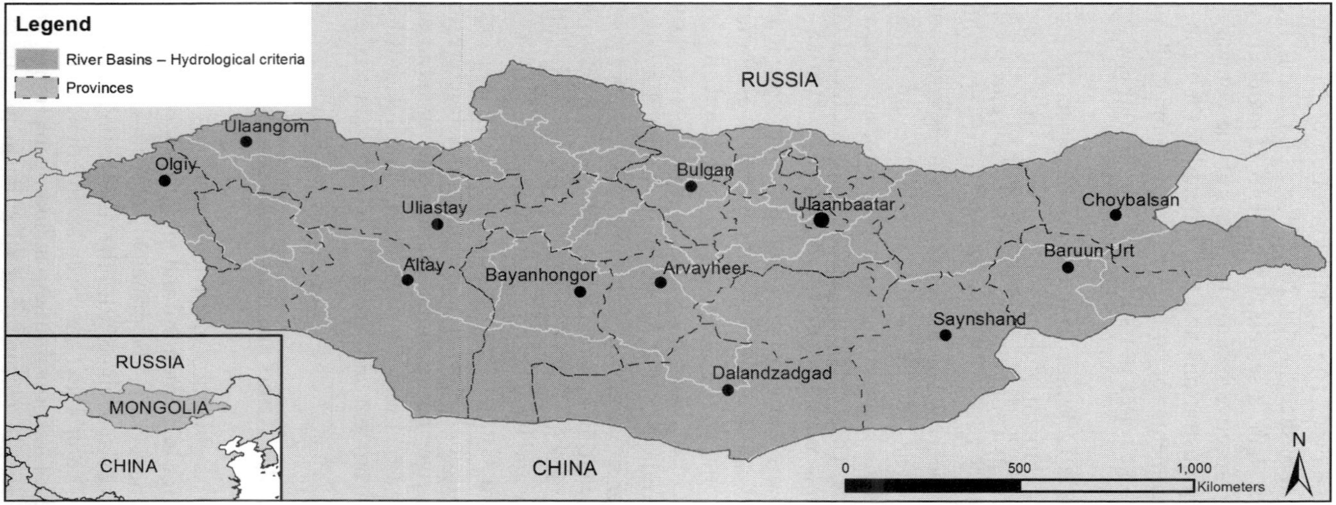

Source: Ron Wunderink, based on the original kindly provided by Professor Basandorj of the Mongolian University of Science and Technology.

Figure 11.3 River basins in Mongolia as defined by hydrological criteria

management and planning. The original limitation of 15 seats in a river basin council was abolished in the 2012 Water Law and the councils can now consist of a chair, a secretary and 31 to 45 members, depending on the size of the respective basin. One-third of the council's members represent public administrations and parliaments of the basin, one-third represent non-governmental organisations (NGOs) and civil society, and the last third represent water users including industrial and agricultural companies.

According to the 2012 Water Law the establishment of a river basin council and the appointment of its members always requires the approval of the Ministry of Environment and Green Development. Before this approval is given, the river basin administration must agree with the local administration on the river basin council's establishment and submit a request to this effect to the Ministry of Environment and Green Development. In practice, the nomination of river basin councils between 2009 and 2012 has varied widely from one basin to another. In the Khovd and Buyant basins the World Wide Fund for Nature (WWF) arranged a broad stakeholder consultation of around 100 people, while in the Kharaa basin the members were nominated after an internal consultation between the local governor, the environmental department and the Ministry of Environment and Green Development. Finally, all river basin council members were appointed by the Ministry of Nature, Environment and Tourism/Ministry of Environment and Green Development. In 2013, all river basin councils were dissolved and they will now be newly constituted.[4]

River basin administrations were created as government authorities in the 2012 Water Law and have been implemented since 2013. The Ministry of Environment and Green Development nominates and dismisses the river basin administration's staff and determines the structure and organisation chart of the river basin administrations. River basin administrations can propose the creation of a river basin council and its members to the Ministry of Environment and Green Development in cooperation with local authorities.[5]

11.3.3 Authority Rules: Actors and their Competencies

At the national level, the Ministry of Environment and Green Development is the key authority for implementing and monitoring water governance in river basins (see section 11.2). The principal actors at river basin level are the river basin councils and, since 2012, the river basin administrations.

Under the 2004 Water Law, the river basin councils were entrusted with a wide range of responsibilities, including planning, monitoring and advising, and the development of river basin management plans.

Except for the power to withhold or suspend water use licences for 14 days, however, their mandate was not to take political decisions, but primarily to make recommendations to local government bodies. Since the 2012 reform, their main tasks are advising the river basin administrations, enabling public participation in water resources planning and implementation, and monitoring water use and pollution. The river basin council may still submit proposals for the suspension of water-user licences. The provincial, district and municipal governors similarly have a range of responsibilities relating to the protection and use of water resources.

While the 2012 Water Law did not change the responsibilities of local parliaments and governors to any major degree, it significantly altered governance at the river basin level. The newly created river basin administrations are now responsible for developing and monitoring the implementation of the river basin management plans, which receive input in the form of recommendations by the river basin council and still need the formal approval of the provincial and district parliaments. The river basin administrations' duties also include maintaining a new annual database on the river basins in cooperation with other public actors. This database feeds information into the national water database coordinated by the Ministry of Environment and Green Development. The river basin administrations grant water licences and set water fees for consumption of between 50 and 100 m^3/day, the provincial and capital city environmental administrations performing this task below that level. Where water consumption exceeds 100m^3/day, the approval of the public authority responsible for water is needed. Mining permits are also issued by the river basin administrations; however, they need the approval of the Water Council (see above).

River basin administrations are thus responsible for rule making relating to the coordination of water demand, water pricing and some production decisions. However, two factors limit the scope for action of the river basin administrations: they do not have a mandate for enforcement (infringements are taken to the court), and the implementation of these rules also depends on the river basin management plans which are drawn up by the river basin administrations and the river basin councils, but are to be implemented by a variety of institutions.

11.3.4 Aggregation Rules

Huitema and Meijerink (Chapter 1, this volume) refer to the aggregation rules as the 'type of rationality' or 'normative motivation' behind collective decisions, which also includes specific decision-making rules in individual institutions. The normative motivation for water governance is laid down in Article 1 of the 2012 Water Law, which defines its aim as being 'the regulation of water resources, the protection of watersheds, the sustainable optimal use and the restoration of water resources'. However, while the river basin administrations are expected to achieve this objective at river basin level and, in so doing, to consider the advice of the river basin councils, decision-making rules within and between these two institutions are not well developed. There are not sufficient regulations detailing how potential disagreements between river basin councils and river basin administrations or between local governors or local parliaments and river basin administrations are to be settled. Only Article 20.4.9 gives river basin councils the right to propose the suspension of activities or policies decided by river basin administrations to the Ministry of Environment and Green Development or a legal authority.[6] According to consultants to the Mongolian government, a by-law regulating these issues is currently being drawn up (as of 2013).[7]

11.3.5 Information Rules

Provisions on the sharing of information among the various actors involved in the assessment and management of water resources were very sparse in the 2004 law, but the situation has improved under the 2012 Water Law. It now provides for increased regulation of information sharing through the development of a national water database. While the Ministry of Environment and Green Development's data assessment unit is the key agency for data collection, river basin administrations are required to assess data annually at river basin level in cooperation with local institutions, governors and water users, and submit them to the Ministry of Environment and Green Development. The law also requires the water unit in charge of data collection at the Ministry of Environment and Green Development to share data and information, but it does not elaborate on the access conditions.

At river basin level, the river basin administration collects data on water resources from the provincial environmental administrations and is also responsible for communicating relevant information to the general public and to the public administration each year. However, the Water Law does not say precisely what information this is and how it should be communicated. The river basin councils basically have two tasks concerning information sharing: communicating violations of the Water Law to the competent public authorities, and assembling and communicating to the river basin administrations citizens' proposals for the protection and restoration of water resources and forests.

11.3.6 Pay-Off Rules and Financing

The financing of the river basin council's operation and activities was not properly regulated in the 2004 Water Law since only the chairperson and the secretary of the council were to be remunerated for their work and there was no budget for the RBMP (Horlemann and Dombrowsky 2012). While there is still no funding for the river basin council members, their activities and the implementation of the river basin management plans are supposed to be financed through the local development plans (financed by tax revenues) and by revenues created by the collection of water use fees.[8] However, the decision about the first potential source remains with local governments and revenues from water fees are so far very low due to weak enforcement. In the past, the river basin councils functioned only where external financial support was provided. Given the persisting uncertainty over their funding, their implementation of the multiple monitoring tasks required by the law remains questionable. The functioning of the river basin administrations is now secured through the national budget, but this does not include funding for activities such as the river basin management plan.

Together with two other new laws (the Law on the Use of Natural Resources and the Law on Water Pollution, both 2012), the new Water Law lays out improved legal foundations for strengthening water pricing and the collection of related fees. The Law on Water Pollution includes the following rules on pricing: environmental rangers may fix the rates for water consumption of 0 to 50 m^3 per day; the river basin administration sets the fees for 50 to 100 m^3 per day after consulting with the local governor; at consumption in excess of 100 m^3 per day, the Ministry of Environment and Green Development's water unit decides the rates. The water use and pollution fees are collected by the fiscal administration and transferred to the Environmental Protection Fund (operating under the auspices of the Ministry of Environment and Green Development), which then allocates some of the revenue to water- and environment-related activities at the local level.[9] The mining sector is required to submit an annual report on its water use to the parliament, which then fixes the fees for pollution in consultation with the Ministry of Environment and Green Development.

These new regulations translate the political will for a more effective collection of water fees and closer monitoring of water pollution. However, uncertainties about implementation persist, since the water monitoring institutions are still underequipped and understaffed. Table 11.1 summarises the main features of the various types of rules and of the major changes made under the 2012 Water Law.

Table 11.1 Rules for Mongolian river basin organisations before and after the 2012 reform

Rule-types	Rules before 2012 reform	Rules after 2012 reform
Boundary rules	RBCs are based on basin boundaries following hydrological but also socio-economic and political criteria. Membership of the RBCs based on representation of all relevant stakeholders, limited to 15 members. Subsidiary principle for RBC appointment.	Geographical boundaries unchanged. Seats in RBCs now extended to 31–45 members, still representation of all stakeholders. RBA coordinates nomination of RBC members, MEGD approves.
Authority rules	RBM introduced under Water Authority, RBC main actor at the river basin (RB) level. RBC responsible for public participation and for monitoring use and pollution. RBC responsible for RBMP and advising local governments. Provincial parliaments, to be involved in appointing RBCs, to approve budget for water-related activities and to approve and mainstream RBMP.	River basin management (RBM) introduced under MEGD, RBAs created to complement RBCs at RB level; appointment of RBCs by MEGD. RBC responsible for public participation; monitoring of use and pollution through RBA and others to be determined (by-laws pending). RBA (partially) coordinates activities of other public and private actors in the basin. RBA grants water licences. RBA responsible for RBMP, RBC advises RBA. Provincial parliaments approve RBMP; budget for water-related activities mainly from local tax revenues and water use/pollution fees, precise regulations pending. Donors encouraged to finance implementation of RBMP.
Aggregation rules	RBCs only institution at the RB level but need approval by local governors for implementation of activities.	RBAs dominant actor, cooperate with RBCs. Decision-making rules between and within RBC and RBA pending (by-laws). RBC

Rule-types	Rules before 2012 reform	Rules after 2012 reform
	Decision making within RBC subject to simple majority rule.	may submit proposals for suspending the RBA's decisions.
Information rules	Information on water quality and quantity spread among various institutions, irregular national reports. No effective information sharing. RBC responsible for assembling and communicating citizen's needs and proposals to local government.	National water database and annual reports to gather water quantity and quality information under the MEGD's auspices. RBA responsible for data collection at river basin level. Conditions of access to these data/reports not (yet) defined. RBC to inform public authorities of violations of the Water Law and to collect and communicate citizens' proposals
Pay-off rules	RBCs: only chairperson and secretary are paid. District parliaments and governor to provide budget for activities, partly fee-based. Voluntary foundations to support RBM.	RBCs to have no public funding, may receive support for implementation of activities from local development plans. Environmental rangers set water fees for 0–50 m^3 per day; RBA for 50–100 m^3 per day; MEGD for more than 100 m^3 per day. RBAs receive funds from central government. RBMP to be financed by provincial government and donors.

Notes: RBC: river basin council; RBA: river basin administration; RBMP: river basin management plan; MEGD: Ministry of Environment and Green Development.

11.3.7 Conclusions on the Type of River Basin Organisation

As section 11.3 shows, in Mongolia two types of RBOs co-exist. River basin councils can be seen as coordinating bodies if categorised as suggested by Huitema and Meijerink (Chapter 1, this volume). They have no authority of their own, but were set up as coordinators between organisations that have responsibilities in the field of water management and to water users and citizens. In contrast, river basin administrations will be more in the nature of river basin agencies, with the responsibility to plan and implement river basin management including the coordination of data assessment and sharing, granting permissions of water use, pricing and monitoring of water use and pollution, and the formulation of proposals on the creation of a river basin council and on the nomination of its members. As such, they are 'created to perform a limited number of specialised tasks at arm's length from the government' (Huitema and Meijerink 2012: 15) and are accountable only to the Ministry of Environment and Green Development. The main rationale behind their establishment seems to be the need to strengthen administrative capacity in the water sector at the sub-national level and to provide for coherent data assessment. The first type of RBO based on river basin councils alone favoured the position of non-state actors, and the creation of river basin administrations with a much stronger mandate may therefore be interpreted as a weakening of these actors. However, the politically and financially stronger position of the river basin administrations also offers a more realistic perspective for actual implementation. Given the de facto very weak position of the river basin councils even under the 2004 Water Law, the creation of river basin administrations as public agencies should therefore not primarily be interpreted as a move against non-state actors, but as a commitment to more effective river basin management given their financially and legally stronger status.

If properly set up, the coexistence of river basin administrations and river basin councils makes a great deal of sense. While the river basin administration serves the implementation of public policies primarily coordinated with the Ministry of Environment and Green Development, the river basin council could complement it by being a multi-stakeholder platform with no more than an advisory function. In terms of ensuring coordination with local and national authorities and public participation, river basin councils will perform a very useful function in advising river basin administrations.

The 2012 Water Law also addressed the question of how institutions at provincial and district levels should interact with the RBOs, whose mandate extends over the political and administrative boundaries. It states that all water-related activities including data assessment and sharing, and implementation, are coordinated by the respective river basin administration which, in turn, is backed by national institutions such as the Water Council.

Section 11.3 shows that, while the formal institutionalisation of river basin management has progressed quickly in Mongolia, the financing and implementation of the new laws and institutions remains incomplete.

As section 11.4 will show, there are also differing views on the essential mandate of the two institutions for river basin management. The final mandates and functions of both will very much depend on the by-laws that are still to be developed and will not only further regulate the respective terms of reference of river basin councils and river basin administrations, but also determine their mutual relationship and their relations with other governmental and non-governmental bodies.

11.4 THE POLITICS OF DESIGNING RIVER BASIN ORGANISATIONS: VIEWS OF VARIOUS ACTORS ON INSTITUTIONAL DESIGN

11.4.1 Introduction

The following chapter refers in more detail to the institutional design of the river basin councils than to that of the river basin administrations, since the implementation of the latter is very recent (2013). Also, not all aspects of their institutional design were decided yet at the time of writing this chapter in 2013.

11.4.2 Positions of Stakeholders Involved in the Establishment of River Basin Organisations

Two aspects can be differentiated where the views of stakeholders on the establishment of RBOs (river basin councils and river basin administrations) are concerned: on the one hand, views of stakeholders involved in the establishment of individual river basin councils or river basin administrations; on the other hand, views on the creation of RBOs in general. In the first case, the main contentious issue is the rules on and actual composition of the membership, which will be analysed in section 11.4.5.

In the second case, foreign actors can be said to play an active and important role in the adoption of integrated water resources management and the establishment of RBOs in Mongolia. Their support can be divided into policy consultation, and project financing and implementation. At policy level, international non-governmental organisations (NGOs) and external consultants were involved in lobbying for the integration of integrated water resources management and its formalisation in the Water Laws of 2004 and 2012. For example, programme managers (Mongolian nationals) from the WWF advised the government on the new law and are involved in the implementation of river basin management.[10] Moreover, two water experts financed by the Dutch government worked at the

Mongolian Water Authority on the development of the national integrated water resources management strategy and the establishment of two river basin councils.

However, Mongolian water scientists also have shown great interest in adopting integrated water resources management since 2004.[11] With respect to the 2012 reform, it was argued that it would give the (neglected) water sector greater political weight and so enable new and comprehensive projects to be implemented; growing concerns about the impacts of increasing mining activities on water use and pollution seem to have played a crucial role.[12] Mongolian water professionals, many of whom had been trained in Eastern Europe during the socialist era, joined forces in order to avert the rapidly increasing water crisis in Mongolia.[13] Hence, in contrast to other cases discussed in this book (see, for instance, Warner and Thomas, Chapter 10), it would be wrong in the Mongolian case to attribute the adoption of integrated water resources management and river basin management only to external actors.

Strategies used by key actors promoting the establishment of RBOs include consultancy and lobbying activities of (internationally funded) NGOs, consultants and academics at the Ministry of Environment and Green Development. The creation of RBOs was also seen as a good opportunity for the adoption of more participatory approaches, an agenda pushed by a group of people within the Ministry of Environment and Green Development and from civil society.

At the level of project design and implementation, many activities related to river basin management would probably not have been undertaken without support from donors. To this extent, the involvement of foreign actors in this process was, as far as our interviews showed, welcomed by the beneficiary river basin councils, the Water Authority and the Ministry of Environment and Green Development.[14] However, the reaction of local governors and government to the creation of river basin councils was mixed. In some cases, governors hoped for new dynamics in natural resources management, a field where their own capacities were very limited.[15] In other cases, local authorities feared they would have to finance new activities from their own already limited budgets and were therefore sceptical about the new institutions.[16] In one of these cases, the WWF helped by communicating the benefits of river basin management and identifying overlaps between local development plans and river basin management plans, which facilitated support by local government.[17]

Finally, the creation of river basin administrations as government authorities under the Ministry of Environment and Green Development was probably also influenced by the politically stronger position of the

Ministry of Environment and Green Development after the 2012 elections, where the Green Party became part of the government. This allowed for the creation of river basin administrations including certain prerogatives over existing administrations.

11.4.3 Views on the Legal Form of the River Basin Organisations

The 2004 Water Law did not clearly define the river basin councils' legal status, which hindered their work. As a member of the Khovd river basin council puts it: 'There is a lot in current legislation and law supplementation we haven't understood properly. For example, if the river basin council is a non-governmental organisation or a government entity. This is not defined by law.[18] As a result, registration with the tax authority was reportedly problematic, and many river basin councils therefore registered as government-organised non-governmental organisations.[19]

Interviewees had different opinions on which legal form would be appropriate for the river basin councils. A member of the Khovd river basin council, for example, argued that river basin councils should be governmental organisations for the next 5–10 years to allow for the development and implementation of strict water regulations. In the future, when people's habits or their attitudes towards the environment and environmental legislation had improved, he felt river basin councils could become NGOs.[20] Others also supported the idea of governmental RBOs because this would secure their budgets.[21]

The 2012 Water Law regards river basin councils as NGOs and also makes it clear that they will not receive public financial support. In contrast, the river basin administrations were created and funded as governmental agencies.[22] In the view of many of our interviewees, a governmental body of this kind needed to be set up at river basin level to reinforce river basin management and to guarantee at least a minimum of funding. However, there was also concern that river basin management would lose the cross-sectoral approach which was meant to be guaranteed by the heterogeneous composition of the river basin councils and their non-governmental status. River basin administrations, in turn, were believed to be more bureaucratic and composed of technical experts rather than representing different users.[23]

11.4.4 Views on the Mandate and Capacities of the River Basin Organisations

As mentioned in section 11.3.3, the 2004 Water Law assigned to the river basin councils a very broad mandate. However, interviewees claimed that the conditions for the performance of the tasks assigned were not clearly defined. For example, the 2004 Water Law entitled the river basin councils to suspend water use licences, but did not mention how they should proceed and how related conflicts were to be handled. Others noted that the river basin councils had very little opportunity to recommend the withdrawal of licences, since they had no influence on the final decisions taken in this regard.[24] Several interviewees also felt that the river basin councils would need stronger legal backing if they were to prevent other, superior actors, such as the mining agency, from taking arbitrary action.[25] According to an interviewee who had been involved in the drafting of the 2004 Water Law, the aim of strengthening users' participation through the river basin councils could not be achieved because the river basin councils did not have sufficient capacities.[26]

A WWF representative also noted that the people might have high expectations of the river basin councils, which they could not meet because of their limited mandate and their restricted financial, human and technical capacities. He was afraid that this might cause disappointment and harm the river basin councils' reputation.[27] It was therefore crucial to support the river basin councils and also to make it clear that their mandate and role was to act as coordinators of activities (to be undertaken by the river basin administration and local authorities) and as facilitators of public information, awareness and participation.[28]

As explained above, the 2012 Water Law also modified the river basin councils' mandate in that responsibility for the river basin management plan was shifted to the river basin administrations. This was perceived by some interviewees as a step backwards in the implementation of participatory river basin management. A former Ministry of Environment and Green Development official compared the 2004 Water Law with the 2012 law and noted: 'There was a different spirit in this [the 2004] law. We really wanted to adopt participatory approaches; we wanted to integrate people into water management. The creation of the river basin councils was a response to this key concern for getting people involved.'[29] A representative of the Water authority (WA) commented on the change as follows: 'The work with the four river basin councils had just started working very well, people were aware of river basin management and integrated water resources management, and then this work was suddenly stopped.'[30] She called for new by-laws to increase the interaction between river basin administrations and river basin councils, the population and water users. Generally speaking, however, interviewees were still hesitant to qualify the creation of the river basin administrations and the changing mandate of the river basin councils, since the outcome in terms of integrated water resources management would depend on the specific relationships between river basin councils and river basin administrations. At the time of writing in 2013, the by-laws concerned still had to be drawn up.

11.4.5 Views on Membership and Participation

According to the 2004 Water Law, the river basin councils' main aim was to increase the participation of civil society in the implementation of the management plan. However, membership of and participation in the river basin councils were highly contested in many cases.[31] A key cause of friction in the Khovd case was the limit on the number of council seats to 15, which turned out to be incompatible with the objective of not only all the administrations in a basin's territory but also the various water users being represented. In the end, the Ministry of Environment and Green Development appointed members in consultation with local authorities and NGOs. Another important concern for NGOs in the development of the 2012 law was the initial idea of the river basin administrations appointing the river basin councils' members. The WWF and others succeeded in influencing the design of the law, which now states that river basin administrations may propose river basin council members to the Ministry of Environment and Green Development in accordance with the regulation on the composition of the council.

River basin councils were furthermore criticised for being dominated by governmental institutions and for including no or too few non-governmental stakeholders, such as NGOs, water users and scientists.[32] According to two interviewees, this could favour political over rational decisions on water management, as in the run-up to elections.[33] A member of the Buyant river basin council, on the other hand, considered it a positive feature that most river basin council members are local government representatives, since government had the final say in water issues in any case.[34]

Another interviewee maintained that only the integration of civil society could help to limit the mining sector's influence over water and the environment, since many local authorities were in the pay of the mining industry and did not care about environmental protection.[35] However, he also argued that NGOs would need better training and a stronger legal mandate for their activities.

The mining sector was not directly represented on the river basin councils created by the time this chapter was composed in 2013, but reference was made to its direct and indirect influence on several occasions. Financial support for local authorities (including the financing of new buildings and the provision of vehicles and fuel) and the lack of capacities for environmental monitoring and control, leading to pollution and resource overuse in the mining areas, were two key issues of concern in this context.[36] It remains to be seen if these two factors will also affect the work of the new river basin administrations, which are most likely to benefit from greater political and financial support than the river basin councils had. Political support for improved control of the mining business will most probably remain difficult as the sector is the largest contributor to the country's spectacular gross domestic product (GDP) growth over recent years (*The Economist* 2012).

Another major contentious issue was the difficulty the river basin councils' members had in performing their tasks, since they had neither financial support nor an official mandate to rely on. To attend meetings, participate in field trips or undergo training, they had to make individual arrangements to be released from work.[37]

The creation of the river basin administrations responds in a certain way to the call for stronger government action in integrated water resources management, but it leaves open the question of the effective contribution of civil society and water users to water governance. The larger number of seats in the river basin councils and the flexibility to adapt this number to individual basins[38] also addresses key issues of discontent (although one may wonder whether a river basin council always needs to have more than 30 members). One interviewee also observed that the degree of participation in river basin management cannot be regarded as separate from Mongolia's more general socio-political structures: 'Water management can't be more democratic than the rest of society.'[39] His view was that the participatory process should be an aim rather than an a priori element of institutional design.

11.4.6 Views on Decision-Making Rules

As explained in section 11.3, the rules on RBO decision making are not well developed. Where the river basin councils are concerned, interviewees argued that a clarification of decision-making powers was needed.[40] According to a number of stakeholders, clear rules are needed for decision making both within and between the institutions involved in river basin management. Within the river basin councils, for example, the heterogeneity of members' backgrounds and of the interests they represent was seen as a potential obstacle to the adoption of common positions and decisions.[41] This was confirmed by a member of the Tuul river basin council, who claimed that 'the members are busier fighting each other than developing any water-related activity.'[42]

Several interviewees also called for the adoption of precise rules on the relationships between different institutions involved in river basin management. The WWF (which advises the government on the design of the new by-laws) maintained that Article 20.4.9 of the 2012 Water Law was of major importance for securing the river basin councils' influence on river basin management. According to its representative, that provision required that any critical decision to be taken by a river basin administration be discussed and approved by the relevant river basin council, but further detail would need to be developed in by-laws.[43] The possibility of the river basin council to protest against decisions or activities of the river basin administration by complaining to the Ministry of Environment and Green Development is the only rule explicitly addressing conflict resolution.

11.4.7 Views on the Financing of River Basin Organisations and their Activities

Almost all interviewees were deeply concerned about the inadequate funding of the river basin councils. Stakeholders noted that proper financing would be needed to train personnel, to employ more staff and to acquire more equipment and vehicles, and also for water infrastructure and afforestation.[44] Before the new Water Law was passed, various interviewees held that the main source of funding for river basin management should be the government.[45] They argued that, whenever river basin management plan activities fell within the local authorities' jurisdiction, the latter should finance them.[46]

The 2012 Water Law clearly states that river basin councils are not to benefit from public funding, but the budget for the implementation of the river basin management plan was to stem from local governments or external funding sources. While all stakeholders interviewed welcomed the creation of the river basin administrations as purportedly more professional and better-equipped bodies, there was also some concern about how funding for research needed to design the river basin management plan, and for its implementation, would be obtained.[47] The passing of a new budget law in 2012 and the consequently higher budgets of local authorities were seen as improving the financing of river basin management, since the river basin management plan is to be financed from the Local Development Fund. Nevertheless, two major concerns were raised in this respect: first, both the precise rate of transfer from national to local budgets and the share of the budget to be allocated to the environment and river basin management were still uncertain. Second, the capacities of local authorities for handling these considerably larger budgets were thought to be very limited.[48]

Given the inadequate funding for river basin councils, the perhaps only active councils appear to be those which have received external support.[49] Nevertheless, it is interesting to observe that, despite a lack of funding, other river basin councils have also been set up in response to local mobilisation. Their actual work is, however, very limited,[50] even if the river basin council format has apparently provided an opportunity for establishing a formalised interest group.

11.5 THE PROSPECTS OF RIVER BASIN MANAGEMENT FURTHERING DEMOCRATISATION

11.5.1 Introduction

Based on the analysis of the institutional design of river basin management in Mongolia, this section evaluates the democratic functioning of the RBOs, the criteria being the mobilisation of funding, coordination, accountability, and dialogue, legitimacy and support.

11.5.2 Prospects for the Mobilisation of Financing

As explained before, Mongolian RBOs have difficulty mobilising the necessary funding for their tasks. The creation of the publicly funded river basin administrations is progress in this regard, but their scope of action will depend on funding for and political backing of their activities. Funding for the functioning of the river basin councils is still not secured in spite of the tasks assigned to these bodies in the 2012 Water Law. Given these difficulties, financing RBOs from external sources remains an alternative. In view of the growing pressure on and the resulting competition for water resources, and of the dynamic evolution of the legal and institutional setting towards integrated water resources management, the prospects for obtaining funds from donor institutions are considered to be good. However, this will probably affect only selected basins and not support the overall homogenous and continuous evolution of river basin management in the country. While the Ministry of Environment and Green Development attempts to harmonise river basin management planning, and while binding guidelines on the development of subsequent plans is being developed, major differences in financial support for river basin management from one basin to another may have negative effects on water resources, lead to competition between different basins and discourage initiatives in underequipped basins.

11.5.3 Prospects for the Functioning of Coordination

From the Water Laws of 2004 and of 2012 it can be inferred that coordination is a key goal of river basin management in Mongolia. The 2012 law places even greater emphasis on this aspect in that it assigns the river basin administrations a prime role in the collection and sharing of data on water resources and in the coordination of water-related activities of different sectors and institutions. So far, the river basin councils' performance in coordination has been poor, partly because of their very limited capacities and mandate, as outlined above. While it is still too early to assess the river basin administrations' contribution to improved coordination, the institutional set-up and the analysis of the overall situation in Mongolia point to three major obstacles in the achievement of this objective.

First, the challenges for both vertical and horizontal interplay in the Mongolian water sector and, more generally, the environmental sector are enormous (Horlemann and Dombrowsky 2012; Tortell et al. 2008). The difficulties of coordinating water monitoring and use at river basin level are partly linked to similar problems at the national level. Besides the large number of public and private sector institutions involved in water use, the unequal political power of the various sectors and their ministries is a cause of concern. At both national and local levels, the mining sector plays a major role in water use and pollution, which has so far largely defied regulation. This is partly due not only to inadequate legislation on decision making, but also, in most cases, to the de facto dominance of informal decision making. However, recent developments, such as the placing of the Ministry of Environment and Green Development and the Water Committee under the direct responsibility of the Prime Minister, as well as the creation of the Water Council, may have positive effects by improving coordination at both national and local levels. Nevertheless, interviews also show that recent institutional innovation has led to new problems of overlapping responsibilities, for example, between the river basin administrations and the newly created public company Mongol Us.[51]

A second challenge to the RBOs' coordinating function is poor data collection and information sharing both in and outside the water sector. While the new Water Law lays down relatively precise rules on responsibilities for communicating data on water, the exact type of data to be collected and communicated is yet to be regulated in future by-laws. Moreover, data collection capacities are considered to be low, and the rules on access to information for the various players are not yet clear. As interviews showed, data sharing is a highly sensitive issue in the Mongolian water sector, and even when working on similar issues, existing institutions have difficulty cooperating in this respect.

The third challenge concerns the relationship between the river basin councils and the river basin administrations, which has to be further defined. As a means of improving coordination, it might be useful for responsibilities to be shared by river basin councils and river basin administrations, with the former responsible for facilitating exchange, information and advice (between the river basin administrations and the various stakeholders) and the latter focusing on the regulation of water resources monitoring, use and restoration.

The development and implementation of the river basin management plan as a major tool for planning and coordinating water use at river basin level illustrates these challenges. In most basins, data availability and collection for river basin management is uncertain, financing of implementation remains unclear, and no rules have yet been adopted on how emerging trade-offs are to be handled and related conflicts resolved. However, the new by-laws to be developed may improve the legal framework for coordination.

11.5.4 Prospects for the Functioning of Accountability

According to Huitema and Meijerink (2012: 28), the evaluation of RBOs' accountability assesses 'the degree to which democratically legitimised bodies (or the polity) can monitor and evaluate organisational behaviour, and enable such bodies to influence organisational behaviour so that it is consistent with their preferences'. As has been said of participation above, the overall context of transition in Mongolia has a major impact on RBOs' accountability. Democratic decision making and decentralised governance have still not been fully implemented, and corruption is reportedly widespread (ADB and OECD 2006; Gelegpil et al. 2009; Asia Foundation 2010). It can therefore be assumed that informal governance also plays an important role in river basin management. However, from a formal perspective, the current Water Law provides some insights into the potential for and limits to democratic control. A key problem emerging from the 2012 Water Law is that it does not provide for enforcement capacities for the RBOs and does not contain precise rules on decision-making processes. For example, while the government has far-reaching rights with respect to water policy and the allocation of water between sectors, it is not clear how it is to integrate the RBOs and local authorities into its decision making on these matters. The provincial and district parliaments can monitor and control the RBOs, since they are responsible for adopting the laws, strategies and plans that concern water-related issues. However, it is not clear if they also have a say in potentially far-reaching decisions by the river basin administrations on such matters as mining licences and the implementation of the river basin management plan.

Interestingly, the river basin councils also have the important mandate of monitoring and controlling the river basin administration's activities, and may help to improve accountability, even if they are not elected bodies. However, the impact of their work will crucially depend on their financial, technical and human capacities. Changes due to the new budget law and possibly larger local budgets for river basin management may have positive or negative effects on the RBOs' accountability. Accountability could increase, since the budget allocated to river basin management would be monitored by the locally elected bodies. However, unfavourable experience in the past with 'Soum development funds' (funds for community development), which were reported to have increased non-transparent spending at district level, is an indication of the challenges to be overcome.[52]

In general, transparency and accountability at river basin level, as at other levels in the water sector, may also be strengthened by the establishment of the new national water database. If this database provides adequate and accessible information at basin level, monitoring and control by the river basin administrations may also increase. Decisions on river basin management, including the granting of licences, could then be considered and discussed on the basis of shared data, which could help in the monitoring of the river basin administrations. Information sharing therefore appears to be a crucial factor in river basin management accountability.

11.5.5 Prospects for the Functioning of Dialogue, Legitimacy and Support

In Mongolia, public awareness of water issues has grown considerably with the rapid development of the mining sector and related water use and pollution problems. The function of river basin management as a means of strengthening the dialogue on water, however, remains very weak owing to the comparatively recent creation of RBOs and to the difficulties encountered by the river basin councils. Nevertheless, our interviews also revealed that the river basin councils have helped to increase public awareness of water issues, such as in the Khovd basin.[53] In the same region, the president of a district parliament stressed the need for information:

> The establishment of the river basin council is very positive. Its activities have been very good so far, and first results are already visible. We need more posters and handbooks and other promotion material to convince citizens. They have to understand that problems will arise if the River Khovd dries up. So far, we have been doing only verbal education by visiting families.[54]

However, local administrations and water users also clearly indicated that while awareness and dialogue on water are on the increase, the river basin is not a common point of reference in this context. In some cases, only its members knew of the river basin council's existence. It is therefore uncertain whether the RBOs are the most appropriate bodies for promoting dialogue on water, or whether they should rather coordinate related activities, which are then undertaken by bodies at the provincial or district level. In this context, the river basin councils have an important role to play in ensuring the legitimacy of river basin management, since they can act as a platform for the promotion of dialogue on water management and for the communication of related strategies at river basin level.

With respect to the river basin administrations, input legitimacy – the way in which institutions deal with the feedback from the wider public – will most probably depend not only on the elected bodies but also, and perhaps to an even greater extent, on the role of the river basin councils. The river basin councils are meant to communicate the general public's view and the interests of the different water users to the river basin administrations, and they can also perform a controlling function by proposing that the river basin administration's decisions should be suspended.

Output legitimacy – the provision of goods and services by RBOs – is so far very limited, since in most cases the river basin councils have not been provided with appropriate resources. Even in cases where support for the river basin councils has existed, their prime task has been the establishment of the river basin management plan, which can be considered a service provided by an RBO but is rarely perceived as such by the population and is therefore unlikely to contribute to the RBOs' legitimacy.

As regards systematic legitimacy, or the 'broader feeling of support for the institutional set-up', support for the institutional set-up of the river basin councils was mostly high when the interviewees were themselves represented on the river basin council and low when they were not (as in the case of NGOs). However, even river basin council members who supported the existence of the institution as such, as well as many other stakeholders, were not satisfied with the limited scope of its action.

Given the recent evolution of river basin management in Mongolia and the major difficulties faced by the RBOs, it is still too early to evaluate the RBOs' performance in terms of their effectiveness in achieving goals such as ecological sustainability. More generally, it can be said that RBOs have in some cases helped to raise awareness of water-related problems and acted as a first conduit for exchanges between rather different stakeholders; by and large, however, the impact on the status of water resources is not yet tangible.

The present evaluation of the RBOs' performance will also depend on whether their institutional design and actual functioning can be seen as a true shift from centralised towards decentralised and more participatory governance. As stated in the introduction, this question is intimately linked in Mongolia to the overall evolution of governance and the context of transition. As other studies confirm, political processes outside the water sector such as political decentralisation play a major role in the establishment and functioning of river basin management (Houdret and Dombrowsky 2013; Kemper et al. 2010). Moreover, as detailed above, crucial issues relating to both the institutional design and the practice of basin management are still uncertain.

11.6 CONCLUSIONS

River basin management in Mongolia is still in the making. The present idea is to govern water at the river basin level through river basin administrations, which can be regarded as river basin agencies, and river basin councils, which can be described as coordinating fora for multi-stakeholder participation. Hence, river basins will be managed in future through two different types of RBOs with complementary functions. River basin administrations have specific functions in data management, river basin planning and monitoring of the implementation of measures. In contrast, river basin councils have important functions in increasing the dialogue on water issues among various stakeholders, and as such, the legitimacy of and accountability for river basin management in Mongolia.

However, both organisations also face constraints. River basin administrations' terms of reference do not currently include the enforcement of legislation, thus their room for manoeuvre will remain restricted. Moreover, given the present deficiencies in planning for water management at sub-national level, it will be important for the river basin administrations to be provided with enough funds. With respect to river basin councils, it remains to be seen whether, in the absence of public financial support, they will operate as envisaged and provide for the democratisation of water governance in Mongolia.

The current dynamics in the implementation of river basin management in Mongolia are positive overall, and reflect a political will for improved water governance. Considering the power struggles between the environmental and the industrial and mining lobbies, the creation of river basin administrations under the auspices of the Ministry of Environment and Green Development may be interpreted as a reappropriation of policy-making in a context where local politics were increasingly influenced by private companies. The choice of the Ministry of Environment and Green Development to create river basin administrations aims at improving water governance through a new institution which would be independent from existing administrations, often perceived as underequipped, inefficient and influenced by the mining lobby. Donors continue to support river basin management, although, at the time of writing this chapter in 2013 their influence seems to be diminishing given that major related projects ran out in 2012 and 2013. In comparison to other countries, however, there is strong internal backing for integrated water resources management by Mongolian policymakers and scientists who have supported the implementation so far and, given the growing problems of water supply and pollution, will most probably continue to do so.

At the same time, the high speed of the process may also threaten its long-term sustainability. As mentioned above, rapid institutional and legal change has led to new problems of overlapping competencies between institutions. Furthermore, it is increasingly difficult to recruit qualified personnel for the newly created institutions. As one interviewee suggested, it may have been more helpful, in a first phase, to concentrate the implementation of river basin management on those basins which are of particular environmental concern.[55]

The Mongolian case study places considerable emphasis on the role played in the establishment of river basin management by the overall political system and its transition. While the transition context has opened windows of opportunity for setting up river basin management, it seems that progress in this area will very much depend on the future evolution of political and fiscal decentralisation, including the successful implementation of the increased budget allocations to the local level.

As the Mongolian context of transition is marked by a number of governance problems, such as a lack of transparency, partial decentralisation and the persistence of informal rules (Gelegpil et al. 2009; Asia Foundation 2010; UNDP 2010), expectations regarding the democratic functioning of the generally young and underequipped RBOs should not, however, be pitched too high. Against this background, it can be stated that while the implementation of river basin management in Mongolia has been relatively fast compared to other countries, the impact and the sustainability of this approach in the longer term will depend not only on further by-laws and their effective implementation, but also on the evolution of the overall governance context.

NOTES

1. See http://www.un.org/esa/agenda21/natlinfo/countr/mongolia/inst.htm (accessed 13 April 2012).
2. Renamed Ministry of Environment and Green Development in 2012.

3. Interview with representative of the Water Authority, Ulaanbaatar, 12 September 2012.
4. Interview with member of the Tuul river basin administration, Ulaanbaator, 22 April 2013.
5. Decree on the creation of a river basin council and its activities, Ministry of Environment and Green Development, draft version, April 2013.
6. 2012 Water Law, Article 20.4.9; and interview with a representative of WWF, Ulaanbaatar, 18 September 2012.
7. Interview with representative of WWF, Ulaanbaatar, 18 September 2012.
8. Interview with representative 1 of the Ministry of Environment and Green Development, department for policy implementation, Ulaanbaatar, 19 April 2013.
9. Interview with representative of the National Water Committee, Ulaanbaatar, 12 September 2012.
10. WWF Mongolia is funded by its German, Dutch, Swedish and Austrian branches and by bi- and multilateral development organisations.
11. Interview with representative 1 of the Water Authority, Ulaanbaatar, 24 April 2010.
12. Interview with scientist, former Ministry of Environment and Green Development official, 17 September 2012; representative of the Ministry of Environment and Green Development, 12 September 2012, both Ulaanbaatar.
13. Interview with representative 1 of the National Water Committee and representative 1 of the Water Authority, both Darkhan, 18 May 2012.
14. See note 13 and interview with representative of the Tuul river basin council, Ulaanbaatar, 12 September 2012; representative of the Kharaa river basin council, Tunkhel, 22 May 2012.
15. Interview with representative of the Asia Foundation, Ulaanbaatar 20 September 2012; and representative of the environmental administration of Zuunkharaa, 21 May 2012.
16. Interview with representative of the Asia Foundation, Ulaanbaatar 20 September 2012; and representative of the environmental administration of Zuunkharaa, 21 May 2012.
17. Interview with representative of the WWF, Ulaanbaatar, 18 September 2012.
18. Interview with representative of the Khovd river basin council, Khovd, 30 August 2010.
19. Interview with representative of the WWF, Ulaanbaatar, 12 May 2009.
20. Interview with representative of the Khovd river basin council, Khovd, 30 August 2010.
21. Interview with representative of the Buyant river basin council, Buyant, 10 September 2010.
22. From 2013 several river basin administrations will be created each year and their respective budgets included in the national budget. For 2013, eight river basin administrations shall be created. Source: Interview with a representative of the Ministry of Environment and Green Development, Ulaanbaatar, 17 April 2013.
23. Interview with representative of the Tuul river basin council, Ulaanbaatar 13 September 2012; and representative of the Water Committee, Ulaanbaatar, 17 September 2012.
24. Interview with representative of the WWF, Ulaanbaatar, 12 May 2009.
25. Interview with representative of the WWF, 16 September 2010; and Board Member Green Party, 15 September 2010, both Ulaanbaatar.
26. Interview with scientist, former Ministry of Environment and Green Development official, Ulaanbaatar, 17 September 2012.
27. Interview with representative of the WWF, Ulaanbaatar, 18 September 2012.
28. Interview with representative of the WWF, Ulaanbaatar, 18 September 2012; and interview with representative of the Asia Foundation, Ulaanbaatar 20 September 2012.
29. Interview with scientist, former Ministry of Environment and Green Development official, Ulaanbaatar, 17 September 2012.
30. Interview with representative 2 of the Water Authority, Ulaanbaatar, 13 September 2012.
31. Interview with representative of the Khovd river basin council, Khovd, 30 August 2010; representative of the Buyant river basin council, Buyant, 10 September 2010; representative of the Kharaa river basin council, Tunkhel, 22 May 2012; representative 2 of the Water Authority, Ulaanbaatar, 13 September 2012.

32. Interview with representative of the Kharaa river basin council, Tunkhel, 22 May 2012; and consultant 2 to the Water Authority, Ulaanbaatar, 24 May 2012.
33. Interview with representative of the Coping with Desertification project, SDC, 17 September 2010; and consultant 1 to the Water Authority, 2 October 2010, both Ulaanbaatar.
34. Interview with representative of the Buyant river basin council, Buyant, 10 September 2010.
35. Interview with representative of the NGO United Mongolian Movement for Rivers and Lakes (UMMRL), Ulaanbaatar, 19 September 2012.
36. Interview with representative of the UMMRL, Ulaanbaatar, 19 September 2012.
37. Interview with representative of the Khovd river basin council, Khovd, 30 August 2010.
38. Interview with representative of the WWF, Ulaanbaatar, 18 September 2012.
39. Interview with consultant 3 to the Water Authority, Ulaanbaatar, 13 September 2012.
40. Interview with representative of the Khovd river basin council, Khovd, 30 August 2010; and consultant 3 to the Water Authority, Ulaanbaatar, 13 September 2012.
41. Interview with representative of the Khovd river basin council, Khovd, 30 August 2010.
42. Interview with representative of the Tuul river basin council, Ulaanbaatar, 13 September 2012.
43. Interview with representative of the WWF, Ulaanbaatar, 18 September 2012.
44. Interview with representative of the district government, Bayannuur Soum, 2 September 2010.
45. Interview with representative of the Khovd river basin council, Khovd, 30 August 2010; representative of the WWF, Khovd, 31 August 2010; representative 4 of the Water Authority, Ulaanbaatar, 24 September 2010.
46. Interview with representative of the Khovd river basin council, Khovd, 30 August 2010.
47. Interview with representative of the Tuul river basin administration, Ulaanbaatar, 23 April 2013.
48. Interview with representative from the United Nations Development Programme (UNDP), 14 September 2012; and consultant 3 to the WA, 13 September 2012, both Ulaanbaatar. A capacity-building programme was initiated in 2012 in this context, led by the UNDP in cooperation with the Swiss Development Cooperation (SDC) and the World Bank.
49. Interview with representative of the WWF, Ulaanbaatar, 18 August 2012. The United Nations Environment Programme (UNEP), in cooperation with the Russian Federation, is supporting the development of an integrated water resources management plan for the Selenge River; the Khar Lake Khovd river basin council was supported by the WWF with external funding (mainly Dutch and Swiss); integrated water resources management in the Kharaa is supported by the Model Region Mongolia project (Integrated Water Resources Management in Central Asia – Model Region Mongolia, funded by the German Federal Ministry for Education and Research), the Tuul and Orkhon river basin councils were supported by the Dutch government.
50. Interview with representative 4 of the Water Authority, Ulaanbaatar, 24 August 2010.
51. Interview with representative of the National Water Committee, Ulaanbaatar, 23 April 2013.
52. Interview with representative of the Swiss Development Cooperation, Ulaanbaatar, 18 August 2012.
53. Interview with representative of the WWF, Ulaanbaatar, 18 August 2012.
54. Interview with representative of the Altantsugts Soum parliament, Altantsugts, 2 August 2010.
55. Interview with representative of the Tuul river basin administration, Ulaanbaatar, 23 April 2013.

REFERENCES

Asia Foundation (2010), *Corruption Benchmarking Survey Mongolia*, Ulaanbataar: Asia Foundation.

Asian Development Bank (ADB) and Organisation for Economic Co-operation and Development (OECD) (2006), *Anti-Corruption Policies in Asia and the Pacific: Progress in Legal and Institutional Reform in 25 Countries*, Manila: Asian Development Bank.

The Economist (2012), 'Booming Mongolia. Mine, all mine', 21 January, available at http://www.economist.com/node/21543113 (accessed 22 January 2012).

Gelegpil, C., H. Hashbat, T. Tseveen, D. Perenlei and H. Oidov (2009), *Millennium Development Goal 9. Indicators and the State of Democracy in Mongolia*, Ulaanbaatar: UNDP Mongolia.

Horlemann, L. and I. Dombrowsky (2012), 'Institutionalising IWRM in developing and transition countries: the case of Mongolia', *Environmental Earth Science*, **65** (5), 1547–1559.

Houdret, A. and I. Dombrowsky (2013), 'The institutionalization of river basin management as politics of scale – insights from Mongolia', *Journal of Hydrology*, 4 December, available at http://dx.doi.org/10.1016/j.jhydrol.2013.11.037.

Huitema, D. and S. Meijerink (2012), 'The politics of river basin organisations: coalitions, institutional design choices and consequences', paper presented at Lund Conference on Earth System Governance, Lund, Sweden.

Kemper, K.E., W. Blomquist and A. Dinar (2010), *Integrated River Basin Management through Decentralization*, Berlin and Heidelberg: Springer.

Parliament of Mongolia (2010), *National Water Programme*, Ulaanbataar: Parliament of Mongolia.

Schlager, E. and W. Blomquist (2008), *Embracing Watershed Politics*, Boulder, CO: University Press of Colorado.

Tortell, P., A.T. Borjigdkhan, and E. Naidansuren (2008), *Institutional Structures for Environmental Management in Mongolia*, Ulaanbaatar, Mongolia and Wellington, Australia: United Nations Development Programme.

United Nations Development Programme (UNDP) (2010), *Governance, Advocacy and Leadership for Water, Sanitation and Hygiene Volume 2*, Ulaanbaatar, Mongolia and New York: United Nations Development Programme.

12. Interplay between new basin organisations, pre-existing institutions and emerging environmental networks in the Mae Kuang watershed, northern Thailand

Santita Ganjanapan and Louis Lebel

12.1 INTRODUCTION

Water management in urbanising and peri-urban regions is complex as it involves many actors with divergent and shifting interests. Effective water governance in these contexts depends on vertical and horizontal coordination, management of conflicts, flexibility to adapt to changing conditions and meaningful public participation (Pahl-Wostl et al. 2012). In many parts of the world integrated water resources management is being promoted through reforms that create river basin organisations (RBOs) to help deal with these inter-sectoral issues, but not without significant challenges (Biswas et al. 2005; Molle 2008). Newly introduced RBOs, for instance, must do their work alongside pre-existing resource management institutions, raising questions of interplay (Young 2002; Mollinga et al. 2007).

One potentially important role for RBOs is to convene and support multi-stakeholder platforms. Multi-stakeholder platforms are often promoted as a way to support deliberations. In favourable circumstances multi-stakeholder platforms can help to build mutual understanding, resolve disputes, build consensus and inform difficult negotiations (Warner 2006; Dore 2007). Deliberation may be challenging but particularly important for dealing with cross-scale issues (Dore and Lebel 2010). In unfavourable contexts such as those of high social inequity, disorganised stakeholder groups and low technical and financial capacities, multi-stakeholder platforms have less predictable influence (Faysse 2006).

In Thailand the drive for the introduction of RBOs came, in part, from multilateral agencies such as the Global Water Partnership, Asian Development Bank and World Bank. RBOs were promoted alongside integrated water resource management and other reforms in the water sector. Conditions placed on loans to the agriculture sector by the Asian Development Bank following the 1997–1998 financial crisis, for instance, included guidelines and requirements for the introduction of river basin organisations, and the creation of so-called apex regulatory authorities under new water laws (Abonyi 2005).

Within Thailand there was also growing support for reorganisation of water management. In 1996 the Office of the National Water Resources Committee was established. In 2000 it released a National Water Vision that stated: 'by the year 2025, Thailand will have sufficient water of good quality for all users through efficient management and an organisational and legal system that will ensure equitable and sustainable use of water resources, with due consideration for the quality of life and the participation of all stakeholders' (Sethaputra et al. 2001: 87). In July 2000 the National Water Resources Committee endorsed RBOs to prepare participatory water management plans. In the Ninth Five-Year National Economic and Social Development Plan starting in October 2001 explicit support was given to establishment of RBOs with participation of all concerned stakeholders. Bureaucratic reforms in 2002 led to the establishment of a new Department of Water Resources in 2003 with responsibilities for establishing and supporting a hierarchy of RBOs across the country.

In 1999 the first pilot committees for three of the 25 recognised major basins of the country were established, including the Upper and Lower Ping basins in northern Thailand (Thomas 2006a). The Department of National Parks, Wildlife and Plant Conservation led initial activities in the Upper Ping, whereas in the Lower Ping this role was taken by the Department of Water Resources. In 2004 funding became available for activities, including a project on participatory watershed management funded through a fund managed by the World Bank and led by the Office of Natural Resources and Environmental Policy and Planning (Thomas 2006a). This latter project studied three parts of the Ping River Basin including the Mae Kuang river sub-basin. Mae Kuang was selected because it was the first priority of stakeholders at preparatory meetings. The key task of the project was to develop an organisational model for integrated water management that could be applied in other watersheds by establishing River Sub-basin Working Groups. The River Sub-basin Working Group was expected to serve as an important link between top-down and bottom-up interactive planning and management approaches (Thomas 2005).

This chapter analyses the establishment, activities and impacts of the

Mae Kuang River Sub-basin Working Group, and the body in which it is nested: the Ping River Basin Committee and its secretariat office (hereafter referred to as the Ping RBO). The chapter pays special attention to how the River Sub-basin Working Group and Ping River Basin Committee (or RBO) have interacted with other institutions and platforms in the Mae Kuang watershed including: traditional irrigation or '*muang fai*' water user groups; the Mae Kuang Udom Tara Dam Irrigation Water User Organisation; and emerging local watershed forest and river conservation networks. The case was chosen because the context illustrates issues arising in an urbanising region with strong agricultural interests and natural vegetation, and because the organisations were among the first to be established in Thailand as pilots for a nationwide programme.

The analysis is based on analysis of official documents and meeting reports, observations of discussions at meetings, and more than 30 in-depth interviews with past and current members of watershed committees and working groups as well as others with knowledge and key responsibilities related to water management. The interviews took place over several years (2006–2013) as an effort was made to follow the evolution of the new institutions over time.

This chapter argues that bureaucratic competition, insufficient resources and lack of constructive engagement with pre-existing institutions and emerging networks have limited the influence of the RBO and River Sub-basin Working Group. Various other water user groups and networks continue to be active in addressing water management problems within the Mae Kuang watershed. To add greater value to existing efforts the RBO and River Sub-basin Working Group should expand its engagement to all stakeholders, taking a more explicit role as convener or at least supporter of multi-stakeholder platforms that would help resolve conflicts, explore alternatives, and inform negotiations and decisions affecting the watershed.

12.2 MAE KUANG WATERSHED: WATER RESOURCES AND USES

The Mae Kuang watershed covers 1661 km^2 in four districts of Chiang Mai province and two districts of Lamphun province (Figure 12.1). Forest, mostly deciduous and dry dipterocarp, covers just over half of the watershed and is mostly restricted to higher elevations (Thomas 2006a). About a third of land is devoted to agriculture and 7–8 per cent to residential use. Mae Kuang has a total population of 291 000 of which half are classified as rural. The Mae Kuang watershed sits within the Ping River

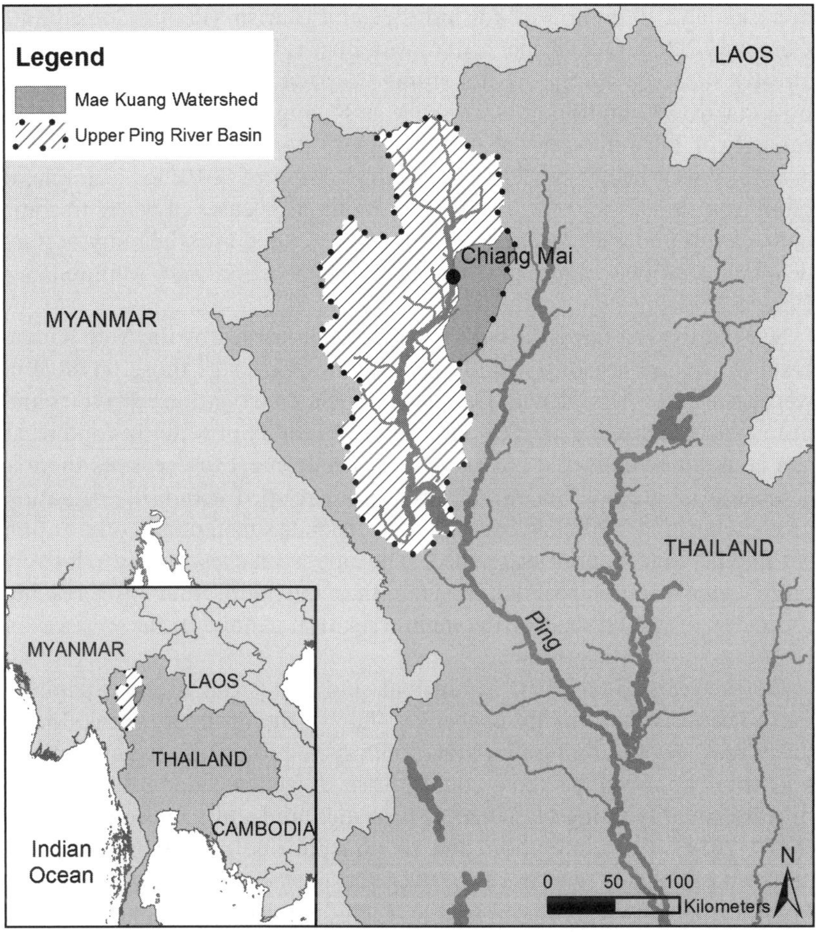

Figure 12.1 The Mae Kuang watershed within the Upper Ping River Basin

Basin and has undergone substantial transformation as the Chiang Mai and Lamphun cities have grown (Lebel et al. 2009).

Understanding the basic annual water budget is important to understanding the interests of different stakeholders in water allocation and other management challenges. Average rainfall in the Mae Kuang sub-basins is just over 1100 mm (Thomas 2006a). Natural surface water yearly averages 564 million m^3, with 470 million m^3 in the rainy season, and 94 million m^3 in the dry season. Several different estimates of total annual water demand have been made with slightly different assumptions. One estimate suggests 489 million m^3 with 459 million m^3 for irrigation, 16 million m^3 for domestic and 10 million m^3 for industry and tourism (Panya Consultants Company 2003). Over 95 per cent of water use is for agriculture, and usage already succeeds storage. Total storage capacities of 68 state-developed projects are 337 million m^3 with the Mae Kuang Dam dominating with a capacity of 263 million m^3. Mae Kuang Dam storage in practice, however, is usually well below capacity. In the dry season of 2010, for example, it plummeted to 17 per cent, resulting in minimal releases of water to some main canals and with resultant effects that extended into the rainy season, delaying wet rice planting (Mae Kuang Operation and Maintenance Project 2013).

Seasonality is a key issue with irrigated areas served by the Mae Kuang Dam in the dry season typically less than a quarter of those possible in wet season. The Mae Kuang Dam and associated irrigation reservoirs and infrastructure raised expectations about availability of water beyond what can actually be delivered. As a consequence in most dry seasons there is a substantial area of land which cannot be cropped within the irrigation area (Lebel et al. 2009). Access to water depends substantially on position within the Mae Kuang watershed. The upper reaches receive relatively more water from the Mae Kuang Dam than the middle and lower reaches especially at the tail-end of the main irrigation canals. In the wet season flooding is a significant issue.

Downstream peri-urban, municipal and industrial areas experience more serious water quality challenges than in the upstream areas. In the lower reaches of Lamphun province, ammonia, biological oxygen demand and total coliform bacteria counts have exceeded standards resulting in water quality being classified as degraded and unfit for consumption (PCD 2011). Contamination due to phosphorus, nitrate-nitrogen and ammonia is higher than in other river sub-basins in the Ping watershed. Pollution is caused by many sources including municipal and industrial effluents, cottage industries, and intensive cropland as well as pig and dairy farms (Thomas 2006a; Chitmanat and Traichaiyaporn 2010).

Finally, groundwater withdrawals are also an important part of water management in the basin. Groundwater recharge averages 40 million m^3 per year with safe yield of 8 million m^3 per year (Panya Consultants Company 2003). The highest rates of groundwater abstraction in the Upper Ping basin are around San Sai and San Kamphaeng, both in the Mae Kuang sub-basin (Margane and Tatong 1999). Perpetual water scarcity due to increasing and diversifying water demand has prompted a number of industries, services, golf courses, housing estates and villages to increasingly extract groundwater.

12.3 INSTITUTIONAL DESIGN

12.3.1 Introduction

According to criteria in Huitema and Meijerink (Chapter 1, this volume), the Mae Kuang River Sub-basin Working Group and Upper Ping RBO are 'coordinating-type' RBOs. Table 12.1 summarises the evidence of key institutional design features. It was not always possible to distinguish the activities and influence of the RBO and River Sub-basin Working

Table 12.1 Institutional design features of the Ping River Basin Organisation and Mae Kuang Sub-basin Working Group

Feature	Explanation
Authority	Roles in planning process, but primarily in compiling rather than integrating. Lacks formal authority: acceptance of role by those with responsibilities has been hard to achieve and remains modest, in part, because competencies with respect to floods or allocation of water are limited.
Aggregation	Planning process involves multiple channels as well as strong editing function by traditional line agencies; as a consequence decision-making role in significant projects is modest. Small, bottom-up proposals, fostered by RBO and River Sub-basin Working Group are filtered to fit national water development and management agendas.
Boundary	Boundaries of River Sub-basin Working Group in theory fit watershed, but in practice activities largely take place outside areas in irrigation schemes. Boundaries of RBO fit better with watershed after merging previously separate Upper and Lower Ping Committees that covered upstream and downstream sections of the same basin. Contrary to earlier visions, committees have become more oriented towards representing interests of different government agencies than non-state actors.
Information	Very modest effort at sharing information with wider public. The RBO and River Sub-basin Working Group do not do their own research but rely strongly on Irrigation and other departments for baseline information. Water engineering consultants have provided key baseline and other assessments.
Payoff	Financial resources insufficient to pursue initial mandate. Budgets remain controlled by line agencies.

Group within the Mae Kuang watershed as they involve some of the same individuals, events and procedures.

12.3.2 Authority Rules

The RBO and River Sub-basin Working Group lack significant formal authority since they were established by the Order of National Water Resources Committee which itself was established by the Order of the Prime Minister's Office, and not by an Act of Parliament. Issues related to authority were contested strongly from the start and as a consequence authority narrowed over time from initial visions promoted. The reasons for this are explored in more detail in section 12.4.

One potential important governance function under authority rules would be to help enable social learning across vertical and horizontal government departmental lines. Evidence for learning, however, was weak. Adjustment by other institutions was limited: organisations continued with former tasks and mandates with little effort to adjust other than recapture authority potentially lost. Furthermore, monitoring and evaluation systems for the RBO and River Sub-basin Working Group were not apparent. There was little evidence that new policy issues were being addressed, with planning tasks being primarily limited to compilation and filtering rather than integration (see section 12.3.3 for details).

Given that the Department of Water Resources is under the Ministry of Natural Resources and Environment, one might have expected an emphasis upon environmental protection, conservation or ecosystem health. As will be seen below this has been the case, in part, with respect to activities in upper tributary watersheds, but in lowland areas where pressures are greatest from agriculture, industry and urbanisation the River Sub-basin Working Group has played a very modest role. Its capacity to implement or shape authority rules has been very limited, and only exercised in areas where it is mostly interacting with other departments (forest, national parks and pollution monitoring) in the same ministry. In 2012, however, representation of stream conservation networks in the River Sub-basin Working Group grew greatly (see section 12.3.4 for details) suggesting that in future the conservation role might be strengthened.

As of April 2014 a national Water Law that would give administrative, policymaking and regulatory teeth to the National Water Resources Committee down the hierarchy to RBOs and River Sub-basin Working Groups had not yet been passed. Meanwhile there has been a movement away from initial emphasis on broad coordinating functions to a narrower subset of administrative, technical, procedures among government departments.

12.3.3 Aggregation Rules

Integration is an example of a normative aggregation rule which nobody opposes – an example of a 'nirvana' concept (Molle 2008). Integrated water resources management plans are typically for five to ten years with shorter-term plans prioritised higher. The annual planning process lasts approximately ten months and will now be described in detail. Most plans considered by the RBO and River Sub-basin Working Group are small-scale local infrastructure projects related to water supply or flood protection.

There are currently two basic pathways to a plan (Figure 12.2). In the first pathway, plans and projects are prepared by line agencies and submitted to the National Water Resources Committee before going to the Bureau of Budget at the end of March. These do not have to pass through a prioritisation process like the second pathway. The second pathway is a bottom-up process where villages propose their own plans and projects. The process starts at declaring water-related needs at village level and the sub-district water working group collects and drafts plans. District, Provincial and Mae Kuang River Sub-basin Working Groups then relate them to watershed and provincial strategies. Another planning subcommittee then considers and screens them so that they fit national government, provincial and watershed strategies and urgent needs, and the secretariat then analyses and screens projects using a system of prioritising criteria. After this the Ping RBO will consider and forward proposals to the Department of Water Resources, National Water Resources Committee, Bureau of Budget, Cabinet, and eventually Parliament.

The planning process for water management projects is complex with multiple levels and multiple centres. In practice coordination across ministry lines remains weak and many procedures remain controlled by top-down mechanisms after the initial period of consultation.

During meetings observed of the Upper Ping Working Group on Integrated Watershed Plans in 2004, representatives from local groups in Chiang Mai negotiated the prioritisation of plans and projects. It was agreed to give opportunities to local representatives at sub-district and district levels to prioritise plans and projects rather than letting higher-level committees prioritise them. They reasoned that local communities are closer to water problems and directly receive both costs and benefits from the projects.

A representative questioned the nature of stakeholder participation saying that government agencies want to organise meetings and take photographs as proof of participation to justify plans, projects and budgets. But, when locally initiated plans or projects are considered at higher levels, they

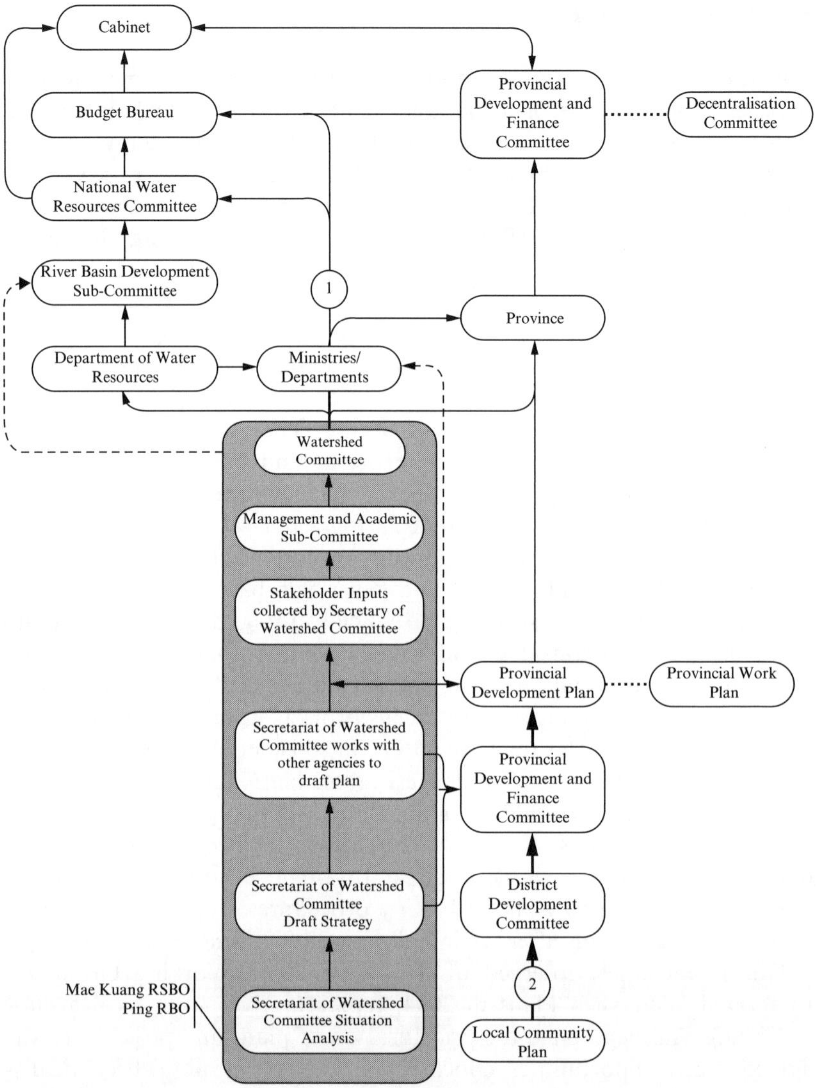

Note: Two planning pathways are highlighted (1 and 2).

Source: DWR (2008), redrawn by authors.

Figure 12.2 Integrated water resources management planning process

are often replaced by those preferred by the line agencies. Local water user groups also complained that they participated in making plans and budgets necessary for solving water problems at the local level, but their bottom-up plans were often changed at higher levels. As a result, key stakeholders remain sceptical, arguing that participation is really about coordination of government agencies and not bottom-up inputs into decision making.

In the 2004 round, most projects were for construction of water sources and distribution systems for agricultural use, domestic use and flood prevention, such as weirs, reservoirs, canal construction and improvement, village tap water, groundwater provision for villages, dykes and drainage. A smaller number of projects focused on improvement, maintenance and restoration of natural and pre-existing water bodies and headwater areas, as well as personnel development. The low priority given to management tasks was surprising given the level of conflicts and governance challenges.

In summary, aggregation rules are deeply embedded in planning structures, procedures and norms. As the basin and integration ideas have gained acceptance in principle these have been adjusted, but in ways which suggest the main purpose has been bureaucratic recapture (of authority and budgets) as much as accommodating the new basin and sub-basin organisations.

12.3.4 Boundary Rules

The Mae Kuang Watershed Working Group established in 2003 is a part of a hierarchy of watershed committees. A sub-national administrative-based hierarchy coexists with a hydrological one. Watershed boundaries and administrative boundaries do not superimpose totally. Mae Kuang sub-basin, for example, physically encompasses some parts of six administrative districts of two provinces, while the other parts of those districts fall into other river sub-basins. The Mae Kuang sub-basin as a management unit includes both the Mae Kuang and Mae Tha watersheds.

Integrated water management based on sub-basin boundary is theoretically convenient, but the state budget is in practice distributed and audited through a hierarchy of line agencies and local administrative organisations. As local administrative organisations have become increasingly important in terms of budget allocation and natural resource management responsibilities within their jurisdictions, it is difficult to plan or manage based on watershed units alone. The hope was that bodies in the watershed and sub-national hierarchies could coordinate their activities and that a flexible and negotiated approach to the design of river sub-basin organisation structure and functions would lead to much better planning.

Boundary rules are also revealed by changes in the structure and

Table 12.2 Changes in membership of the Mae Kuang River Sub-basin Working Group

Group represented	First committee 2003–2004	At end 2008	In mid-2012
Old, existing community organisations	6	0	0
Community forestry networks	3	0	0
Stream conservation networks	3	0	16
Farmers' networks	3	6	2
Ethnic minority groups	3	0	0
Housewives' groups, women's development groups	3	0	0
Local government (Tambon Administrative	6	6	22
Organization, municipalities, Provincial	6	6	2
Administrative Organization)	4	5	12
Kamnan/village headmen			
Government officials (line agencies)			
Local specialists (academics)	4	1	0
Non-governmental organisations	2	6	0
Industry/business	2	6	0
Total membership	45	36	54

composition of the River Sub-basin Working Group and RBO. Initial proposals for membership of the River Sub-basin Working Group were modified following discussions. In the early stages the River Sub-basin Working Group's design was shaped and reshaped repeatedly to reduce state domination and increase representation of diverse local stakeholders. This aimed to include pre-existing groups and evolving networks that had allied among themselves and with some state agencies (Thomas 2005, 2006a). First a group of 45 was established (Table 12.2). Not all positions were filled as planned and some individuals filled more than one role. Two persons concurrently held two positions within the first committee, as agricultural water user on the one hand and as sub-district officer or representative of a people's organisation on the other. A third person held posts in higher- and lower-level working groups and the Ping RBO. A fourth was both a River Sub-basin Working Group member and leader and founder of a recently established network in the lower reaches of the Mae Kuang. Individuals with formal roles in multiple organisations could help to improve information flow among them.

During 2007–2012 changes in governments caused multiple changes in RBOs and River Sub-basin Working Group committees with new rounds of selection. After the 2008 reshuffle the Mae Kuang River Sub-basin Working Group had 36 members distributed widely across stakeholder groups (Table 12.2). In April 2012, a new Mae Kuang River Sub-basin Working Group was established with an increased number of members but decreased diversity. The higher representation from local government was primarily from administrative rather than elected positions, is thus a reassertion of the conventional administrative and planning hierarchy within the Ministry of Interior. Another significant shift in orientation among non-state members was the increase in 2012 of representatives from stream conservation networks relative to other farmer and civil-society organisations (Table 12.2).

The selection process used in 2012 was as follows. A Mae Kuang Sub-River Basin selection committee was established by the Ping River Basin Management and Academics Sub-committee of the RBO with broad representation from different sectors. This selection committee reviewed applications from local farmers and water user groups. In the end, all applicants were selected because there were few people willing to serve. This selection process did not apply to representatives from government agencies who were appointed by respective agencies. The new Mae Kuang Working Committee was subsequently approved by the National Water Resources Committee.

In terms of areas being represented within the River Sub-basin Working Group there were some gaps. For instance, no one applied as a representative of industrial and services water user groups in Lamphun in 2008 or 2012. This is significant, given the proximity of large industrial estates. Neither were there any representatives from academics and sub-district administrative organisations from downstream areas of the Mae Kuang river sub-basin in both years. On the other hand, in 2012, the chairperson of the Mae Kuang Watershed Conservation Network in the upper reaches was included in the selection committee, indicating some level of recognition of 'alternative' local networks.

At the RBO level a new Ping River Basin Committee was appointed in August 2008 that combined the Upper and the Lower Ping Committees (Figure 12.3). Government representation in the Upper Ping River Basin Sub-committee has increased with the changes in organisation in 2008 (Figure 12.3). At the same time, representation of farmers was slashed. The trend at both Mae Kuang and Ping level has been a shift away from wider representation and a greater focus on central and local government agencies and business.

Compared with local leaders and agricultural water user groups who regularly participate in meetings, however, representatives from industrial and service water user groups participate less in meetings. Some admitted that they are too busy to attend and seldom joined meetings; others

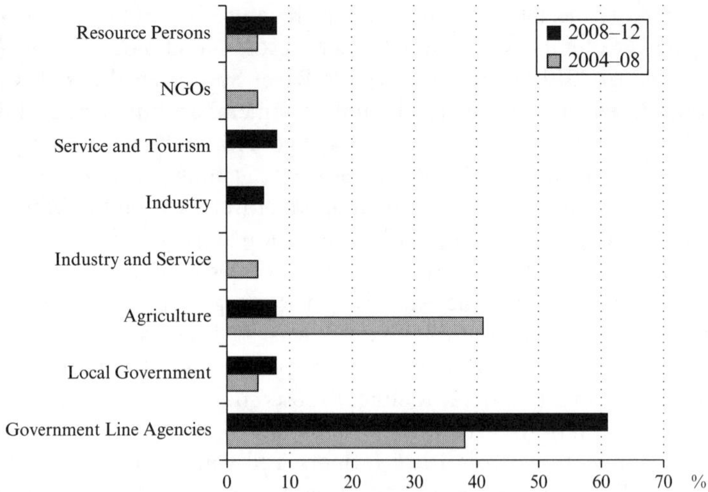

Figure 12.3 Stakeholder composition of the old Upper Ping River Basin Sub-committee and the current Ping River Basin Committee which combine both Upper and Lower Ping Committees into one (as of October 2012)

claimed that they did not know that they were representatives and were never invited to join.

Following similar selection procedures to those described above, a new Ping River Basin Management and Academics Sub-committee was introduced in 2009 with 47 members; just under half were from government. The Basin Committees have four-year terms, so in early 2013 expect another process of selection to start.

The key point is that throughout the establishment period of the hierarchy of watershed management structures there has been a lot of attention to representation on committees, with efforts made to come up with a system that can fit in with existing administrative hierarchies. Entry and exit boundary rules have changed without ever being made explicit. As the basin organisation initiative has aged there has been significant recapture of agendas by state agencies, meaning that non-state interests are less likely to be taken into account.

12.3.5 Information Rules

Within the sub-basin area, the Mae Kuang River Sub-basin Working Group is responsible for: (1) collecting and disseminating information regarding water and other natural resources; (2) submitting local demands for plans and projects on water resource development to the Provincial Watershed Working Group; (3) coordinating state agencies' operation plans regarding development and conservation of water bodies, water allocation, watershed rehabilitation, prevention and control of floods, droughts and water pollution; (4) resolving water management conflicts; and (5) following up and evaluating the performance of government agencies regarding water resources and reporting to the RBO committee.

A non-governmental organisation (NGO) representative interviewed thought that the River Sub-basin Working Group was a good idea but ineffective because it lacked physical and financial resources.[1] With no local branch office there was no clear place to store or share information with the public. With little budget for travel, meetings had to be done in hotels or in parallel with events held at district or provincial level. Consequently, the collection and dissemination of information activities appears to have been largely with other government agencies rather than residents.

The RBO drew substantially on the work of consultants to put together initial datasets for assessments and plans. As the years went by most of the information gathering and analysis capability remained with the irrigation department, while modest collation and dissemination roles were taken up. In terms of information rules the River Sub-basin Working Group and parent RBO depend greatly on views and contributions of technical experts in other agencies. The initial emphasis on stakeholder consultation and representation of non-state actors – which may have brought other sources of knowledge legitimately into planning processes – has largely faded away as the traditional elements of the bureaucracy have reasserted power.

12.3.6 Pay-Off Rules

Analysis of budget allocation gives some indication of the role and influence of stakeholders within the sub-basin. The 2010 yearly budget allocation for state plans within the Mae Kuang sub-river basin went largely to the irrigation department. Out of this budget allocation, most went to infrastructure development and improvement, that is, for flood prevention (55 per cent), reservoirs (14 per cent), canals (12 per cent), weirs (9 per cent), groundwater drilling (4.8 per cent), pipelines (2.7 per cent), water pumping (2.5 per cent) and check dams (0.3 per cent), and reforestation (0.1 per cent).

The responses of government agencies, traditional irrigation groups, and alternative watershed networks to the introduction of the River

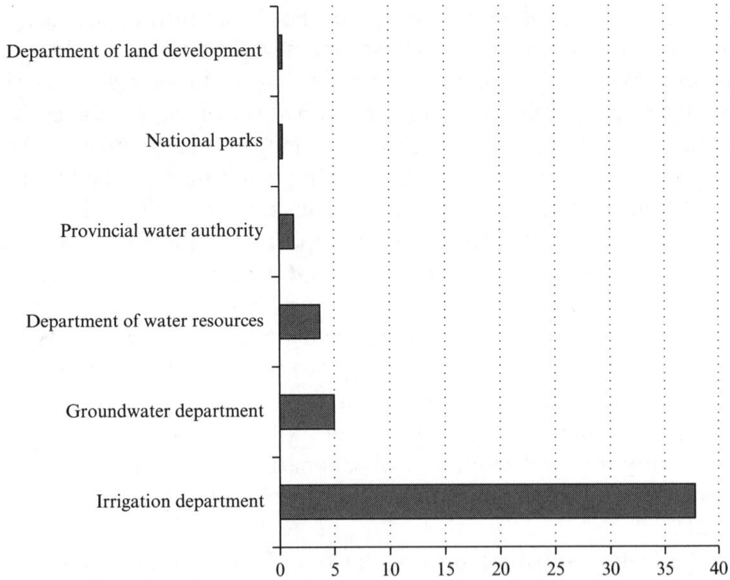

Figure 12.4 Allocation of 2010 budget in the Mae Kuang basin

Sub-basin Working Group have been distinct. Each group has also adopted a mixture of strategies reflecting existing support base, but also needs to build wider coalitions if it hopes to have influence. The way budgets have continued to be allocated suggests that the River Sub-basin Working Group has had only very minor influence.

At the higher level the Ping Basin Committee has a supporting office under the Department of Water Resources. Its budget has been about US$2.7 million per year and is expected to continue at around this level.

In summary, the way in which the RBO and River Sub-basin Working Group structures have emerged and evolved in the Mae Kuang watershed has not matched initial visions or expectations of proponents of a functional organisation helping to facilitate deliberation among stakeholders, resolving conflicts and coordinating activities of relevant government agencies. The first decade has been largely spent filling and refilling committees and working groups which rarely meet and have only very modest budgets, little dedicated office support and limited authority. The role has been narrowed to coordinating plans of different agencies and local government. The explanation for this difficult beginning lies in how these new organisational structures tried and largely failed to integrate their activities with pre-existing institutions and alternative platforms (see section 12.4).

12.4 PRE-EXISTING INSTITUTIONS AND EMERGING NETWORKS

12.4.1 Introduction

This section explains how various actors responded to the introduction of new basin organisations. The Mae Kuang River Sub-basin Working Group and Ping RBO were introduced into a context with a history of conflicts over allocation of water and uncoordinated activities by various agencies and community-based organisations. In the upland areas these are typically called watershed organisations or networks, and are typically involved in the management of land, forests and water resources. In the rural lowlands, these watershed organisations or networks are more commonly known as water user groups or associations, being engaged in allocation and management of water for irrigation purposes. Pre-existing institutions and alternative platforms were already in place or emerging in parallel leading to a set of potential interactions (Figure 12.5). In this

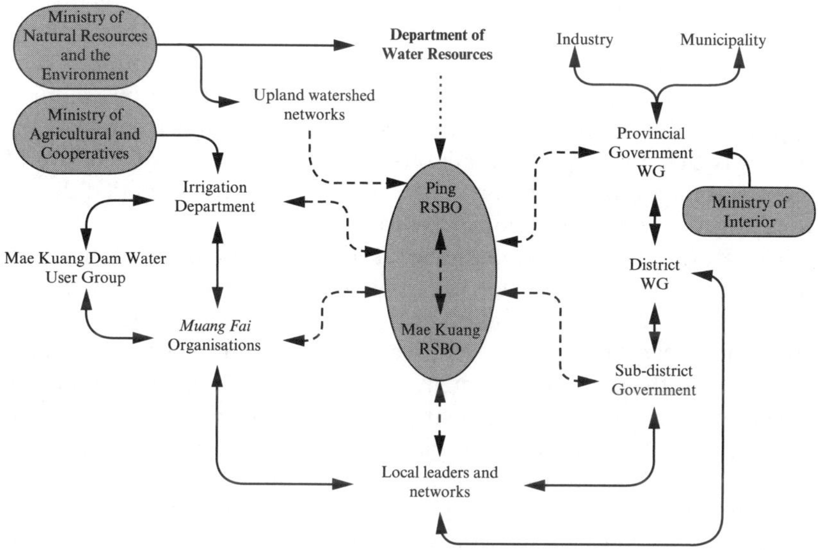

Notes: New organisations are in bold with fine-dotted lines indicating formal relationships. Solid lines indicate pre-existing institutions and relationships and dashed lines some of the important new relationships that had to be developed, but were not always successful. Not all possible links are shown.
WG = working groups.

Figure 12.5 The complex institutional context of the Mae Kuang River Sub-basin Working Group

section we examine the actors supporting, resisting and ignoring the establishment and evolution of the river basin organisations and explore what strategies they pursued, successful and not.

12.4.2 Government Agencies: Coordinating and Competing

A key challenge for the River Sub-basin Working Group and the RBO, according to a committee member we interviewed, is persistent bureaucratic competition.[2] At the national level this starts between ministries (Molle 2007). The relatively new Department of Water Resources in the Ministry of Natural Resources and the Environment tries to establish itself and develop new working mechanisms through water law and watershed committees (Figure 12.2). Other state agencies, in particular, the Royal Irrigation Department, in the Ministry of Agriculture and Agricultural Cooperatives, may follow this pathway or directly submit plans and budgets to the Bureau of Budget and eventually to the cabinet.

Different agencies have their own plans for the Mae Kuang watershed. Negotiating a shared or compromised vision has been tough. The first plans therefore focused on organisational issues and procedures more than objectives (Thomas 2006b) and this trend has continued. The presence of several strong groups made it harder to agree on common approaches during the establishment phase. A common perception at the start was that the separate coalitions or factions, with a sub-culture of accountability to their superiors rather than to local communities, do not work together very well (Thomas 2006a).

The Royal Irrigation Department is the most important actor with respect to allocation of water for irrigation at larger scales in the Mae Kuang watershed. Two meetings a year at Mae Kuang Dam focus on what areas will receive how much water (in cubic metres) during what time (how many days with water and how many days without water). Farmers and the Irrigation Department are both strategic in negotiations and subsequent patterns of water use. Farmers typically overplant relative to agreements, expecting to be able to use the image of crops in the ground as a way to lever additional water at the end of the dry season when shortages are most acute (to prevent disastrous crop losses); the Royal Irrigation Department, on the other hand, is conservative in its recommendations for land uses, aware that in practice it will often have to deal with appeals for special assistance as the dry season unfolds. Location within irrigation systems is also important at finer scales in water competition and negotiation.

The biggest and most controversial solution to water deficit in the Mae Kuang sub-basin is an inter-basin water transfer project proposed by the Royal Irrigation Department. The mega-project is expected to annually divert 147 million m^3 from the Mae Taeng and Mae Ngad rivers, which are tributaries of the Ping River, through 23 km of tunnels into the Mae Kuang reservoir. It is anticipated that for irrigated farmland in the dry season and for urban and industrial purposes available water could be roughly tripled. Public hearings for this controversial mega-project were held and an environmental impact assessment was done. Cabinet has approved the 11 billion baht project (€270 million), and construction is expected to be completed in 2017.

A retired River Sub-basin Working Group committee member (who is also a wet-rice farmer) criticised the River Sub-basin Working Group as ineffective in solving the water scarcity problem in the Mae Kuang sub-basin, in particular by not supporting the inter-basin water transfer project, claiming that both the RBO and the River Sub-basin Working Group have neither power nor resources, and are very slow to take action. The Royal Irrigation Department initiated the project and tried to gain public acceptance by organising several public hearings during 2008–2009 (some RBO and River Sub-basin Working Group committee members attended the public hearings, but they were not convenors). Some local leaders supported the project by organising meetings in 33 sub-districts in Chiang Mai and Lamphun in order to get water users' signatures (which they called 'referendums' when, in fact, these are more like petitions) to be submitted to the government. The act of seeking water users' views through a 'referendum' was frowned upon by some Ping RBO and River Sub-basin Working Group members, claiming that it is not a normal process for project development. The process should start from an annual visioning and planning process for developing projects. Plans and projects must correspond to provincial development strategies. The 'referendum' was later discouraged. There were also criticisms from some water users of the Mae Taeng irrigation system that they have long experienced water shortage and disputes in their own system and that inter-basin transfer would exacerbate these problems.

When water users are faced with water-related problems, they usually ask for help from formal local leaders, district officials and contact the Royal Irrigation Department (RID) directly. They do not contact the RBO or River Sub-basin Working Group because their procedures are lengthy and time-consuming, and the RBO and River Sub-basin Working Group will eventually have to coordinate with line agencies that have direct responsibilities. For instance, water users in six sub-districts in the irrigation area of the Mae Kuang Dam were successful in having water delivered to their dry areas by shifting water from Mae Kuang to the more abundant Mae Faek-Mae Ngad River through direct negotiations with line agencies.

12.4.3 *Muang Fai* Irrigation Organisations: Transforming Traditions

Muang fai irrigation organisations manage common irrigation water. Farmers in these organisations grow rice, vegetables or fruit. *Muang fai* organisations have accumulated experience in effective water allocation and conflict management through the centuries, and have shown potential and flexibility in dealing with new challenges (Cohen and Pearson 1998; Bastakoti et al. 2010). Vanpen Surarerks (1986) pointed out that *muang fai* organisations were typically more effective than state irrigation organisations. Tan-kim-yong (1983) and Tanabe (1981) indicated that these organisations practiced a form of participatory democracy at the grass-roots level. They were granted customary use rights of facilities and water rights in exchange for responsibilities in operation and maintenance according to communal rules and regulations. Members elected an irrigation head along with a committee based on technical skills and leadership ability. As *muang fai* organisations prioritised rice production, only rice landowners have the right to vote.

With the introduction of modern agricultural technology and increasing intensification and diversification of agriculture, multiple cropping became prevalent (Lebel et al. 2009). As a result, water shortage was more pronounced, especially in the dry season and at the tail-end portion of irrigated farmland. Water demand and water allocation became complex because of diverse cash crops with spatially and temporally different water requirements (Ganjanapan 1984; Thangphet 1989).

In the past *muang fai* leaders solved water allocation and conflicts based on their (moral) authority. As urban and industrial water users increase in number and in their water demands, new water conflicts erupt among farmers and non-farm water users. The traditional rules, leadership and processes of *muang fai* organisations could not cope effectively with the complexity of conflicts (Charoenmuang 1994). Consequently, semi-government officials such as village headmen, provincial governors, or local and national politicians, are increasingly asked to mediate rather than relying on negotiations among irrigation leaders. At the same time a number of *muang fai* leaders initially became members of the Mae Kuang River Sub-basin Working Group or Ping RBO, but in recent reorganisations the representation of farmers has dropped (Table 12.2, Figure 12.3) suggesting reduced influence, and maybe also reduced interest in, these organisations.

Within the context of agrarian restructuring, urbanisation and industrialisation, *muang fai* organisations have to mobilise financial and technical resources from government agencies for system rehabilitation and maintenance, and seek legal status so that they have the right to access development funds and programmes. They restructured their organisations by reducing irrigation assistants and including semi-government village heads in the management committee to facilitate resource mobilisation from government agencies. Overall, the trend is towards greater state control over the allocation of water and the bureaucratic incorporation of *muang fai* organisations (Pearson 1999).

12.4.4 Alternative Watershed Networks: Environmental Agendas

In upland valleys, diverse conservation activities had been going on for several years, with suggestions and support from government agencies, academic institutions, the private sector and international organisations (Thomas 2006b). Management of the headwaters of the Upper Ping watershed are now under the responsibility of the National Parks, Wildlife and Plant Conservation Department. Most projects were short term, including reforestation, slope stabilisation via vetiver grass cultivation, small weirs, information systems that integrate both scientific and local knowledge, and development of sub-basin environmental volunteer networks.

Since 2004 the First Regional Office of Environment has supported more than 20 local groups in activities including tree ordination ceremonies, river life prolongation ceremonies, check dam construction, repair and digging of irrigation canals, tree planting, improvement of landscape around reservoirs, and training in water resource management.

The Office of the Ping River Basin Committee has supported only one local group in the Mae Kuang watershed. Beginning in 2006 it has provided modest financial support to the Huay Kaew Reservoir Conservation Group in the Mae Faek subdistrict, for tree ordination ceremonies in watershed forests as well as reforestation activities. Many other groups and networks work independently of the RBO and River Sub-basin Working Group.

The Village Volunteers for Natural Resources and Environment group in Lamphun province, for example, were informed about water pollution in Mae Kuang river by the First Regional Office of Environment that regularly monitors water quality through seven river water monitoring stations. They have since involved government agencies, NGOs, mass media, religious organisations and civil society in grass-roots pollution abatement and control projects. Other related activities were to raise awareness of deforestation and changing water quality, and to carry out 'river life prolongation' rituals.

This group also coordinated activities with the Mae Kuang Watershed Forest Conservation Committee which has been active since 2001 in the upper reaches of the watershed. The Conservation Committee holds monthly meetings with members from 55 villages. Also present in meetings are government officials and members of local administrative organisations. Members do volunteer work on forest protection and restoration through fire control, tree ordination and river life prolongation ceremonies. Activities are done without allocated budgets from state line agencies. The two networks have considered joining. So far they do not include either industrial or service sector stakeholders, nor coordinate with the River Sub-basin Working Group. In addition, these networks are not included in the formal list of local networks held by either the Office of Ping River Basin Committee or the First Regional Office of Environment though both offices give support to them.

Mae Kuang watershed is covered with a patchwork of partly overlapping layers of institutions: some long-established and others relatively new and covering different parts of the watershed. In the uplands, however, there are already well-established forest and watershed-related networks in place that have continued to do most or all of their work independently. In the lowlands, key planning activities and budgets remain under strong influence of traditional line agencies like the Department of Irrigation.

In summary, pre-existing institutions and emerging watershed networks have had a significant impact on the evolution of roles and functions of the Mae Kuang River Sub-basin Working Group and the Ping RBO as the latter struggled to justify their existence. The resulting interplay has pushed the new River Sub-basin Working Group and RBO to try and work more in upland watershed areas where they can engage in familiar environmental management issues and avoid the allocation and management issues in the lowlands where well-established irrigation and urban-industrial interests predominate. This shift avoids having to address the cross-sectoral issues of 'integration' which were used to justify establishment of RBOs and River Sub-basin Working Groups in the first place.

12.5 PERFORMANCE AND DESIGN

12.5.1 Introduction

In this section we synthesise material from earlier sections and additional evidence to evaluate the performance of the Mae Kuang River Sub-basin Working Group and Ping RBO in Mae Kuang watershed, and where possible relate this to features of the evolving institutional design (Table 12.3). As assessed above in the Mae Kuang the two organisations have had primarily coordination roles with most activities related to coordination, information gathering and dissemination.

Table 12.3 Key performance issues

Dimension	Observations
Coordination	Irrelevant and ineffective as a multi-stakeholder platform. Compiled rather than integrated plans. Key government stakeholders do not support watershed perspective.
Legitimacy and accountability	Modest as viewed as having limited capacity and authority. Other organisations with greater legitimacy present. Seem more concerned with coordinating government stakeholders than meaningful public participation.
Water management	Avoid difficult allocation issues. Peripheral to major water infrastructure decisions. Unable to leverage financing for their tasks.

12.5.2 Coordination

RBOs have made a modest contribution to coordination, mostly on paper. This is evident by the lack of engagement in planning-related conflicts over the allocation and quality of water in the Mae Kuang watershed. There are several cases of controversial water projects in which one would expect the basin organisations to have got involved, but they largely have not, including: the Mae Taeng–Mae Ngad–Mae Kuang water transfer; the construction of Ping water gates to replace 300-year-old weirs on Ping River which is adjacent to the Mae Kuang sub-basin; water competition between farmers and the Provincial Water Authority within the Mae Kuang sub-basin; and water pollution in the lower Mae Kuang sub-basin near industrial estates. In the first two cases the topic was put on the meeting agenda and members of the Ping RBO and Mae Kuang River Sub-basin Working Group participated in the related public meetings, but the organisations did not themselves organise or convene meetings of stakeholders to try and manage these conflicts.

Conflict management still rely on local administrative organisations and district officials rather than the RBO and River Sub-basin Working Group. They would like the greater authority that would be granted to them if the Water Bill was passed. It is not clear, however, whether this would actually enhance their roles or lead to even more serious turf wars. Key government stakeholders do not fully accept or concur with the watershed perspective, nor do they see the RBO and River Sub-basin Working Group as providing a platform for multiple stakeholders to consider alternatives to controversial projects, policies and decisions.

The coordination roles of the Mae Kuang River Sub-basin Working Group and Ping RBO are limited to planning activities and this is mostly about compiling other agencies' plans and projects. Each fiscal year, the RBO and River Sub-basin Working Groups consider plans and projects from water-related government agencies and local administrative organisations. Due to limited time, limited personnel and voluminous tasks, all plans and projects are put together with only limited consideration of redundancies and not much integration. Although some plans, projects and budget allocation are prepared on a watershed basis, many are also proposed via local and supra-local administrative organisations. Budget allocation, critically, is still based on conventional administrative jurisdictions.

12.5.3 Legitimacy and Accountability

The legitimacy of the RBO and River Sub-basin Working Group in Mae Kuang is modest. Many key government agencies do not fully accept their roles as water managers, or believe they are not capable of carrying out mandated tasks. Local representatives have been given a chance to belong to higher-level committees, but planning processes have largely filtered out bottom-up proposals in ways that are not transparent to local water users. The shift in membership since 2008–2009 towards much greater representation of government in the Mae Kuang River Sub-basin Working Group, along with similar shifts in other parts of the multi-level hierarchy, would seem to imply fewer opportunities for public participation and deliberation and thus even less chance that it can be effective as a multi-stakeholder platform. Given these trends the wider public may withdraw what little support it has given. On the other hand, networking efforts appear to continue especially at the RBO level.

Working with other local organisations the Ping RBO may be improving its legitimacy. It now tries to link with around 100 diverse groups including *muang fai* organisations, forest conservation groups, youth groups, housewives' groups, and so on. The most cited are two *muang fai* organisations in Chom Thong district, one Mae Wang forest and water conservation network, one Wiang Nong Long Water Development for Life Group in Lamphun, and one Mae Sa Watershed Restoration Group in Mae Rim district, Chiang Mai. All of them are outside Mae Kuang river sub-basin, and four of them are further away from urban Chiang Mai. The nature of such networking is that the Ping RBO gives small financial, technical and material support to projects that are initiated by those groups. These efforts are likely to increase social acceptability of the RBO and River Sub-basin Working Group.

But when it comes to water allocation in lowland areas other organisations still have greater legitimacy. *Muang fai* irrigation groups in the peri-urban interface linger and still have potential to manage water resources and manage conflicts, but they need support from government agencies. *Muang fai* organisations have allocated irrigation water effectively for a long time, but in many places have grown weaker due to lack of labour and money, environmental changes and the activities of state agencies. Recently the Royal Irrigation Department has strengthened links with water user and old *muang fai* irrigation groups through a project called Irrigation Volunteers which serve as liaison between irrigation groups and department officials. These volunteers are elected to be in charge of irrigation canals and water gates. The Ping RBO and related River Sub-basin Working Groups have learnt this lesson, and are reaching out to create networks with them. Their networks, however, are still limited in number, and do not include other active networks in the Mae Kuang sub-basin that have developed in parallel to the RBO and River Sub-basin Working Group. Those networks are working with other government agencies.

In a study of the Mae Sa River Basin Management Working Group, another River Sub-basin Working Group-like structure established in 2005 within the Upper Ping River Basin, the structure was also dominated by state agencies that barely interacted in the working group meetings observed (Kanjina 2008). Local government representatives were all from administrative positions and not the elected council. The observed passive participation of stakeholders, apart from the convening office with responsibility for the pilot project, can be seen as a form of bureaucratic resistance to the reforms.

12.5.4 Water Management

In the Mae Kuang watershed the River Sub-basin Working Group and RBO have struggled to influence let alone enhance water management. Farmers felt they were irrelevant to water allocation problems compared to alternative platforms available, such as those supported by *muang fai* and the Mae Kuang Dam water user organisations. Irrigation officials view the River Sub-basin Working Group as irrelevant to water allocation because it works outside the Mae Kuang irrigated areas, and see their own work as having been largely successful at dealing with allocation challenges.

The inability to adequately finance tasks is both a cause and a consequence of performance. The RBO and River Sub-basin Working Group in Mae Kuang watershed have been plagued with budgetary problems and long delays in appointments of committees. National-level political instabilities have been one factor: committees are appointed by Order of the Prime Minister's Office, which has changed often in the last six years.

Most River Sub-basin Working Group decisions continue to be about process and scope, rather than allocation or integration. Even basic visions are still in dispute and barely formed. This is in contrast to other platforms in the area which are longer-established and have much-better-defined agendas and decision-making sets. From the point of view of most stakeholders the River Sub-basin Working Group has not performed well on issues that matter most: reducing conflicts, building consensus or otherwise informing critical decisions on water allocation and infrastructure.

12.6 CONCLUSIONS

Several platforms co-exist in the Mae Kuang River watershed with varying scope and engagement by different stakeholder groups. The efforts of the Department of Water Resources to introduce a River Sub-basin Working Group appears to have met with little success for a variety of reasons, including bureaucratic competition, lack of resources, and a failure to adequately take into account pre-existing platforms and institutions within the sub-river basin. The River Sub-basin Working Group has become just a working group that meets irregularly. At the larger scale, Ping RBO is not faring much better with limited resources and a still unclear role beyond the compilation of plans. More powerful organisations, with the support of local coalitions, work through alternative platforms and channels; but problems with cross-sectoral allocation and coordination persist.

For the medium term it seems highly unlikely that a comprehensive platform will emerge from these basin organisations to address the diverse water allocation problems in the Mae Kuang watershed. The Ping RBO and Mae Kuang River Sub-basin Working Group, if they are to contribute to integrated water resource management, need to pay much more attention to, and learn how to work with, pre-existing platforms and institutions which themselves are also changing.

ACKNOWLEDGEMENTS

Funding for this work came from the Asia Pacific Network for Global Change Research, and the Challenge Programme on Water Grants to Project PN67 (European Commission) implemented by M-POWER (www.mpowernetwork.org).

NOTES

1. Interview with Mr. Chaisin Rinkaew, 14 February 2013. Previously Mae Kuang Working Group and now Ping RBO member.
2. Interview with committee member from environmental NGO, 1 July 2008.

REFERENCES

Abonyi, G. (2005), 'Policy reform in Thailand and the Asian Development Bank's Agricultural Sector Program Loan', ERD Working Paper Series No. 71, Manila: Asian Development Bank.

Bastakoti, R.C., G.P. Shivakoti and L. Lebel (2010), 'Local irrigation management institutions mediate changes driven by external policy and market pressures in Nepal and Thailand', *Environmental Management*, **46**, 411–423.

Biswas, A.K., O. Varis and C. Tortajada (eds) (2005), *Integrated Water Resources Management in South and Southeast Asia*, Oxford: Oxford University Press.

Charoenmuang, T. (1994), 'The governance of water allocation problems in Thailand: four case studies from the upper northern region', in Thailand Development Research Institute (ed.), *Water Conflicts*, Bangkok: Thailand Development Research Institute, pp. 111–147.

Chitmanat, C. and S. Traichaiyaporn (2010), 'Spatial and temporal variations of physical-chemical water quality and some heavy metals in water, sediments and fish of the Mae Kuang river, northern Thailand', *International Journal of Agriculture and Biology*, **12**, 816–820.

Cohen, P.T. and R.E. Pearson (1998), 'Communal irrigation, state, and capital in the Chiang Mai Valley (Northern Thailand): twentieth-century transformations', *Journal of Southeast Asian Studies*, **29**, 86–110.

Department of Water Resources (DWR) (2008), 'Office of Water Resources Policy and Planning Report', 8 June, Bangkok: Department of Water Resources.

Dore, J. (2007), 'Multi-stakeholder platforms (MSPS): unfulfilled potential', in L. Lebel, J. Dore, R. Daniel and Y. Koma (eds), *Democratizing Water Governance in the Mekong Region*, Chiang Mai: Mekong Press, pp. 197–226.

Dore, J. and L. Lebel (2010), 'Deliberation, scale and the governance of water resources in the Mekong Region', *Environmental Management*, **46**, 60–80.

Faysse, N. (2006), 'Trouble on the way: an analysis of the challenges faced by multi-stakeholder platforms', *Natural Resources Forum*, **30**, 219–229.

Ganjanapan, A. (1984), *The Partial Commercialization of Rice Production in Northern Thailand (1900–1981)*, New York: Cornell University.

Kanjina, S. (2008), 'Participatory water resource management in Thailand: where are the local communities?', presented at the 12th Biennial Conference of the International Association for the Study of Commons, University of Gloucestershire, available at http://iasc2008.glos.ac.uk/ (accessed 14 May 2013).

Lebel, L., P. Garden, N. Subsin and S.N. Nan (2009), 'Averted crises, contested transitions: water management in the Upper Ping River Basin, northern Thailand', in D. Huitema and S. Meijerink (eds), *Water Policy Entrepreneurs: A Research Companion to Water Transitions around the Globe*, Cheltenham, UK and Northampton, MA, USA: Edward Elgar, pp. 137–157.

Mae Kuang Operation and Maintenance Project (2013), 'Statistics on levels from various years', Chiang Mai: Mae Kuang Operation and Maintenance Project, available at http://www.maekuang.net/ (accessed 18 May 2013).

Margane, A. and T. Tatong (1999), 'Aspects of the hydrogeology of the Chiang Mai-Lamphun Basin, Thailand that are important for the groundwater management', *Zeitschrift für Angewandte Geologie*, 45, 188–197.

Molle, F. (2007), 'Irrigation and water policies: trends and challenges', in L. Lebel, J. Dore, R. Daniel and Y.S. Koma (eds), *Democratizing Water Governance in the Mekong Region*, Chiang Mai: Mekong Press, pp. 9–36.

Molle, F. (2008), 'Nirvana concepts, narratives and policy models: insights from the water sector', *Water Alternatives*, **1**, 23–40.

Mollinga, P., R. Meinzen-Dick and D. Merrey (2007), 'Politics, plurality and problemsheds: a strategic approach for reform of agricultural water resources management', *Development Policy Review*, **25**, 699–719.

Pahl-Wostl, C., L. Lebel, C. Knieper and E. Nikitina (2012), 'From applying panaceas to mastering complexity: Towards adaptive water governance in basins', *Environmental Science and Policy*, **23**, 24–34.

Panya Consultants Company (2003), 'Integrated plan for water resources management in the Ping River Basin', *Final Report, Volume 2: Executive Summary*, Bangkok: Panya Consultants Co.

Pearson, R. (1999), 'A political economy analysis of the impact of agrarian change and urbanisation on communal irrigation systems in the Chiang Mai Valley, Northern Thailand', PhD, Sydney: Macquarie University.

Pollution Control Department (PCD) (2011), *Annual Report of the Office of Water Management 2011*, Bangkok: Pollution Control Department.

Sethaputra, S., S. Thanopanuwat, L. Kumpa and S. Pattanee (2001), 'Thailand's water vision: a case study', in L. Ti and T. Facon (eds), *From Vision to Action: A Synthesis of Experiences in Southeast Asia*, Bangkok: Food and Agricultural Organisation of the United Nations, United Nations Economic and Social Commission for Asia and the Pacific, pp. 71–98.

Surarerks, V. (1986), *Historical Development and Management of Irrigation Systems in Northern Thailand*, Chiang Mai: Department of Geography, Chiang Mai University.

Tan-kim-yong, U. (1983), *Resource Mobilization in Traditional Irrigation Systems of Northern Thailand: a Comparison between the Lowland and the Upland Communities*, New York: Cornell University.

Tanabe, S. (1981), *Peasant Farming in Thailand: A Comparative Study of Rice Cultivation and Agricultural Technology in Chiang Mai and Ayutthaya*, London: School of Oriental and African Studies, University of London.

Thangphet, S. (1989), 'Intensive, diversified cropping and irrigation management: a case study of the *fai tachumpa* system in Northern Thailand', unpublished MS thesis, Manila: Ateneo de Manila University.

Thomas, D.E. (2005), *Participatory Watershed Management in Upper Northern Region of Thailand*, Chiang Mai: ICRAF.

Thomas, D.E. (2006a), *Participatory Watershed Management in Ping Watershed: Final Report*, Bangkok: Office of Natural Resources and Environmental Policy and Planning, Ministry of Natural Resources and Environment, Thailand.

Thomas, D.E. (2006b), *Results Measurement Framework for Pilot Sub-basins, Bangkok: Office of Natural Resources and Environment Policy and Planning*, Bangkok: Ministry of Natural Resources and Environment, Thailand.

Warner, J.F. (2006), 'More sustainable participation? Multi-stakeholder platforms for integrated catchment management', *Water Resources Development*, **22**, 15–35.

Young, O.R. (2002), 'Institutional interplay: the environmental consequences of cross-scale interactions', in T. D.E. Ostrom, N. Dolsak, P.C. Stern, S. Stonich and E. Weber (eds), *The Drama of the Commons*, Washington, DC: National Academy Press, pp. 263–291.

13. The evolution of river basin management in the Murray–Darling Basin

Andrew Ross and Daniel Connell

13.1 INTRODUCTION

The Murray–Darling Basin (MDB) occupies 1 043 000 km² in south-eastern Australia. The MDB has a population of 2.1 million people and supplies about 40 per cent of the gross value of Australian agricultural production. The MDB spans four Australian states (Queensland, New South Wales, Victoria and South Australia) and the Australian Capital Territory. River basin organisation (RBO) in the MDB offers an example of federal multilevel governance (compare de Loë and Morris, Chapter 3 in this volume). The Australian Government, state governments and corporations, catchment management organisations and local governments are involved in basin management together with water users and other interests. Under Australia's federal system of government, the primary right to own or to control and use water is vested with the states and territories (Lucy 2008), and the states manage water allocation, planning and implementation. The Australian Government plays a major role in water policy development and financing.

The MDB is widely considered to be a success story in river basin management (Kemper et al. 2005). The MDB institutions have been put forward as a model for other countries. In many respects the management of the MDB has been a success story, but in other respects it has been less successful and new challenges are continuing to emerge.

Coordinated management of the Murray River began in 1915 with the establishment of the River Murray Agreement and the River Murray Commission. The Murray–Darling Basin Agreement, together with the Murray–Darling Basin Ministerial Council and the Murray–Darling Basin Commission, were established more recently, during the period between 1987 and 1992. The MDB Agreement has now been subsumed into the Australian Government Water Act 2007,[1] which provides for a

Murray–Darling Basin Plan to be prepared by a newly established Murray–Darling Basin Authority, which supersedes the MDB Commission.

This chapter focuses on the period since 1987 when the first interstate agreement on the MDB was signed.[2] In the terms of the typology of RBOs introduced in this book, the recent changes in MDB institutions and organisations can be broadly characterised as a shift from an interstate coordinating body to an Australian Government agency. However, the MDB states continue to play an important role in the development and implementation of basin policy and planning. The chapter is based on an analysis of official and academic documents coupled with previous research by the authors, which included interviews with government officials and other stakeholders.

The following analysis of the MDB management begins with a brief introduction to the MDB, and describes the evolution and reform of management institutions over the last 20 years. Then, groups and organisations which have promoted or resisted reforms are identified, together with the strategies that they have used. The performance of MDB institutions is analysed in terms of overall outcomes, coordination, accountability, legitimacy and financing, before the final section of the chapter draws conclusions.

13.2 CONTEXT, INSTITUTIONAL DESIGN AND DEVELOPMENT

13.2.1 Socio-Economic and Biophysical Context

The MDB boundary is defined by the catchment areas of the Murray and Darling rivers and their many tributaries, together comprising 23 major river valleys (Figure 13.1). The basin covers one-seventh of the Australian mainland, extending over 1 million km^2 of south-eastern Australia, and including three-quarters of New South Wales, more than half of Victoria, significant portions of Queensland and South Australia, and all of the Australian Capital Territory.[3]

The basin has a dry and highly variable climate, with annual average rainfall varying from under 200 mm in arid western regions to over 1000 mm in some eastern upland areas. It is estimated that the MDB receives an average annual rainfall of about 530 000 gigalitres (GL), most of which evaporates. Run-off is estimated to average 32 800 GL/year, and groundwater recharge is estimated to average 26 500 GL/year under natural conditions. Inflows to rivers have ranged from 118 000 GL in 1956 to less than 6740 GL in 2006.

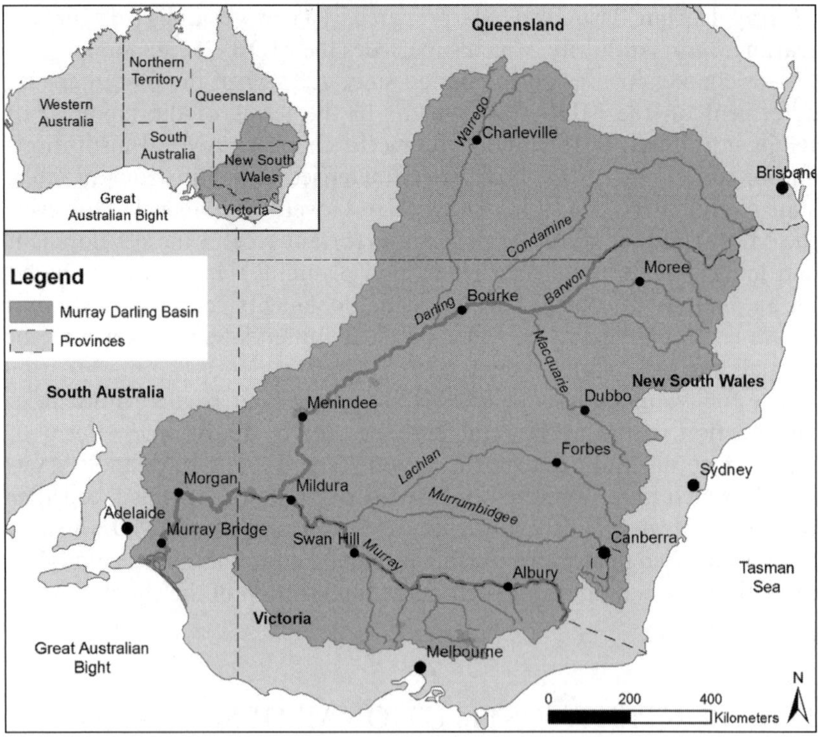

Source: Haisman (2004).

Figure 13.1 The Murray–Darling Basin

Current surface water diversion limits in the MDB total 10 940 GL, including irrigation, urban supplies, stock water, domestic supplies and industry. Groundwater diversion limits total 1790 GL.[4] The share of surface water and groundwater varies substantially between catchments. Groundwater supplies about 16 per cent across the MDB rising to 27 per cent in the Northern Basin (CSIRO 2008). It is estimated that an average of 2700 GL of run-off is intercepted by farm dams and forestry plantations (ABS 2008; MDBA 2010).

Water use in the basin has expanded substantially owing to growth in irrigated agriculture, from about 4000 GL a year in the mid-1950s to over 12 000 GL in the 1990s. However, it fell to an average of 8100 GL between 1999/2000 and 2008/2009, and 4600 GL in 2006/2007–2008/2009, owing to dry weather conditions during the last decade (MDBA 2010; MDBC various).

Substantial water storage and regulating structures have been built across the MDB to cope with the inter-annual variability of streamflow and to enable longer-term storage and re-release of water in drier years. The total public storage capacity in the basin's larger surface water storages (>10 GL) is approximately 22700 GL (MDBA 2010). However, storage falls to as low as 20 per cent of capacity during a major drought (ABS 2008). Assessments of future climate scenarios suggest that the extent and frequency of exceptionally hot years, and exceptionally low rainfall and low soil moisture years, are likely to increase (Henessey et al. 2008).

13.2.2 Policy and Institutional Development

The institutional arrangements in the MDB have evolved through three major stages (Blomquist et al. 2004):

1. The 1915 River Murray Agreement which allocated water flows of the Murray River and provided for the construction and operation of infrastructure on the River, benefitting the three River Murray States (South Australia, New South Wales, and Victoria).
2. The 1992 MDB Agreement which extended the scope and structure of the intergovernmental arrangements to the Darling River.[5]
3. Further development of integrated water resource management in the MDB culminating in the Australian Government's Water Act 2007 and the 2008 Intergovernmental Agreement on Murray–Darling Basin Reform.

Both the 1992 MDB Agreement and the 2008 Intergovernmental Agreement on MDB Reform were signed by the leaders of the basin jurisdictions. This represented a commitment by the jurisdictions to implement the Agreement, and to pass legislation or regulations required to give effect to the Agreement in their jurisdictions.

Until the late 1980s most of the emphasis in water management in the MDB was on rural community development: building infrastructure and water supply capacity, with an emphasis on surface water delivered at low cost through large highly regulated delivery systems. By the 1980s there were increasing concerns about pressures on water resources, water pollution and salinity (Smith 2001). In response a cap on surface water use in the MDB was established under the 1992 MDB Agreement. Available surface water is allocated to the MDB jurisdictions on the basis of the formula set out in the 1992 MDB Agreement but groundwater use was not capped and continued to increase (MDBC 2006).

During the 1990s and the following decade the traditional goal of supplying water for community development was supplemented by two additional goals. Firstly, there was a new emphasis on economic efficiency, and the establishment of water markets to allocate scarce water resources, embodied in the 1994 Council of Australian Governments Agreement on water reform. Secondly, there was a growing emphasis on environmentally sustainable development and integrated catchment management. The three goals of economic efficiency, environmental sustainability and community development were embodied in the 2004 National Water Initiative.[6] The National Water Initiative established an ambitious agenda for comprehensive, integrated and participative water reform in Australia (Connell et al. 2007).

The National Water Initiative addressed overuse and overallocation issues by requiring the comprehensive development of catchment-based water plans (surface and groundwater), restoration of overallocated (sub-) basins to 'environmentally sustainable levels', provision for the supply of environmental water and further development of water markets. These reforms were extended by the Water Act 2007, the 2008 Intergovernmental Agreement on MDB Reform and the preparation of a draft MDB Plan.[7]

13.3 THE CHANGING DESIGN OF MURRAY–DARLING BASIN MANAGEMENT INSTITUTIONS SINCE 1987

13.3.1 Introduction

MDB management since 1987 can be divided into two periods; from 1987–2006 during the development and implementation of the 1992 MDB Agreement, and from 2007 to the present following the Water Act 2007 and the 2008 Intergovernmental Agreement on MDB Reform.

In the following sections institutional arrangements during these two periods are discussed under the following headings:

- Authorities and accountability: the roles, responsibilities and mandate for action given to the MDB executive and administrative bodies, who they are accountable to and by what means.
- Aggregation: how the views of individual governments and other stakeholders are weighted and how collective choices and decisions relevant to MDB management are made. In a federal system with relatively complex multilevel governance, aggregation rules and procedures are particularly important.

Table 13.1 Major changes in Murray-Darling Basin management since 2007–2008 reform

Rule types	Before reform	After reform
Authority	Joint national and state powers to manage MDB water resources	Australian government authority to make and enforce the Basin Plan
Aggregation	Consensus decision making	Greater centralisation of powers with the Australian government
Boundaries	Physical boundaries clearly established. National and state governments participate in decision making; some consultation with water users and third parties	No substantial change
Information	Information sharing between jurisdictions and their representatives	Greater centralisation of information collection and provision with Australian government agencies
Financing	Intergovernmental cost sharing formulae	Greater proportion of funding from the Australian government

- Boundaries: which organisations and representative groups participate in MDB management and its implementation, and who else (for example, governments, water users, civil society) typically influences the decisions of these bodies.

Reference is also made to information collection and provision, financial arrangements and incentive mechanisms. Authorities, accountability and aggregation are discussed in sections 13.3.2 and 13.3.3, the participating organisations are discussed in section 13.4.

There were major reforms in MDB management institutions and rules in 2007–2008. The rules before and after these changes are summarised in Table 13.1 and discussed in the remainder of this section.

13.3.2 Basin Management under the 1992 Murray–Darling Basin Agreement

The purpose of the MDB Agreement was 'to promote and co-ordinate effective planning and management for the equitable, efficient and sustainable use of the water, land and other environmental resources of the Murray–Darling Basin' (Commonwealth Parliament 1993). The Agreement included details on the authorities and processes of the MDB Ministerial Council, Commission and Office, distribution of (available) water between the contracting states, investigation and monitoring, construction and maintenance of works. Each of the contracting governments agreed to provide for the execution and enforcement of the Agreement in their jurisdiction.[8]

Authorities

An RBO was established to coordinate activities and organisations in order to implement the Agreement. In the Australian federal system neither the Australian Government nor an RBO can have absolute authority because under the Australian Constitution the state governments (and their agencies) have plenary legislative power to govern resources in their state. Moreover, section 100 of the Constitution restricts the Australian Government from abridging the right of a state or residents therein to the reasonable use of waters or rivers for conservation or irrigation (Gardner et al. 2009).[9]

The river basin organisation included four parts: the Murray–Darling Basin Ministerial Council (the political or policy-making body), the MDB Commission and its Office (executive and administrative bodies), and the Community Advisory Committee (stakeholder consultative body); see Figure 13.2 (MDBC 2008).

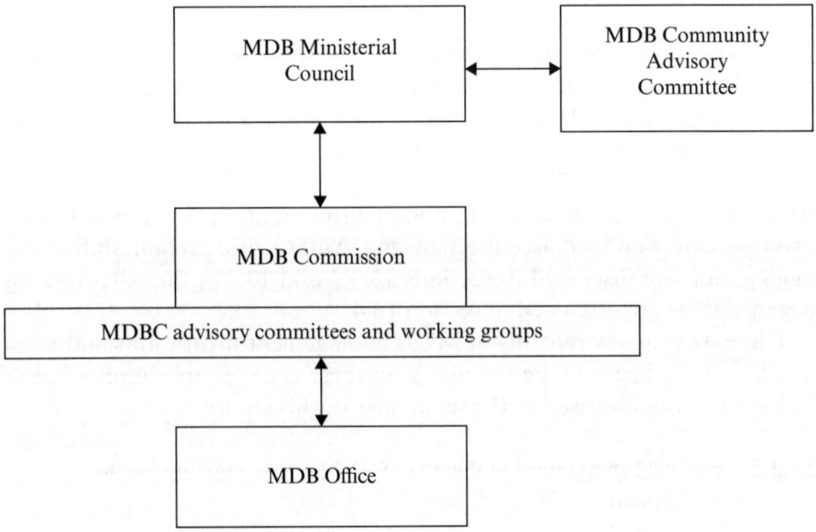

Figure 13.2 The Murray–Darling Basin organisation under the 1992 Murray–Darling Basin Agreement

The authority of MDB Ministerial Council under the MDB Agreement included:

1. Decisions on major policy issues of common interest to the contracting governments concerning effective planning and management for the equitable efficient and sustainable use of the water, land and other environmental resources of the MDB. Management and regulation of groundwater were not included in the Agreement.
2. Authorisation, where appropriate, of measures for the equitable, efficient and sustainable use of such water, land and other environmental resources. These measures included research on water quality, aquatic and riverine environments and the impacts of groundwater management on surface water (MDB Agreement, Section 9).[10]

The MDB Ministerial Council included up to three ministers from each of the state and Australian governments and a representative from the Australian Capital Territory (ministers responsible for water, land and the environment), and was chaired by one of the Australian Government ministers. The MDB Ministerial Council had a broad mandate to take decisions and develop measures concerning planning and management for the equitable efficient and sustainable use of water, land and other environmental resources of the MDB, to authorise related capital works and to agree on amendments to the Agreement. The Council took a consensus approach in order to achieve unanimous agreement on actions; participating governments each had an effective veto (MDB Agreement, section 12).

The MDB Commission was the executive body for developing, coordinating and implementing measures authorised by the Council. It also advised the MDB Ministerial Council on basin conditions and resource development, planning and management. Over time the Commission's activities expanded into the management of salinity and native fish, interstate water trade and water policy development. The Commission was an unincorporated joint venture of the MDB jurisdictions, chaired by an independent president and included representatives (commissioners) from each basin government. The Commission was advised by five high-level committees (River Murray Water, Natural Resource Management, the Living Murray, Commission Strategy, Audit and Compliance). The committees and associated working groups linked the policy development and on-ground project implementation of Council and Commission initiatives (Craik and Cleaver 2008). The MDBC worked with state governments to monitor the quantity and quality of the River Murray system and to assess and carry out works.

The MDB Commission was supported and advised by the MDB Office.

The Commission and the Office coordinated intergovernmental management of the MDB, worked with state authorities to construct and maintain infrastructure, and worked in partnership with other water programmes, organisations and research bodies. The Office also coordinated monitoring and research on the Murray–Darling Basin and commissioned important studies on salinity, environmental assets and groundwater as well as producing an annual report on the implementation of the cap and water use in the MDB.[11] The results of this monitoring and research were shared among participating jurisdictions. This work provided the basis for the development of MDB Commission programmes such as the Basin Salinity Strategy and the Living Murray, which operated under a federal state shared financing arrangement.

Accountability and enforcement

The MDB Commission and its Office were accountable to the MDB Ministerial Council, which was in turn responsible to participating governments. Council approval was required for works in excess of A$2 million, Commission approval was sufficient for works of less than A$2 million. The MDB Commission was accountable to the Council, and to each contracting government for the monies it received. It was required to report annually on its activities and the achievement of its objectives, policies and plans.

The MDB Ministerial Council and the MDB Commission were advised by a Community Advisory Committee that included representatives of local communities and stakeholder groups from throughout the basin. However, the MDB Ministerial Council was not bound to take account of input from the Community Advisory Committee. A MDB Commission commissioned assessment of community consultation procedures suggested that while the general public were usually satisfied with access to information on the management of the MDB, there was a lower level of satisfaction with opportunities to influence the MDB Commission's agenda and processes. People felt they were being told about decisions made previously rather than participating in decision making (Chenoweth et al. 2002). One assessment of the MDB Commission concluded that the Commission lacked experts from outside government, and community understanding of basin-wide natural resource management issues and decision-making processes was relatively poor (Blomquist et al. 2004).

The MDB Agreement is a contract, and the approving legislation of each contracting state provides authority to implement the Agreement in its territory, including by the authorisation of MDB Commission activities. However, the approving legislation did not express any obligations on citizens or government authorities to implement the decisions of the Council or the MDB Commission (Gardner et al. 2009).

Aggregation and outcomes

Council decisions could only be carried by the unanimous vote of all ministers present who constituted a quorum.[12] Decisions of the Commission could only be made by resolutions carried by a unanimous vote of the MDB Commission's president and all of the commissioners (two from each jurisdiction). However, advice by the Commission to the Ministerial Council could be determined on the basis of a majority rather than a unanimous vote.

Jurisdictions funded their own commissioners and jointly funded MDB Commission programmes, which were accompanied by ongoing and reliable measurement, development of scientific knowledge, and an integrated and participatory approach. The MDB Commission was empowered to appoint committees, to employ staff (including public servants from participating jurisdictions) and consultants. The MDB Commission acquired a high level of technical expertise from its employees, participating jurisdictions and the committee system (Blomquist et al. 2004; Craik and Cleaver 2008).

Under its enabling legislation the MDB Ministerial Council and the MDB Commission could deal with any issue, provided all members agreed. In practice that was rare, but some important cross-policy and cross-jurisdictional programmes were introduced. The culmination of the achievements of the 1980s pioneer generation of MDB policymakers was the cap on water extractions introduced in 1996. Implementation of the Basin Salinity Management Strategy and the Living Murray programme also required significant sacrifices by certain states. The educational programme Special Forever, which involved hundreds of primary schools, showed awareness of the fundamentally values-based nature of water reform. In the same way, the settlement negotiated with indigenous communities that ended the long-running Lake Victoria dispute in 2002 revealed a capacity to deal effectively with environmental justice issues.

While the MDB Ministerial Council's authorities gave it considerable potential for action, in practice the Council's capacity to act was severely limited for two reasons. Firstly, the MDB Ministerial Council members and MDB commissioners represented jurisdictional positions rather than acting in the best interests of the basin as a whole (Scanlon 2006). This made it difficult for them to take any action that harmed the interests of individual states, such as collective action to deal with the overallocation of water use entitlements.

Secondly, the capacity of the Council for decisive action was constrained by the requirement for unanimous voting to pass its resolutions. This requirement resulted in a cautious incremental approach by the Council. Hard decisions were avoided (Watson 2008).

Moreover, contracting states did not fully implement all the decisions made by the Council and the Commission, and the Australian Government lacked powers to remedy failures of action by the state governments (Gardner et al. 2009). There were continuing weaknesses in implementing the MDB Agreement and some provisions of the National Water Initiative, especially slow progress in returning over allocated catchments in the MDB to sustainable levels of allocation, and in ensuring the health of river and groundwater systems (NWC 2009).

These weaknesses coupled with the limitations of the MDB Commission's capacity to act quickly in response to drought conditions after 2002 led to proposals for reform including majority voting, greater openness in Council and Commission proceedings and an enhanced Community Advisory Committee (Connell 2007).

Financial arrangements

The costs of basin infrastructure works, monitoring and research and other collective action under the 1992 MDB Agreement were shared between participating jurisdictions under various cost-sharing formulae. In 1997 the MDB Ministerial Council agreed on a formula for sharing operating and maintenance costs between the states of 40 per cent New South Wales, 36 per cent Victoria and 24 per cent South Australia (Haisman 2004). Capital costs were shared equally between the Australian Government, New South Wales, Victoria and South Australia. In 2003 MDB jurisdictions agreed on new funding of A$500 million over five years to recover and manage water for environmental purposes. The Australian Government provided A$200 million with the remainder shared between the states (New South Wales and Victoria 38 per cent each, South Australia 22 per cent and the Australian Capital Territory 1 per cent).[13]

13.3.3 Basin Management under the Water Act 2007 and the 2008 Intergovernmental Agreement

In his speech introducing the 2007 Water Bill, Minister for Environment and Water Resources Malcolm Turnbull noted that water use in the MDB had increased from 2000 GL in the 1920s to over 10 000 GL at the end of the twentieth century. This increase in water use has underpinned massive agricultural development, but has also led to increased salinity, substantial degradation of rivers and wetlands, and losses of native fish.[14] In addition the provisions of the 1992 MDB Agreement proved insufficient to guarantee water supplies during severe drought conditions such as 2006–2007 when volumes in major basin water storages fell below 15 per cent.

Authorities, accountability and enforcement

The Intergovernmental Agreement on Murray–Darling Basin Reform[15] was signed on 3 July 2008 by the Australian Government, Queensland, New South Wales, Victoria, South Australia and the Australian Capital Territory. The 2008 Intergovernmental Agreement on MDB Reform embodied a new phase of Australian Government–state relations in the management of the MDB. The provisions of the 2008 Intergovernmental Agreement on MDB Reform are reflected in the Water Amendment Act 2008.[16] The Water Act 2007 and the 2008 Intergovernmental Agreement are included as schedules to the 2008 Act.

The most important new provisions in the 2008 Act and Agreement are the requirements for the preparation of a MDB Plan, the allocation of responsibility for the plan to the Australian Government, the establishment of the Commonwealth Environmental Water holder, and the extension of the role of the Australian Competition and Consumer Commission to regulate markets and charges for water in the basin. State governments will make plans for specific water resource areas consistent with the Basin Plan. The key objectives of these reforms are to improve the efficiency of water use and to return water sources to environmentally sustainable levels of extraction. The revised post-2007 RBO arrangements are summarised in Figure 13.3.

Under the Water Act 2007 the Basin Plan is required to address environmental outcomes, water quality and salinity, long-term sustainable diversion limits and trading in water access entitlements. It requires the identification of risks to the condition or continued availability of basin water resources including climate change, but does not specifically cover drought or flood management.

The Australian Government Minister (responsible for water) is

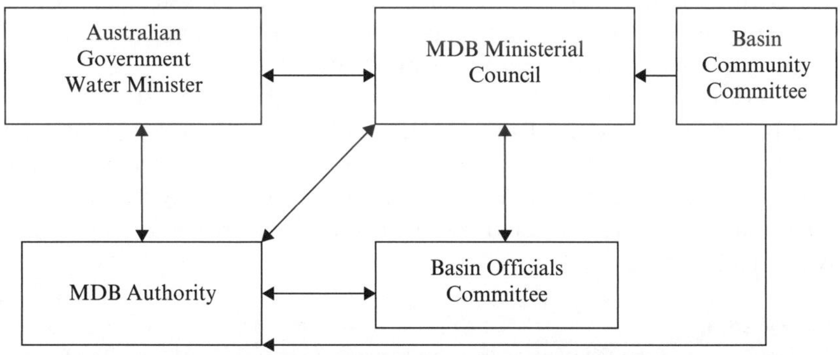

Figure 13.3 The Murray–Darling Basin organisation after 2007

responsible for deciding whether or not to adopt the MDB Plan, and for the MDB Authority. The Minister is accountable to the Parliament of Australia.

The MDB Ministerial Council, chaired by the Australian Government Minister and comprising one minister from each jurisdiction, advises the Australian Government Minister on the draft Basin Plan, and considers other major policy issues not addressed in the Basin Plan. The Council also approves the MDB's annual corporate plans and budgets and agrees on amendments to the Intergovernmental Agreement on MDB Reform. The MDB Ministerial Council receives advice from a basin officials committee, chaired by an appointee of the Australian Government, and including one representative of each basin government. The basin officials committee has a lesser role than the MDB Commission in the sense that it has a more limited role in deciding and implementing works and measures. The accountabilities of the Council and basin officials committee are similar to those of their predecessors.

The MDB Authority is responsible for the preparation, implementation and monitoring of the Basin Plan and for implementing decisions made by the Ministerial Council and officials committee. The Authority takes responsibility for the activities of the former MDB Commission, including the protection of water and environmental resources. The MDB Authority also manages River Murray operations and carries out investigations and monitoring responsibilities conferred by the MDB Agreement for which it is accountable to the MDB Ministerial Council and basin officials committee. The Authority is generally accountable to the Australian Government Minister, in particular for activities related to the Basin Plan.

The Commonwealth Environmental Water Holder is established by the Water Act 2007 to manage the water entitlements that the Australian Government is purchasing to use for environmental conservation. The Water Act requires that the Australian Government Minister responsible for water be given a report each financial year on the management of the Australian Government's environmental water.[17]

A Basin Community Committee with a chair and up to 16 other members provides advice to the MDB Authority and to the Ministerial Council. The Water Act 2007 (Part 8) sets out detailed mechanisms for the appropriate Australian Government authorities to enforce the Basin Plan. This represents a switch of enforcement powers from the states to the Australian Government.

Aggregation

The reconstituted MDB Ministerial Council operates on the basis of consensus. When an issue affects the interests of a contracting government, each minister representing that government is entitled to vote. A unanimous vote of those ministers is required to pass a resolution. The powers of the Council are less than the pre-2008 Council in two respects. First, the Council can only agree policy within the constraints defined by the Basin Plan, which has to be ultimately approved by the Australian Government Minister. The Minister has to take account of comments by the Council but does not have to implement them. Secondly, amendments to the Intergovernmental Agreement by the Council can only take effect on the registration of legislative instrument in Australian Government law. This may be disallowed by the Parliament of Australia.

The decision-making process in the basin management committee varies depending on the issue being addressed involving Queensland and/or the Australian Capital Territory. A quorum for a meeting is a member from each contracting government, if issues relevant to that government are being considered.

Information collection and provision

Two new sources supplement information collected by state governments and provided to the MDB Authority. The Water Act 2007 provides for the National Bureau of Meteorology to collect up-to-date, accurate and comprehensive water information across Australia. This includes a national water account and reports on resource use and availability. The Act also provides for the National Water Commission to undertake periodic audits of the Basin Plan and related state water resource plans and to report to the Minister.

Financial and market arrangements

The Australian Government has taken on major new financial responsibilities under the Intergovernmental Agreement on MDB Reform including for the purchase of water entitlements and financing state infrastructure projects. The government has purchased 1094 GL of entitlements at the cost of A$2.3 billion, and committed A$3.3 billion to state infrastructure improvement projects. The government has also committed an additional A$1.8 billion over the next ten years to acquire additional water and address water delivery constraints.[18]

The Australian Competition and Consumer Commission provides advice to the Commonwealth on water charge and market rules to apply within the MDB and is responsible for enforcing these rules. The rules ensure that charges are set efficiently and that participants in the water market act competitively. The Australian Competition and Consumer Commission determines bulk water charges.

The cooperative funding model for MDB Authority infrastructure, monitoring and research has continued under the post-2007 arrangements. In 2011 funding for the MDB Authority was sourced from the Australian Government (A$20 million) Victoria (A$31 million) New South Wales (A$32 million), South Australia (A$26 million) with small additional amounts from Queensland and the Australian Capital Territory.

In 2012 the long-standing cooperative financing arrangement started to unravel. New South Wales and South Australia announced large cuts in their financial contributions; New South Wales from A$32 million to A$9 million (over two years) and South Australia from A$26 million to A$12 million. Owing to these cuts the financial future of the RBO is uncertain.[19] If Victoria and the other basin jurisdictions follow suit the joint programme budget will be drastically cut, and the remaining funding may not be enough to maintain existing infrastructure.[20]

13.4 STAKEHOLDERS AND THEIR STRATEGIES TOWARDS REFORM

13.4.1 Introduction

The main stakeholders who have influenced the post 2007 MDB management reforms are Australian Government and state government ministries and departments; the MDB Ministerial Council, Commission and Office and their successor the Murray–Darling Basin Authority; water user groups (especially irrigators); and third parties including environmental groups. These stakeholders are heterogeneous groups with differing views on particular issues. Also the positions of some stakeholders, especially governments, have changed over time.

13.4.2 The Australian Government

In recent years both major parties in Australian federal politics have supported the Australian Government takeover of primary responsibility for MDB policy and planning. The process was begun by the Howard Liberal–National coalition government and Minister Malcolm Turnbull, and continued by the Rudd and Gillard Labour governments. Australian Government control of the basin has also been supported by the current leader of the opposition Tony Abbott.

In response to slow progress with aspects of the Council of Australian Governments 1994 water reforms and the National Water Initiative, the Howard government initiated the reforms in the Water Act 2007. The Australian Government sought to rely on the ‘referrals’ power in the

Constitution (s51, xxxvii), and to refer state powers over river management to the Australian Government in order to enact the Water Act 2007. The government was forced to change this approach and use a range of trade and commerce, corporations, external affairs and territories powers because of Victoria's initial opposition to the reform (Kildea and Williams 2011).

The Australian Government takeover was supported by the findings of enquiries by the Australian Senate and the Productivity Commission. In 2009 the Senate Rural and Regional Affairs and Transport References Committee Inquiry into 'Implications for the long term sustainability management of the Murray–Darling Basin System' (2009: xii) recommended that: 'The Australian Government work towards a full and unconditional referral of powers relevant to the management of the MDB and, in the absence of full referral, consider pursuing other options to provide for complete federal management'. In the report of its 2009 inquiry into Market Mechanisms for Recovering Water in the Murray Darling Basin, the Productivity Commission found that:

> The referral of powers to create the MDBA and to give it the power to develop and enforce a Basin-wide plan for the use of water resources should improve coordination in water planning across the Basin. This will principally be achieved by requiring that all entities manage their Basin water resources in accordance with water resource plans accredited under the Basin Plan. (Productivity Commission 2010: 208)

It is noteworthy that the Basin Plan focuses only on the use of water resources; powers to manage land and enforce property rights are separate, and remain with the states. This is an important decision that limits the scope of the RBO. Thus, water recovery and allocation actions by local environmental managers will need to be consistent with these plans. Cooperative arrangements will still be needed with other levels of government to manage all of the inputs required to achieve environmental outcomes.

The main strategies that the Australian Government has used to pursue its reform goals have been the referral of constitutional powers and consequent enabling legislation, as discussed above, together with financial incentives. The Australian Government's Water for the Future programme includes A$6 million to support improvements in water efficiency and A$3 million for water entitlement buybacks.[21] This strategy includes collaboration with state governments but has been criticised for not sufficiently engaging regional and local land and water management organisations and groups. It has destroyed goodwill and social capital built up over many years' work on collaborative natural resource management by trying to impose a plan developed in isolation by experts (Byron 2011).

13.4.3 State Governments

State government positions in relation to MDB reforms have been informed by self-interest, politics and pragmatism. The positions and strategies of state ministries and departments in relation to MDB reform have changed over time, but a consistent theme has been their concern with maintaining State autonomy and existing water allocation arrangements. South Australia, the downstream state in the MDB, has sought to maintain downstream flows while the other states have supported diversions, especially for irrigation. But these divisions between upstream and downstream states have been blurred by rivalries among the downstream states and the increasing role of the Australian Government including through financial assistance to the states.

Prior to the negotiation of the River Murray Waters Agreement, South Australia was mainly concerned about maintaining river flows for navigation, while Victoria and New South Wales were concerned that any Australian Government action to ensure river navigability could potentially supersede their interests in using water for irrigation (Kildea and Williams 2011).

During negotiations of the MDB cap, South Australia wished to maintain flows along the River Murray for both environmental and consumptive use reasons, and supported restrictions on consumptive use and increased allocations for the environment. New South Wales and Queensland objected to restrictions on consumptive use. New South Wales has the most severe overallocation problem among the Murray–Darling Basin states, and feared that it would have to introduce substantial rollbacks in entitlements to comply with a strict cap. Queensland has been a relative latecomer to irrigation development and did not want to restrict future development (Heinmiller 2007). One important factor in persuading the states to accept the MDB cap was that they wanted to receive favourable financial treatment from the Australian Government including Australian Competition and Consumer Commission payments and infrastructure investments.

In 2007 the Labour premiers of South Australia and Queensland both supported the proposal to refer basin management powers to the Australian Government because the agreement would unlock over A$10 billion worth of Australian Government funding for infrastructure and voluntary buyback while maintaining their access to water. New South Wales stood to benefit from the Australian Government taking responsibility for water buyback and helping to solve the chronic overallocation of water licences by previous New South Wales governments, but wanted assurances that the Australian Government could pay compensation to New South Wales irrigators who had to leave the industry.

Victoria was the only state to oppose Australian Government proposals. Victoria had two concerns. One was that the Australian Government would remove restrictions on interstate water trading that would result in a flow of entitlements from Victoria to New South Wales, and Victoria has obtained the right to veto any change in state shares under the Water Act. The other was that the Australian Government does not have sufficient skill and knowledge to get involved in state water management. Victoria finally agreed to support a modified Australian Government proposal on the basis that it had an extra five years until 2019 to bring its water plans into compliance with the new Basin Plan, and that the Australian Competition and Consumer Commission would oversee water markets.

State governments have found more common ground in relation to the proposed Basin Plan since the labour governments of Victoria, New South Wales and Queensland were replaced by liberal coalition governments in 2010, 2011 and 2012 respectively. These governments have adopted three common criticisms of the proposed plan:[22]

- The environmental benefits of the proposed plan are uncertain and their scientific basis has not been sufficiently demonstrated.
- The proposed plan will require substantial adjustment in some communities and it is not clear how this adjustment will be supported.
- There is no mechanism to ensure equitable impact on states and territories of downstream water requirements, and to take account of steps that states have already taken to improve water use efficiency and, in the case of Victoria, to return water to the environment.

Also the states argue that environmental watering requirements could be met by improved distribution and infrastructure, and improved water use efficiency rather than Australian Government purchase of water use entitlements. These arguments are contradicted by economic analysis which shows that the purchase of water entitlements is much more cost-effective than infrastructure investment (Grafton and Jiang 2010).

The division between the positions of the Australian Government and South Australia on one hand, and Queensland, New South Wales and Victoria on the other, is leading to relations between governments becoming increasingly strained. South Australia is challenging the constitutional validity of Victoria's trading on water entitlements, and New South Wales has imposed its own trading cap. South Australia has also threatened High Court action against the Basin Plan because the proposed additional 2750 GL allocation for environmental purposes is inadequate. Queensland has threatened to withdraw support for the referral of MDB management powers to the Australian Government if the Basin Plan is not amended in accordance with its requests.

These challenges and threats may partly reflect strategic positioning in the context of Basin Plan negotiations, but if they are realised they will be problematic not only because of the delay and expense involved in any litigation, but because it is not clear that litigation based upon the Australian Constitution is capable of producing outcomes consistent with the best management of Australia's river systems (Kildea and Williams 2011).

13.4.4 The Murray–Darling Basin Commission

The Commission and its Office coordinated intergovernmental management of the MDB under the direction of the MDB Ministerial Council. The Office was able to influence policy development by establishing and maintaining itself as a source of scientifically sound and neutral technical advice. The Office also administered a number of programmes to improve the MDB management, as discussed above.

The states were uncomfortable with some of the reforms and programmes promoted by the Commission but went along with them because of growing public demand for action to reverse continuing degradation. The Australian Government took the view that the Commission (that is, the Office) was usurping its role in the field of natural resource management and environmental restoration. The Government reduced the influence of the Commission by channelling new natural resource management programme money through the Natural Heritage Trust and the National Action Plan on Salinity and Water Quality directly to state agencies rather than through the Commission.[23]

13.4.5 Major Water Uses

The main water user in the MDB is the agriculture sector (85 per cent) followed by the water supply industry (13 per cent).[24] Development of agriculture in the Murray–Darling has been substantially assisted by the supply of cheap water through public infrastructure. Recently farmers have been granted individual transferable water use entitlements.

Irrigators are generally opposed to any changes in current arrangements that could result in any reduction in their water use entitlements or loss of access to cheap water. However, there are some divergences among irrigators. Producers with permanent plantings such as vines or orchards have favoured a more restrictive policy on the issue of water entitlements and in that sense shared some common ground with environmental interests; see below (Heinmiller 2007). Some crop farmers (cereals, rice) have been prepared to accept temporary reductions in access to water, providing that they can have full access to water entitlements in normal to wet years (Blomquist et al. 2004).

Farmer and irrigator groups are sceptical about the referral of MDB management powers to the Australian Government. In relation to a pledge by Tony Abbott, leader of the federal opposition, to take full control of the Murray–Darling river system, Laurie Arthur, chairman of the water committee for the National Farmers Federation, commented: 'The issues are so complex, and for Mr Abbott to come out and say that he can fix the river system by the Australian Government taking full control. Well, I think it's a bit simplistic' (Madden 2010). Danny O'Brian, chief executive officer (CEO) of the National Irrigators Council, said it does not support a full federal takeover because there was no evidence shifting management from one group of politicians and bureaucrats to another would improve the operation of the system (Madden 2010).

Agricultural and irrigator organisations have strongly lobbied governments to maintain their access to water, and have organised public meetings to press their case. Some meetings have achieved widespread media coverage, such as the consultation meeting in Griffith, when copies of the consultation draft guide to the Basin Plan were burned. More generally the influence of agricultural sector lobbying has been enhanced by the significant number of marginal seats in rural and peri-urban Australian Government and state electorates.

The largest cities in the MDB jurisdictions are on the coast, and outside the MDB catchment areas. Urban water suppliers and municipal authorities in the MDB have relied on consumer demand restrictions to manage water shortages when water storages fall to low levels (Productivity Commission 2008). Urban suppliers can accommodate reduced supplies providing this does not jeopardise their overall financial situation (Blomquist et al. 2004).

13.4.6 Environmental and Scientific Groups

Environmental groups have supported Australian Government intervention to maintain water allocations for the environment and to restrain further allocations for irrigation and other extractive uses. Environmental groups have formed an alliance with independent experts such as the Wentworth Group, and the Australian Government, in order to press their case for increased environmental allocations.

Environmental and scientific groups generally rely on media, lobbying and submissions to advance their views. For example, the Wentworth Group[25] statement on the draft MDB plan criticised the lack of specification of the amount of water required to deliver a healthy working river, the lack of modelling of the effects of increased groundwater extraction on surface water, and the lack of assessment of the ability of the plan to cope with prolonged severe drought. Following recent state elections, environmental and scientific groups are in a weaker position, and the Wentworth Group's comments on the draft Basin Plan have not been fully addressed in subsequent drafts.

13.5 PERFORMANCE OF THE PRE- AND POST-2007 MANAGEMENT ARRANGEMENTS

13.5.1 Core Governance Approach and Culture

The MDB initiative introduced in the mid-1980s and later codified in the 1992 MDB Agreement (and in parallel legislation in the MDB states) was fundamentally based on the concept of comity. The Water Act 2007 and the 2008 Intergovernmental Agreement on MDB Reform introduced a very substantial change in the philosophy shaping integrated water resource management in the MDB. This is in essence a shift to a contractual or purchaser provider model.

Comity is a long-established concept in international law whereby all parties agree to work for mutually beneficial solutions and avoid beggar-your-neighbour or zero-sum strategies (Wanna et al. 2009). At the core of the MDB initiative there was a belief that cultivating trust between the many different stakeholders was the key to progress with water reform. This led to substantial progress on some difficult policy issues despite frequent and intense conflicts between jurisdictions and competing interests.

In order to encourage participants to work together through persuasion rather than to compete, all decisions in the first MDB Ministerial Council had to be unanimous. Subsequent history exposed both the merits and limitations with this approach. Coordination was relatively comprehensive across sectors and levels of government, but some contentious issues were not dealt with. On the one hand there have been significant achievements in cooperative action such as the generally effective implementation of the surface water cap, the salinity and drainage strategy, the Living Murray and community education programmes. On the other hand there was little progress on issues where one or more states resisted change, such as a reduction of overallocated entitlements or cross-border water marketing.

Perceptions of the MDB initiative are unavoidably coloured by the situation during its final years in which policy and cross-border management were in gridlock and apparently unable to deal with ongoing disputes about institutional reform during the worst drought in a century. This policy failure, together with political opportunism, led the Australian Government to promote an Australian Government-led basin governance model.

The purchaser–provider model at the heart of the 2008 water reform package is very different and more coercive than the previous approach. Under the 2008 legislation the Australian Government is developing an overarching Basin Plan that in principle can take account of all issues of importance to water management. Then, if a state fails or refuses to develop a satisfactory implementation plan, the Australian Government is empowered to act in lieu of the state. The Australian Government has the required financial leverage to push through reforms. In addition to the legal powers of enforcement contained in the Act there is the inducement to cooperation provided by the A$12 billion Water for the Future fund which can be accessed by complying states. The Fund finances improvements in water efficiency and buyback of water entitlements where they have been overallocated by management authorities.

The purchaser–provider model creates a relationship where the purchaser is contractually dependent on the provider and vice versa. The contracting purchaser, the Australian Government, can reject what is provided, but this may be difficult when urgent or continued service provision is at stake. The contracted providers, state governments, may have the option to reject the offered deal but this can be difficult when the deal comes with attractive financial incentives, coupled with the threat of further Australian Government intervention in their affairs if they do not agree. As a result the states tend to express agreement in principle with the new system, and work to defend or promote their interests in other ways. Their current reactions are in some ways like travellers taking shelter from a passing storm: it remains to be seen how much fundamental change will occur, and whether the results of the new system will be very similar to the old.

13.5.2 Coordination

Planning and management in the MDB faces both horizontal and vertical coordination challenges arising from the horizontal separation of policy sectors at each level of government and the vertical separation of policy and administrative functions between different levels of government. Coordination can be further broken down into tasks of varying levels of difficulty ranging from communication between sectors and governments through to joint plans and strategies (Ross 2008). Effective coordination can be defined as a situation in which the policies and programmes of governments are characterised by minimal redundancy, incoherence and gaps (Peters 1998).

The complicated multilevel process for policy development, policy coordination and decision making in the MDB is summarised in the following assessment:

> Policy development [in the MDB] now involves complicated interactions between a large number of individuals, groups, organizations and institutions including governments. The Australian Government and the States are focal points around which contending interests arrange themselves, moving from one to the other as their members make strategic decisions about alliances and how to best promote their goals or block those of others. In practice, decisions are not made through a top-down process but are the product of complex cycles of interaction in which the participants have varying degrees of influence but no single one is dominant. (Connell 2007: 180)

The Murray–Darling Commission achieved coordination in some areas such as management of the River Murray. In some areas such as basin salinity management, the Commission achieved some positive results but did not achieve overall policy coordination. In some areas such as the overallocation of water entitlements the Commission did not make progress. It remains to be seen whether stronger top-down coordination led by the Australian Government Minister will be able to achieve more effective coordination.

The Australian Government adopted a more coercive approach to coordination in response to the gridlock among the states over sustainable diversion levels and the need to deal with overallocated resources. It remains to be seen whether the top-down approach can achieve greater horizontal or vertical coordination. For example, both irrigators and environmentalists have strongly criticised the draft Basin Plan, and the states have canvassed options for resisting the current draft, ranging from delayed implementation to possible legal action against the Basin Plan. Certainly effective cooperation between the Australian and state governments will be required in order to implement the plan.

It is going to be difficult to implement the Basin Plan, with its top-down planning process and rigid requirements for state coordination, in the turbulent pluralist political environment of the MDB. However, the likelihood of the Australian Government acquiring sufficient water entitlements to achieve by itself most of the environmental targets of the Basin Plan renders the timetable for developing the state sub-plans to implement the Basin Plan less essential. When the Commonwealth Environmental Water Holder's purchasing programme is complete, it will hold about a quarter of all water entitlements in the MDB, and will be largely unfettered by legal constraints beyond the requirements of the Environmental Watering Plan to be developed by the MDB Authority in consultation with the states. It is likely therefore that in coming years the Commonwealth Environmental Water Holder will emerge as one of the most important water management institutions in the MDB, working with representatives of other jurisdictions and other policy sectors.

13.5.3 Accountability

Under the MDB Initiative, the Australian Government and the states worked together to develop policy and were less able to disclaim responsibility. The Ministerial Council introduced audits for the cap, the Basin Salinity Management Strategy and the environmental condition of the rivers. These activities complemented the auditing processes undertaken to help implement the 1994 Council of Australian Governments water reforms as part of National Competition Policy. The Australian Government supplied the bulk of project funds to a variety of recipients but usually relied on indirect processes of accountability to influence implementation. The MDB Commission and its Office provided a reasonably good balance of discretionary powers and accountability, but accountability was somewhat fragmented across different programmes and levels of government.

Under the new governance framework objectives, performance requirements and monitoring and evaluation processes are spelt out in some detail in the Basin Plan legislation. The imposition of a technocratic solution acceptable at higher levels of government comes at the cost of less accountability to regional and local levels, and risks increasing the 'implementation deficit'. Perhaps reflecting a sense of new-found policy primacy during the early years of the current system there was reduced effort by the Australian Government to work with the states. As a result the various drafts of the Basin Plan have been strongly contested by the states and powerful stakeholders such as the irrigation industry and environmental interests.

Moreover, the strong legislative framework does not remove the potential for the Australian Government to blame future failure on ineffective implementation by the states. Legislated requirements for the Basin Plan performance to be assessed and reported on by the MDB Authority, and compliance to be audited by the National Water Initiative, may increase the likelihood of the Australian Government blaming the states for any implementation deficit. The states in turn will probably claim that the cause is bad policy and legislation developed by the Australian Government.

In response to the new approach encompassed in the Basin Plan some of the states, notably New South Wales, have adopted a minimalist strategy whereby they commit to do what is funded by the Australian Government but not much more. In addition the power of the Australian Government to pressure uncooperative states is proving rather illusionary in that all governments are elected by the same voters. Any Australian Government that 'punished' a state is likely to suffer losses in the following federal election.

13.5.4 Legitimacy

The high-level coercive approach may have produced a more comprehensive and integrated Basin Plan, but this has come at the cost of reduced (input) legitimacy at regional and local levels. In the current Australian Government there appears to be growing disenchantment with the benefits of a strong purchaser–provider model. Its ministers are increasingly working to establish mechanisms that promote cooperation, rather along the lines of the old MDB initiative.

These are still very early days for the Basin Plan. Full implementation is not due to commence until 2019 and even then there will be a five year phase-in. It is unclear whether output legitimacy can be achieved under the current model. If output legitimacy means balancing productive and environmental goals of basin management, it is unclear whether long-term environmental goals will be achieved under the new Basin Plan. The implementation of the plan will certainly require a restoration or establishment of trust between key stakeholders across sectors and at different levels of implementation. It remains to be seen how this will be achieved.

13.5.5 Financing

Since 1915 MDB jurisdictions have managed basin assets and programmes on a cooperative basis. This arrangement has succeeded because the basin jurisdictions have agreed to share management costs. It is unclear whether this successful financing model will continue, following large funding cuts by New South Wales and South Australia.

13.5.6 Impact of Institutional Changes on Effectiveness of the River Basin Organisation

It is quite likely that the Basin Plan will not be implemented as envisaged in the Water Act 2007. Implementation will depend on a very high degree of support from the basin states which the Australian Government could have difficulty attracting, given the political strength of irrigation communities. The Basin Plan is to be implemented through sub-plans developed by each of these states, but there has been a long history of reluctant cooperation by the Australian states with such national water reform programmes (Connell 2007). This pattern does not appear to be changing and it is hard to see why it will change.

Given these potential problems with implementation, the Basin Plan is unlikely to end up as the centrepiece of the MDB's high-level policy and management system in the medium term. However, the Water Act contains within it the potential for a set of arrangements that would focus on the Commonwealth Environmental Water Holder. The Australian Government is likely to acquire sufficient water entitlements under its direct control, purchased under the Water for the Future programme and lodged with the Commonwealth Environmental Water Holder, to be able to achieve most of the environmental targets of the Basin Plan even if the States withhold their support. The Commonwealth Environmental Water Holder has a A$3 billion budget, sufficient to acquire up to a quarter of all entitlements and is now more than halfway to that target. The Commonwealth Environmental Water Holder has the potential to play a major role in securing and coordinating environmental water supplies leading to the maintenance and improvement of the condition of water-based ecosystems.

13.6 CONCLUSIONS

Australian experience suggests that there is no ideal set of structures, strategies and processes capable of driving integrated river basin management. Leadership and cultural change have key roles to play, and persistence, evaluation and feedback are more important than the precise model adopted.

Over the last 25 years some promising results have been achieved through integrated catchment management, separation of land and water title, and development of water markets, but the effectiveness of new initiatives has been reduced because initiatives are overtaken or abandoned before they have had time to prove themselves or before they have been properly evaluated (Ross and Dovers 2008). It remains to be seen how much fundamental change will occur, and whether the results of the new system will be very similar to the old.

In the case of the MDB it remains unclear how much central steering and what incentives are needed to bring about effective coordination, accountability and legitimacy. Now the Australian Government has taken power over MDB governance, but the Australian Government cannot implement the new Basin Plan on its own, it requires cooperation from the states. The structures of RBO can only provide part of the solution for river basin governance; much depends on relationships between key stakeholders, the effectiveness of multi-scale coordination, the inclusiveness of participation and the flexibility of institutions in response to changing conditions and knowledge.

If implementation of the Basin Plan process is blocked by political, financial and legal disputes and the Commonwealth Environmental Water Holder becomes a significant organisation, there may be further evolution of the water governance system in the MDB.

NOTES

1. The MDB agreement is schedule 1 of the Water Act 2007.
2. In 1985, South Australia, New South Wales and Victoria and the Australian Government began negotiations on a new Murray–Darling Basin Agreement. In 1987 they signed an initial version of the Agreement as an amendment to the River Murray Waters Agreement. In 1988 the MDB Commission was established. In 1992 the above states signed a new Murray–Darling Basin Agreement.
3. Available at http://www.mdba.gov.au/explore-the-basin/maps (accessed 25 July 2012).
4. Diversion limits comprise plan limits for areas with accredited plans and estimated current use where there is no accredited plan.
5. In 1992, New South Wales, Victoria and Queensland signed the Agreement; Queensland signed in 1996 and the Australian Capital Territory formalised its participation in 1998 through a Memorandum of Understanding.
6. Available at http://nwc.gov.au/__data/assets/pdf_file/0008/24749/Intergovernmental-Agreement-on-a-national-water-initiative.pdf (accessed 13 September 2013).
7. These developments paralleled and gave practical effect to the international concept of integrated water management.
8. Available at http://www.austlii.edu.au/au/legis/nsw/consol_act/mba1992209/ (accessed 25 October 2013).
9. Gardner argues that Australian Government law which promotes an environmentally sustainable use of river water would not violate this provision, since this would be intended to provide for the reasonable use of river water.
10. See paragraph 9 of Murray–Darling Basin Agreement 1992, available at http://www.austlii.edu.au/au/legis/nsw/consol_act/mba1992209/ (accessed 6 August 2012).
11. Available at http://www.mdba.gov.au/what-we-do/managing-rivers/the-cap/cap-archives (accessed 25 October 2013).
12. Each jurisdiction nominated one minister to vote. The votes of each nominated minister were required to constitute a quorum.
13. Available at http://archive.coag.gov.au/coag_meeting_outcomes/2004–06–25/index.cfm#nwi (accessed 1 March 2013).
14. Available at http://www.openaustralia.org/debates/?id=2007–08–08.12.1 (accessed 6 August 2012).
15. Available at http://www.coag.gov.au/node/277 (accessed 7 August 2012).
16. Available at http://www.comlaw.gov.au/Details/C2008B00243 (accessed 7 August 2012). The Water Amendment Act 2008 includes the Water Act 2007 (as a schedule),

and a few amendments to reflect subsequent additions in the Intergovernmental Agreement on MDB Reform.

17. Available at http://www.environment.gov.au/about/publications/annual-report/10–11/index.html (accessed 13 September 2013).
18. Available at http://www.environment.gov.au/water/basin-plan/faq.html (accessed 28 February 2013).
19. Available at http://www.theaustralian.com.au/national-affairs/state-politics/burke-stunned-by-murray-funds-cut/story-e6frgczx-1226541430053 (accessed 18 February 2013).
20. Available at http://www.mdba.gov.au/media_centre/media_releases/Statement-from-MDBA-Chair (accessed 18 February 2013).
21. Available at http://www.climatechange.gov.au/~/media/Files/minister/previous per cent20minister/wong/2008/major-speeches/April/sp20080429.pdf (accessed 7 August 2012).
22. Available at http://www.mdba.gov.au/what-we-do/basin-plan/development/steps-in-the-development-of-the-basin-plan (accessed 25 October 2013).
23. Cullen, P. (2004), interviewer D. Connell, oral history collection Australian National Library, Canberra.
24. Available at http://www.abs.gov.au/ausstats/abs@.nsf/mf/4610.0.55.007 (accessed 8 August 2012).
25. The Wentworth Group of Concerned Scientists is an independent group comprising leading Australian scientists, economists and business people with conservation interests. Available at http://www.wentworthgroup.org/about-us (accessed 19 September 2012).

REFERENCES

Australian Bureau of Statistics (ABS) (2008), *Water and the Murray–Darling Basin – A Statistical Profile, 2000–01 to 2005–06*, Canberra: ABS.

Blomquist, W., A. Dinar and K. Kemper (2004), 'Comparison of institutional arrangements for river basin management in eight basins', Washington Research Working Paper 3636, Washington, DC: World Bank.

Byron, N. (2011), 'What can the Murray Darling Basin Plan achieve? Will it be enough?', in D. Connell. and R. Quentin Grafton (eds), *Basin Futures*, Canberra: ANU Press, pp. 385–398.

Chenoweth, J., S. Ewing and J. Bird (2002), 'Procedures for ensuring community involvement in multijurisdictional river basins: a comparison of the Murray–Darling and Mekong River Basins', *Environmental Management*, **29** (4), 497–509.

Commonwealth Parliament (1993), *Murray–Darling Basin Agreement (1992) Act 38*, Canberra: Commonwealth Parliament.

Connell, D. (2007), *Water Politics in the Murray Darling Basin*, Annandale: Federation Press.

Connell, D., L. Robins and S. Dovers (2007), 'Delivering the national water initiative: institutional roles, responsibilities and capacities', in K. Hussey and S. Dovers (eds), *Managing Water for Australia: The Social and Institutional Challenges*, Canberra: CSIRO, pp. 127–140.

Craik, W. and J. Cleaver (2008), *Institutions and Incentives for Promoting Better Policy Outcomes*, Canberra: Murray–Darling Basin Commission.

CSIRO (2008), 'Water availability in the Murray–Darling Basin: a report from CSIRO to the Australian Government', Canberra: CSIRO.

Gardner, A., R. Bartlett and J. Gray (2009), *Water Resources Law*, Chatswood, NSW: LexisNexus.

Grafton, R.Q. and Q. Jiang (2010), 'Economics of drought, water diversions, water recovery and climate change in the Murray–Darling Basin', Environment and Policy (CWEEP) Research Paper 10–01, Canberra: Centre for Water Economics.

Haisman, B. (2004), 'Murray–Darling Basin, Australia, case study background paper', available at http://siteresources.worldbank.org/INTSAREGTOPWATRES/Resources/Australia_MurrayDarlingFINAL.pdf (accessed 8 July 2013).

Heinmiller, T. (2007), 'The politics of cap and trade policies', *Natural Resources Journal*, **47**, 445–467.

Hennessy, K., R. Fawcett, D. Kirono, F. Mpelasoka, D. Jones, J. Bathols, P. Whetton, M. Stafford Smith, M. Howden, C. Mitchell and N. Plumme (2008), *An Assessment of the Impact of Climate Change on the Nature and Frequency of Exceptional Climatic Events*, Canberra and Melbourne: CSIRO and the Australian Bureau of Meteorology.

Kemper, K., A. Dinar and W. Blomquist (2005), *Institutional and Policy Analysis of River Basin Management Decentralisation*, Washington, DC: World Bank.

Kildea, P. and G. Williams (2011), 'The constitution and the management of water in Australia's rivers', *Sydney Law Review*, **32**, 595–616.

Lucy, J. (2008), *Water Regulation: The Laws of Australia*, Sydney: Thomson Reuters Australia.

Madden, J. (2010), 'Abbott mauled for talk of Murray–Darling takeover', *Australian*, 16 January, available at http://www.theaustralian.com.au/news/nation/abbott-mauled-for-talk-of-Murray-Darling-takeover/story-e6frg6nf-1225820177750 (accessed 9 September 2013).

Murray–Darling Basin Authority (MDBA) (2010), *Guide to the Proposed Basin Plan: Volume 1: Overview*, Canberra: MDBA.

Murray–Darling Basin Commission (MDBC) (1996–2008, various years), *Water Audit Monitoring Report*, Canberra: MDBC.

Murray–Darling Basin Commission (MDBC) (2006), *Murray–Darling Basin Agreement*, Canberra: MDBC.

Murray–Darling Basin Commission (MDBC) (2008), *Review of Implementation: Report of the Independent Audit Group 2006–07*, Canberra: MDBC.

National Water Commission (NWC) (2009), *Australian Water Reform 2009: Second Biennial Assessment of Progress in Implementation of the National Water Initiative*, Canberra: National Water Commission.

Peters, G. (1998), 'Managing horizontal government: the politics of coordination', Research Paper Number 21, Ottawa: Canadian Centre for Management Development.

Productivity Commission (2008), 'Towards urban water reform: a discussion paper', Melbourne: Productivity Commission.

Productivity Commission (2010), *Market Mechanisms for Recovering Water in the Murray Darling Basin*, Melbourne: Productivity Commission.

Ross, A. (2008), 'Australia', in A. Jordan and A. Lenschow (eds), *Innovation in Environment Policy: Integrating the Environment for Sustainability*, Cheltenham, UK and Northampton, MA, USA: Edward Elgar, pp. 289–310.

Ross, A. and S. Dovers (2008), 'Making the harder yards: environmental policy integration in Australia', *Australian Journal of Public Administration*, **67** (3), 245–60.

Scanlon, J. (2006), '100 years of negotiation with no end in sight – where is the Murray Darling Basin Initiative leading us?', Keynote Address, Environment Institute of Australia and New Zealand Conference, Adelaide.

Senate Rural and Regional Affairs and Transport References Committee (2009), 'Implications for the long term sustainability management of the Murray–Darling Basin System', 25 June, Australian Parliament.

Smith, D. (2001), *Water in Australia*, Melbourne: Oxford University Press.

Wanna, J., J. Phillimore, A. Fenna and J. Harwood (2009), 'Common cause: strengthening Australia's cooperative federalism', Final Report to the Council for the Australian Federation, Melbourne.

Watson, A. (2008), 'The Howard–Turnbull National Plan for Water Security of January 2007: rescue or rejection?', *Economic Papers*, **27** (2), 146–159.

14. Institutional design, politics and performance of river basin organisations

Sander Meijerink and Dave Huitema

14.1 INTRODUCTION

The founding of river basin organisations (RBOs) is an important prescription in modern water resources management. The core idea is that institutions for managing water resources need to fit the scale at which many water problems manifest themselves, which is the scale of a hydrological (sub-)basin. As was explained in the introductory Chapter 1 in this volume, the foundation of RBOs may be conceived as a shift in governance from the nation state to other jurisdictional levels, independent agencies, markets and/or to civil society. One of our objectives in collecting 11 in-depth case studies of the institutional design and performance of RBOs was to better understand these shifts in governance in river basin management. In addition, we wanted to learn more about the relationship between the institutional design and performance of RBOs, and to formulate recommendations for those who are involved in processes of RBO formation or reform. In this final chapter we want to take up the research questions, which we formulated in the introductory chapter:

1. What are institutional design characteristics of river basin organisations and how were their geographical boundaries chosen?
2. Who supported and who resisted the founding or reform of river basin organisations?
3. How well do the river basin organisations function in a democratic sense and in terms of realising coordination and environmental effectiveness, and to what degree can this performance be attributed to the institutional design of these organisations and their geographical delineation?
4. What are the implications of these findings for those who are involved in institutional design or reform processes in the water sector?

In Chapter 1 we used Elinor Ostrom's institutional analyses and development (IAD) framework to develop an ideal typology of river basin organisation, distinguishing between autonomous, agency, coordinating and partnership-type RBOs. For each ideal type we specified the corresponding authority, aggregation, boundary, information and pay-off rules. The chapter authors have all used these rule types to describe their respective cases of RBO formation or reform, and to characterise the RBOs, which they have studied. In section 14.2, we will use the results of their analyses to explore the relationship between the establishment of RBOs and various shifts in governance which were discussed in Chapters 1 and 2: the devolution of power from the state to regional and local governments (decentralisation, devolution), shifts in responsibilities from the government to non-governmental organisations (participatory and self-governance) and the private sector (privatisation), and from the state to independent bodies (agencification). Do the examples of RBO formation and reform discussed in this book provide evidence for either one or more of these shifts in governance or does the national government continue to dominate the water policy domain in these cases? In addition, we discuss the usefulness of our typology in specifying and analysing institutional similarities and differences between the RBOs.

In section 14.3 we will discuss the triggers for and the main actors propagating or resisting RBO formation and reform. In this section we will test our analytical claim that the establishment and reform of RBOs is a highly political process in which parties with different interests try to promote or hinder the establishment of RBOs or to influence their institutional design. Section 14.4 continues with a discussion on the performance of the RBOs studied. We will compare their democratic performance, their contribution to realising coordination and their environmental effectiveness. Finally, section 14.5 addresses the question of the extent to which the performance of RBOs may be attributed to their institutional design, and, based on these findings, presents some recommendations for those who are involved in (re)designing RBOs.

14.2 INSTITUTIONAL DESIGN CHARACTERISTICS OF RIVER BASIN ORGANISATIONS

14.2.1 Applying the Ideal Typology

Table 14.1 gives an overview of the countries and RBOs studied and a characterisation of these RBOs with the use of the ideal typology, which we presented in Chapter 1. As is to be expected with ideal types, none

Table 14.1 Characterisation of the RBOs

Country	Chapter	Name of RBO	Type of RBO
Canada	3	Mackenzie River Basin Board	Coordinating
United States	4	Oregon Watershed Enhancement Board	Agency
UK	5	Westcountry Rivers Trust	Partnership
Germany	6	Erftverband	Agency
Portugal	7	River Basin District Authorities	Autonomous
South Africa	8	Breede–Overberg Catchment Management Agency	Agency
Ukraine	9	Western Bug River Basin Administration	Agency
		Western Bug River Basin Council	Coordinating
Afghanistan	10	Lower Kunduz and Taloquan River Basin Agencies	Agency
		Lower Kunduz and Taloquan River Basin Councils	Coordinating
Mongolia	11	River Basin Councils	Coordinating
		River Basin Administrations	Agency
Thailand	12	Ping River Basin Committee	Coordinating
		Mae Kuang Sub-basin Working Group	Coordinating
Australia	13	Murray–Darling Basin Authority (MDBA)	Agency
		Murray-Darling Basin Ministerial Council (MDBMC)	Coordinating
		Basin Community Committee	Coordinating

of the RBOs discussed in this book was exactly identical to the types we had identified. To mention just a few examples: the Oregon Watershed Enhancement Board is an agency-type RBO as it was given a specific mandate by the state of Oregon, but it still shares some characteristics with both the coordinating and partnership-type RBOs. The German case study of the Erftverband is an example of an agency-type RBO but, because of its fairly independent position and far-reaching competencies, it also shares some characteristics with the ideal-type autonomous RBO. The same goes for most other RBOs which have been analysed in this book. Indeed, almost all authors demonstrate how 'their' RBO shows the most similarity to one ideal type, but is different in some respects.

Although the case studies presented in this book are far from a random sample of the total population of RBOs – instead cases were chosen that authors were already familiar with – it is worth reflecting on the types of RBOs that have been discussed in this collection. Table 14.1 provides an overview.

As can be seen, agency and coordinating-type RBOs (with some characteristics of either one or more of the other types) are well represented, but only one example of an autonomous RBO was presented. The limited number of autonomous RBOs, which are characterised by their independent position vis-à-vis other government organisations, is perhaps not surprising as shifts in governance in the direction of this type are rare (Schlager and Blomquist 2008), despite the fact that the academic literature on RBOs shows a clear preference for it (Huitema et al. 2009). A second observation is that in quite a number of cases we found more than one RBO in the same basin; often the presence of more than one RBO takes the form of a combination of an agency and a coordinating-type RBO. In these cases the agency is given a specific mandate by the national government, and the coordinating agency aims to broker between subnational governments, policy sectors and/or different societal interests. Following the account by Jaspers and Gupta (Chapter 2), we can conclude that such institutional complexity or layering is not unique for the cases that we have included in this book. They refer to the 'increasing density of the organisational framework that operates at multiple levels of governance and has multiple influences from different social actors'. Although the typology has proven useful for a basic characterisation of the RBOs, and to position them in the institutional landscape, the case studies display a rich institutional variety. Each RBO has unique institutional design features, and understanding these features is essential for understanding their functioning and effectiveness.

14.2.2 Comparing Rule Types

Introduction

For a fine-grained institutional analysis we need to learn more about the five rule types, which we derived from Ostrom's IAD framework. Table 14.2 summarises the authority, aggregation, boundary, information and pay-off rules for each RBO.

Authority rules

In comparing the authority rules of the RBOs we can observe that a distinction can be made between RBOs with a narrow and a broad mandate. The Oregon Watershed Enhancement Board, for instance, was founded only for the restoration of salmon habitat (Lach and Calvert, Chapter 4). But the German Erftverband has a broad mandate which

Table 14.2 Summary of authority, aggregation, boundary, information and pay-off rules

Name of RBO	Authority rules	Aggregation rules	Boundary rules	Information rules	Pay-off rules
Mackenzie River Basin Board	Delegated mandate focuses on coordination, facilitation, dispute resolution and information gathering and sharing. Pertains to ecological integrity and sustainability.	Decisions require the consent of two-thirds of the members of the Board.	Geographical boundaries based on hydrological criteria. Membership is limited to representatives of governments, and to Aboriginal peoples.	Emphasis on science and traditional knowledge. Meeting minutes not public.	Budget provided by the 'basin jurisdictions'.
Oregon Watershed Enhancement Board	Delegated mandate for salmon restoration. Coordinating and non-regulatory roles. Mobilising resources and managing grants. Reporting of results.	Majority voting in the appointed Board.	Partnerships at the watershed level (all within Oregon). Board members represent different geographical areas of the state.	Science and experiential knowledge valued. Progress reports public.	Budget provided by state lottery money and specific federal grants.

Westcountry Rivers Trust	No formal authorities delegated. Aimed at leveraging resources, building collaborative relations, and fulfilling advisory and brokerage roles.	Consensus.	Boundaries defined on the basis of hydrological criteria, and water quality and habitat issues. Voluntary and locally inclusive participation.	Scientific and local/experiential knowledge welcomed.	Costs and benefits shared between participants.
Erftverband	Broad delegated mandate, includes water quality management, flood management, sewerage and waste water treatment, and water supply.	Majority voting; representation according to the level of contributions.	Boundaries set on the basis of hydrological criteria. Mandatory membership for water users and polluters.	Emphasis on science.	Water charges/ tariffs levied to members.
Portuguese River Basin District Authorities and Councils	Financial and administrative autonomy, coordination, monitoring, provision, dispute resolution.	Decision making in councils is consensual.	Boundaries based on hydrological and ecological criteria. Municipalities and users associations represented.	Emphasis on science. Much emphasis on information exchange between government layers.	Water charges/ tariffs levied to users.

Table 14.2 (continued)

Name of RBO	Authority rules	Aggregation rules	Boundary rules	Information rules	Pay-off rules
Breede–Overberg Catchment Management Agency	Mandate for water allocation, water resources protection, water use, development, conservation, management and control.	Mix of elite and consensus decision making. Strong focus on egalitarianism.	Boundaries based on natural hydrological borders and financial viability. Inclusive decision making.	Emphasis on scientific knowledge, stakeholder input solicited.	Water charges/tariffs levied to users, and seed funding from the Ministry of Water and Environmental Affairs
Western Bug River Basin Administration (RBA); Western Bug River Basin Council (RBC)	Delegated mandate to collect data, and to develop water management plans including specific measures.	RBC takes decisions on the river basin plan by majority rule. State Agency for Water Management has veto power.	Boundaries based on hydrological boundaries of surface waters. Members are regional and local authorities, non-governmental organisations (NGOs), industrial representatives and scientists.	Emphasis on science, NGO input solicited.	State funding for RBA staff, and implementation. Additional funds from local government, charges/tariffs, and international donors.

Lower Kunduz and Taloquan River Basin Agencies; Lower Kunduz and Taloquan River Basin Councils	Delegated mandate to deal with a variety of water issues. (In practice addressing water allocation only.)	Water users as key decision makers. (In practice: use of traditional *abandâz*.)	Boundaries nominally based on hydrological criteria. Formally very inclusive decision making	Emphasis on science.	Reasonable upstream–downstream equity
Mongolian River Basin Councils (RBCs); River Basin Administrations (RBAs)	Delegated mandate to coordinate activities of public and private parties in the basin, to draft river basin management plans and to grant water licences. RBCs responsible for public participation, and monitoring of water use and pollution (through RBAs and other organisations).	RBAs prepare river basin management plans which need formal approval of provincial and district parliaments. RBCs may submit proposals for suspending the RBA's decisions.	Boundaries based on hydrological, socio-economic and political criteria. Stakeholder-based representation in the RBCs.	Emphasis on science, stakeholder input solicited. Sharing of data not yet fully regulated.	Charges/tariffs, state funds, provincial funds and international donors.

Table 14.2 (continued)

Name of RBO	Authority rules	Aggregation rules	Boundary rules	Information rules	Pay-off rules
	RBCs formulates recommendations to RBAs.				
Ping River Basin Committee; Mae Kuang Sub-basin Working Group	Limited delegated mandate for coordination and planning.	Joint decision making but with a strong editing function of traditional line agencies.	Boundaries based on hydrological considerations. State and non-state actors have access.	Emphasis on science. Few requirements to share information publicly.	State funding through line agencies.
Murray–Darling Basin Authority (MDBA); Murray–Darling Basin Ministerial Council (MDBMC); Basin Community Committee	Delegated mandate for water allocation through planning for and advising to the government in the field of water allocation.	The Australian Government Minister (who is accountable to the Parliament of Australia) decides on the Basin Plan. Decision making in MDBMC and basin community committee is based on consensus.	Boundaries based on hydrological considerations. Governments decide, non-state actors consulted.	Emphasis on science, stakeholder input solicited. Data centrally managed.	Government funding. Emphasis on water markets.

covers groundwater and surface water management, water quality and water quantity, sewerage, wastewater treatment, planning and monitoring (Hüesker and Bernhardt, Chapter 6). It should, however, be noted that the formal mandate that RBOs have, is not always consistent with the issues that they focus on. The Afghan river basin agencies described in this book, for instance, have a very broad mandate on paper, but in practice focus on allocation issues only (Warner and Thomas, Chapter 10).

We can also conclude that the authorities acquired by RBOs can derive from multiple sources, and these are different for all types. The only autonomous RBO we encountered in this book (in Portugal) derived its authority directly from legislation adopted by parliament and we hypothesise that this is typical for this type. In the case of coordinating RBOs, the mandate depends on the willingness of the contracting parties to transfer power to the RBO (for example, in case of the Mackenzie River Basin Board: de Loë and Morris, Chapter 3). For agency-type RBOs the mandate depends on the willingness of the 'principal' (see Waterman and Meier 1998) to delegate a certain number of authorities; for examples see the Ukrainian (Hagemann and Leidel, Chapter 9) and Mongolian (Dombrowsky et al., Chapter 11) river basin administrations, and the Murray–Darling Basin organisation (Ross and Connell, Chapter 13). In the case of partnership-type RBOs, it is the founding members who decide on the scope of activities and decision-making power themselves; see, for example, the Westcountry Rivers Trust (Cook et al., Chapter 5).

Based on our 11 case studies we cannot observe a repeated or consistent pattern of development, such as from single-issue towards more holistic RBOs, or from local to more global scales (see Jaspers and Gupta, Chapter 2). Our observations rather confirm Jaspers and Gupta's statement that there is nothing linear in RBO development, and that each evolution has a space- and time-specific character.

The overview of authority rules that are relevant in the RBOs that have been analysed in this book provides food for thought on the connection between the foundation of RBOs and the shift from government to governance (see Huitema and Meijerink, Chapter 1). Our book contains quite a few cases in which a river basin agency or authority receives its mandate from a national ministry or government, and we have also seen multiple cases where river basin councils were essentially filled with representatives of regional and local government agencies; see case studies on Ukraine (Hagemann and Leidel, Chapter 9), Afghanistan (Warner and Thomas, Chapter 10) and Mongolia (Dombrowsky et al., Chapter 11). Although these are cases of agencification and decentralisation, we cannot conclude that the state has fully disappeared from the picture; in fact in many cases the state continues to be a powerful actor. This is because the state defines the specific mandate of the agency (sometimes pushed by international donor organisations, see also section 14.3.3), and either final approval of plans and policies rests with the nation state, or the nation state has veto power. Interestingly, the developments in the Australian Murray–Darling Basin as described by Ross and Connell (Chapter 13), and the latest reform decisions in South Africa (Meissner and Funke, Chapter 8), and reform proposals in Portugal (Thiel and de Brito, Chapter 7) reveal a swing in the pendulum. As national ministries managed to recapture power, the national government is gaining influence at the cost of the earlier-established RBOs and sub-national governments.

Aggregation rules

In international RBOs the decision-making rule is unanimity in most cases. Because of the absence of central authority, sometimes called 'international anarchy' (Sebenius 1992), basin states need to reach a consensus in order to conclude international agreements or decide on plans and policies. Table 14.2 shows that consensus is also frequently used as an aggregation rule in the (national) cases discussed in this book; see, for instance, the Westcountry Rivers Trust, a partnership type of RBO (Cook et al., Chapter 5), and most coordinating RBOs. However, other aggregation rules are applied too: majority voting was found in a number of cases: the Mackenzie River Basin Board (de Loë and Morris, Chapter 3), the Oregon Watershed Enhancement Board (Lach and Calvert, Chapter 4), and the Western Bug River Basin Council (Hagemann and Leidel, Chapter 9).

Analytically distinct but related to the voting rules of RBOs is the question of whose vote counts (see also the boundary rule discussion). In the German Erftverband case, we find majority voting linked to a principle that assigns greater voting powers to those that pay more to the RBO (see Hüesker and Bernhardt, Chapter 6). Also, harking back to our point about state dominance of several RBOs, the Ping River Basin Committee and the Mae Kuang Sub-basin Working Group are platforms in which line agencies dominate and essentially are the only relevant voters; their voting power is underpinned by their financial resources, which they can choose not to spend in line with RBO policies (Ganjanapan and Lebel, Chapter 12).

Boundary rules

In our analysis of boundary rules we have made a distinction between the rules describing the geographical delineation of the RBO, and the inclusiveness of the RBO's decision-making processes. When comparing the geographical boundaries of the RBOs analysed for this book, the difference in scale of the basins studied stands out. The surface area of the basins studied ranges from the small Mae Kuang watershed in Thailand (1661 square kilometres, Chapter 12) to the massive Mackenzie River Basin (1.8 million square kilometres, Chapter 3). Calling all these organisations RBOs means that we are not applying an often-used distinction, which discerns between watersheds, catchments and river basins (in ascending order of size). This is because we believe that our typology (autonomous, coordinating, agency and partnership-type RBOs) is more fundamental for the analysis of RBOs. This is, however, not to say that scale does not matter: it actually appears to have an effect on the available options in terms of institutional design. For example, organising a meeting of the basin board in the Mackenzie basin, let alone a full-blown multi-stakeholder process, is a laborious process if only because of the immense travel distances. Indeed, the partnership-type RBO discussed in this book is exclusively found at the level of relatively small basins. For the larger basins, such as the Mackenzie and Murray–Darling basins, coordinating or agency types of RBOs would appear to be the most feasible type.

In accordance with Schlager and Blomquist's (2008) and Molle and Wester's (2009) contention that the definition of the RBO's geographical scope and basin boundaries is not a straightforward technical exercise, but a process in which other factors often also play a role, we have found ample evidence that the geographical scope of the RBOs is determined by criteria in addition to hydrological criteria. In the Mongolian case study, next to hydrological criteria, economic and political factors played a role in defining the geographical scope of the RBOs. 'Economically relevant rivers were granted the status of discrete river basins even when, from a hydrological perspective, they actually belonged to larger catchments' (Dombrowsky et al., Chapter 11). In the Afghan case study, the geographical delineation of the RBOs does not match hydrological borders either. Where sub-basins either proved too big or were considered too heterogeneous in terms of land use, they were split. Other relevant factors, which played a role in the delineation made by the sponsors were the transport costs to attend meetings, the number of council seats and the anticipated income through fee collection (Warner and Thomas, Chapter 10). Apart from these considerations Warner and Thomas conclude that in practice the 'problemshed' turns out to be more important than the watershed. They argue that for solving allocation issues it may indeed be pragmatic to reduce complexity and to discuss issues with a limited set of main water users only. In the South African case study (Meissner and Funke, Chapter 8), next to hydrological criteria, the potential for a catchment management agency to achieve financial viability in the medium to long term already was an important criterion for delineating the geographical boundaries of the RBOs and has recently become even more important.

In addition to the use of political and social-economic criteria, the diverging boundaries associated with the surface water and groundwater basins can be a complicating factor in defining the basin boundaries. Examples of this can be found in the case studies on the Erftverband (Hüesker and Bernhardt, Chapter 6), and the Western Bug River Basin Administration (Hagemann and Leidel, Chapter 9). As Dombrowsky et al. (Chapter 11) explain, in the southern part of Mongolia, the Gobi Desert, there is no surface water, and since 'not enough was known about the groundwater aquifers for the catchment areas to be determined, the river basins were defined by reference to provincial boundaries'. On the other hand, in some other cases such as the Mackenzie River Basin Board, the definition of the boundaries of the basin has not been an issue at all (de Loë and Morris, Chapter 3). This may be due to the low population density and the lack of political controversies in the border areas of the basin. Although, next to hydrological criteria, other criteria have proven important for defining the geographical scope of most RBOs studied, we have not really found evidence for power struggles or political conflict over the geographical boundaries of the RBO jurisdictions. It rather seems that parties are pragmatic enough to deviate from hydrologically defined basin boundaries where this is considered useful or necessary.

The second type of boundary rules concerns the inclusion (or exclusion) of parties in decision-making processes. The findings on this rule type are directly related to the type of RBO studied. The Westcountry Rivers Trust, as an example of a partnership, is characterised by voluntary and broad participation of societal stakeholders (Cook et al., Chapter 5). On the other hand, an interstate coordinating RBO, such as the Mackenzie River Basin Board, includes governmental and First Nations representatives only (de Loë and Morris, Chapter 3). Even though decision making in the coordinating RBOs may seem pretty inclusive, the analyses presented in this book demonstrate that more often than not considerable power remains with the state (see also the discussion of authority rules above). In the South African case study this power is reflected in the state's competency to decide on the composition of the catchment management board. It is argued, however, that this state power is probably needed to manage the challenge of power differences between vested economic interests and marginalised parties (Meissner and Funke, Chapter 8). If we confront the findings on authority and boundary rules with the shifts in governance, which we discussed earlier, we do not really see evidence of a shift in power from the nation state to sub-national governments or civil society. Even in the partnership case of the Westcountry Rivers Trust, the authors argue that the trust is complementing rather than replacing existing, traditional, hierarchical governmental institutions in place (Cook et al., Chapter 5). It would thus seem safe to conclude that in Thelen's (2004) terminology, the foundation of RBOs is normally not an example of institutional 'displacement', which is associated with power shifts in the here and now. Instead, the foundation of RBOs is usually an example of 'layering': the foundation of a new institution on top of existing ones, and a process that is associated with existing systems that are actually 'intractable for change' in the short term and can only be changed in the long term – perhaps as a consequence of the new institution (Comodomos and Ferrer 2011). If anything, most of our cases demonstrate how extremely cautious existing institutions are, and how careful they are to curtail the powers of the new RBOs by negotiating institutional designs limiting their tasks and competencies.

Information rules

Information rules regulate the use of specific types of expertise, data and information as well as the openness of data and information on the status of the basin and effectiveness of basin policies to the public. Based on our 11 case studies we conclude that scientific expertise (which may relate to all natural sciences including ecology, but also to engineering) really takes centre stage in the design of almost all RBOs as it is considered the most relevant source for decision making, monitoring and reporting. Conversely, experiential knowledge (be it local, indigenous or professional knowledge) was hardly ever considered relevant by those that set up and designed RBOs, although many RBOs have to meet a requirement to solicit knowledge from groups that have such knowledge (farmers, fishermen, inhabitants). These conclusions are largely in line with the observation that Jaspers and Gupta (Chapter 2) make: water management is normally considered a highly technical field. Indeed, the several instances described in this book of (often Western) non-governmental organisations (NGOs) and donor countries trying to help establish RBOs, show that such RBOs are expected to work almost exclusively on the basis of scientific information, whilst largely ignoring the traditional and local knowledge embedded in the existing institutions for water management; see the chapter on Afghanistan that reveals how the Afghan *abandâz* system was hardly recognised by the new institutional architects (Warner and Thomas, Chapter 10).

The exceptions to the almost exclusive emphasis on scientific expertise that we encountered in this book show that greater attention to other forms of knowledge is more likely in partnership-type RBOs (see Cook et al., Chapter 5 and Lach and Calvert, Chapter 4); experiential expertise and narratives about 'success stories' play a key role in the functioning of the Oregon Watershed Enhancement Board (Lach and Calvert, Chapter 4). The Canadian example of the Mackenzie River Basin Board shows the relevance of clear requirements to incorporate traditional (Aboriginal) ecological knowledge (de Loë and Morris, Chapter 3).

Finally, and supportive of our point on the layering of institutions, we could observe that even if the exchange of information across levels and sectors is often a main objective of coordinating-type RBOs, such exchange may be hindered by previously existing institutional practices. According to Hagemann and Leidel (Chapter 9), cross-sectoral information exchange in Ukraine is hampered because 'historically, communication is organised on a vertical axis. This means that data travels up from lower-level to higher-level authorities, but not horizontally between different sectors'. So the foundation of the RBO was supposed to change existing information rules, but this may only become visible in the longer term.

Pay-off rules

The pay-off rules determine how costs and benefits are shared amongst participants. Again, the case studies display a rich variety of pay-off rules. These relate to the various ways in which the staffing of the RBO is financed, the financing of the policies and measures proposed by the RBO, and the distribution of costs and benefits amongst RBO members. It is here that we see quite a few attempts by existing organisations to keep the powers of the new RBOs in check, as most RBOs are expected to operate on a shoestring or need to request others to spend money in line with their plans and policies.

There are various different mechanisms and procedures for setting budgets. The budget of the Mackenzie River Basin Board is provided by the basin jurisdictions according to a clear formula specified in the Master Agreement (de Loë and Morris, Chapter 3). For the management of the Australian Murray–Darling Basin the jurisdictions rely on a negotiated intergovernmental cost-sharing formula as well, which is supplemented by funding from the Australian Government (Ross and Connell, Chapter 13). The Oregon Watershed Enhancement Board receives a fixed percentage of the revenues from the state's lottery (Lach and Calvert, Chapter 4). In still other cases the user-pays and polluter-pays principles are used as a means to allocate costs and benefits amongst water users. In the German Erftverband members generally contribute according to the amount of water used or polluted (but there are exceptions for mining companies; Hüesker and Bernhardt, Chapter 6). In the Westcountry Rivers Trust 'payments for ecosystem services' (PES) is practiced as an alternative to the polluter-pays principle (Cook et al., Chapter 5). Two or more parties voluntarily reach an agreement on the provision and payment for ecosystem services, such as farmers providing water quality services to drinking water companies. This specific case provides a good example of self-governance and self-organisation. Within the substantially different context of Afghanistan, upstream and downstream water users sometimes manage to voluntarily reach an agreement on the temporary release of water by upstream water users to benefit parties downstream, and by that to achieve a reasonable upstream–downstream equity (Warner and Thomas, Chapter 10).

The case studies of developing countries reveal the crucial importance of financial resources provided by international donor organisations for the functioning of RBOs. The problem in these cases is often that resources for staffing the RBOs are provided and the planning stage is covered, but sufficient resources for realising policy measures are lacking (Warner and Thomas, Chapter 10).

14.3 INSTITUTIONAL DYNAMICS AND THE POLITICS OF RBO FORMATION AND REFORM

14.3.1 Introduction

Now that we have discussed and compared the main institutional features of the RBOs, we move on to the second research question addressing the political dimension of RBO formation and reform. As we have outlined in the theoretical chapter, we expect some actors, such as the international water expert community, environmentalists or downstream actors, to be important advocates for the foundation of RBOs; and other parties, such as general-purpose governments, economic interests or upstream actors, to be less interested in this. In this study we found ample evidence for the highly political nature of RBO formation and reform processes, and the crucial role of change agents therein, but we have also learned that the actions of these agents interact with specific events or triggers, which in the literature are often referred to as 'windows of opportunity' (Kingdon 1995; Huitema and Meijerink 2009). Table 14.3 lists the main windows of opportunity, and the actors pushing or hindering RBO formation and reform for each case study. We found that three (sometimes related) categories of windows of opportunity have played a role: problem windows, political windows and windows related to the adoption of new legislation.

14.3.2 Triggers for RBO Design and Reform

What stands out in most cases is that the institutional reform proposed is a reaction to observed shortcomings of the institutional design in place

Table 14.3 Windows of opportunity, and actors pushing or hindering the founding and reform of RBOs

Name	Windows of opportunity	Pushing actors	Hindering actors
Mackenzie River Basin Board	Problem: drying out of the Peace–Athabasca Delta.	Federal and state governments, but primarily the downstream Northwest Territories.	Upstream provinces refuse to grant more power to the Board.
Oregon Watershed Enhancement Board	Legislation: ESA listing of salmon species.	The Governor of Oregon.	Environmental groups which preferred legal obligations to restore habitat above voluntary action. Republican legislators because of use of funds for land acquisition.
Westcountry Rivers Trust	Problem: failure of existing institutions to effectively solve pollution and habitat loss issues.	Amongst others, Arlin Rickard.	None.
Erftverband	Problem: issue of distribution of groundwater resources between municipalities and mining companies. Later, new challenges such as water pollution.	State of North Rhine-Westphalia enabled the foundation of Special Law Water Boards.	The Registered Society of Rhenish Brown Coal Plants considered establishment of the Erftverband unnecessary.
Portuguese River Basin	Legislation: the European Water	Influential group of Portuguese	Ministry of Finance which

Name	Windows of opportunity	Pushing actors	Hindering actors
District Authorities and Councils	Framework Directive.	water experts, and a minister who was part of this group.	feared greater costs. Regional Development and Coordination Commissions. 'Technocrats' within the national Water Institute.
South African Catchment Management Agencies	Political: the country's democratic transition.	Minister of Water Affairs, Kadar Asmal.	'Technocrats' within the Department of Water Affairs and Forestry and systems modellers opposing decentralisation.
Western Bug River Basin Administration and Council	Political: the country's transitioning process after independence.	International (donor) organisations, for example, European Union (EU) through project TACIS.	Existing (hierarchical) institutional structures from Soviet times.
Lower Kunduz and Taloquan River Basin Agencies and Councils	Political: fall of the Taliban.	Donor organisations: Germany's overseas agency for technical cooperation and Landell Mills consultancy, both hired/ funded by the EU.	Existing institutions, such as the Governor and the traditional *mirab* and *abandâz* system.
Mongolian River Basin	Political: decentralisation	World Wide Fund for Nature.	Some local authorities

Table 14.3 (continued)

Name	Windows of opportunity	Pushing actors	Hindering actors
Administrations and Councils	process after peaceful democratic revolution. Problem: environmental degradation caused by rapid expansion of the mining sector.	Dutch water experts. Mongolian water professionals trained in Eastern Europe.	which feared they would have to finance new activities. Powerful mining sector.
Ping River Basin Committee; Mae Kuang Sub-basin Working Group	Problem: 1997–1998 financial crisis.	Asian Development Bank, World Bank, Global Water Partnership.	Persistent bureaucratic competition is detrimental to the success of the RBO.
Murray–Darling Basin Authority	Problem: existing arrangements proved insufficient to guarantee water supply during severe drought conditions such as 2006–2007.	The Australian Government. For most issues, South Australia.	State governments other than South Australia. Farmer and irrigator groups.

previously. These shortcomings often became clear after a shock or focusing event, or because of persistent water management problems. In the Canadian case study, concerns related to the impacts of upstream development on the Peace–Athabasca Delta led governments of British Columbia, Alberta, Saskatchewan, Yukon and the Northwest Territories to recognise the need to coordinate, and establish the Mackenzie River Basin Liaison Committee (de Loë and Morris, Chapter 3). In the German case study of the Erftverband, strongly intensified brown coal mining in the 1950s and the related problems of distributing groundwater resources between municipalities and mining companies were the main triggers for legally strengthening the RBO, and when new challenges such as water pollution were recognised later, the scope of activities of the RBO was broadened (Hüesker and Bernhardt, Chapter 6). In the Mongolian case study, environmental degradation, which was caused by the rapid expansion of the mining sector, was a main trigger for RBO formation (Dombrowsky et al., Chapter 11). In the Murray–Darling case study, increased water use has led to increased salinity, substantial degradation of rivers and wetlands, and losses of native fish, and existing arrangements proved insufficient to guarantee water supply during severe drought conditions such as 2006–2007 (Ross and Connell, Chapter 13). For these reasons, the Australian national government established a new Murray–Darling Basin Plan, and a new organisational arrangement to implement it including a new Murray–Darling Basin Authority. Finally, the development of partnerships in the UK, among which are the Westcountry Rivers Trust, was a reaction to the top-down implementation of the European Water Framework Directives by agency-type RBOs, and their failure to effectively solve pollution and habitat loss issues (Cook et al., Chapter 5). In all these cases failure of existing water management institutions to address urgent water management problems served as a main trigger for either establishing new or reforming existing RBOs. Finally, the involvement of donor agencies in Thai water management following the Asian financial crisis formed a trigger for the establishment of RBOs in Thailand (Ganjanapan and Lebel, Chapter 12).

In addition to persistent problems, new regulations or legislation may open a window of opportunity for those advocating institutional change. After the Governor of Oregon had bad experiences with the US Endangered Species Act (ESA) listing of the Northern Spotted Owl, which had resulted in endless litigation, he first tried to prevent a listing of salmon species, later to develop an alternative, bottom-up approach to the challenge of habitat restoration. According to one of the governor's staff, 'You don't recover salmon by issuing orders from Salem (Oregon's capital). But you don't recover them with orders from D.C. either. People on the ground have to want to do it. You can't make them do it' (Lach and Calvert, Chapter 4). In similar vein, the adoption of the European Water Framework Directive was exploited by high-level Portuguese policy entrepreneurs to see through their plans, and to establish river basin district authorities (Thiel and de Brito, Chapter 7).

In still other cases, developments in the broader political environment opened a window of opportunity for institutional innovation. Examples of these are the democratic transition of South Africa (Meissner and Funke, Chapter 8), the fall of the Taliban in Afghanistan (Warner and Thomas, Chapter 10), the decentralisation process after peaceful democratic revolution in Mongolia (Dombrowsky et al., Chapter 11), and the transitioning process after independence in Ukraine (Hagemann and Leidel, Chapter 9).

14.3.3 Actors Pushing or Hindering RBO Formation and Reform

Introduction

Although the triggers or windows of opportunity described above may to some extent explain why initiatives for RBO formation or reform were taken, it is only actors who can interpret events, exploit windows of opportunity and initiate institutional reform processes, or try to block or frustrate these. In our study we came across five categories of actors which either pushed or hindered the founding or reform of RBOs (see also Table 14.3): upstream and downstream parties, parties having vested economic interests, political parties, general-purpose governments, and international donor organisations.

Upstream and downstream parties

Different interests of upstream and downstream water users and related power asymmetries often translate into disagreements about RBO design. In the Mackenzie basin, the Government of the Northwest Territories, which is situated downstream of the other basin jurisdictions, has advocated changes in the Board's membership and funding so as to make the Board more effective, but reform proposals have not succeeded so far (de Loë and Morris, Chapter 3). As the Mackenzie River Basin Board is a coordinating type of RBO, a downstream party such as the Northwest Territories always needs the cooperation of upstream parties, and therefore they are dependent on their willingness for that. In the Australian case study, South Australia, the downstream basin state in the Murray–Darling Basin, has an interest in maintaining downstream flows to ensure water supplies for the state, while the other states have supported diversions, especially for irrigation (Ross and Connell, Chapter 13). Likewise, in the Afghan case study the relationship between upstream and downstream water users is crucial for understanding the functioning of the RBO. Interestingly, the traditional *abandâz* system, which was essentially disregarded by foreign donor organisations, was developed exactly for the purpose of solving allocation issues between upstream and downstream water users, making the decision to ignore the system look even more perplexing (Warner and Thomas, Chapter 10). Clearly, hydro-politics does not only exist in the international context, but is also ubiquitous within states.

Parties with vested economic interests

As in other cases of institutional reform, the case studies presented in this book demonstrate that parties with vested economic interests are often more interested in maintaining the status quo than in institutional reform.

For example, in the German case study, the Registered Society of Rhenish Brown Coal Plants, representing the interest of the coal industry, considered the establishment of the Erftverband unnecessary (Hüesker and Bernhardt, Chapter 6). The mining industry is an important actor in the Mongolian case study as well. Dombrowsky et al. (Chapter 11, referencing *The Economist*) state that 'Political support for improved control of the mining industry [to be achieved in part by the foundation of an RBO] will most probably remain difficult as the sector is the largest contributor to the country's spectacular gross domestic product (GDP) growth over recent years'. In the Australian case, farmer and irrigation groups are sceptical about the referral of Murray–Darling Basin (MDB) management powers to Australian Government (Ross and Connell, Chapter 13). The reason for this may be that they question the effectiveness of a more centralised river basin management, but also because they perceive such centralisation as detrimental to their interests.

Political parties

The case studies produce some evidence for the role of party politics as well. In the case study of the Oregon Watershed Enhancement Board, Republican legislators opposed the use of funds for the expropriation of agricultural land, so as to protect agricultural interests (Lach and Calvert, Chapter 4). In Australia, both main political parties on the national level supported the centralisation of water policies, but on the state level, the liberal coalitions of Victoria, New South Wales and Queensland voiced similar criticisms in response to the Basin Plan that had been developed. One of the criticisms was around the uncertain environmental benefits of the Basin Plan (Ross and Connell, Chapter 13). The study of the transformation of Portuguese RBOs illustrates how elections and changes in government have played a crucial role in the institutional reform processes, which has meant that the water governance system of Portugal has oscillated between two competing visions for years (Thiel and de Brito, Chapter 7). Finally, the Mongolian case study suggests that the increased influence of the Green Party in that country may have been conducive to RBO formation (Dombrowsky et al., Chapter 11).

General-purpose governments

RBOs are an example of special-purpose governments (Skelcher 2006). Unlike general-purpose governments, such as the nation state, provinces or municipalities, RBOs are designed to address water management issues specifically. The founding of special-purpose governments, such as RBOs, may come at the cost of the competencies of general-purpose governments. As Jaspers and Gupta (Chapter 2) argue in their analysis of global discourse on RBOs: 'RBOs embody a power struggle – a shift of power from the state to an administrative body at both international and national level – and this power struggle has often led to minimising the mandate of the RBO'. Very true words, as the case studies analysed in our book show. The Australian case study exemplifies this power struggle: the states reluctantly accepted an increase of federal powers in return for increased federal contributions (Ross and Connell, Chapter 13). In the Ukrainian, Afghan and Mongolian case studies, existing (hierarchically operating) government agencies are hindering effective RBO formation. In Portugal the Regional Development and Coordination Commissions opposed the founding of river basin administrations (Thiel and de Brito, Chapter 7). Although the tension between existing general-purpose governments and newly created RBOs stands out in most cases, perhaps because RBOs are seen as a threat of institutional displacement, some RBOs such as the Oregan Watershed Enhancement Board (Lach and Calvert, Chapter 4) and the Westcountry Rivers Trust (Cook et al., Chapter 5) do not so much rival as complement existing institutions, which may explain why the struggle over competencies has been less prominent in these cases. This is an indication that an introduction strategy explicitly aimed at layering (Thelen 2004) meets with less resistance.

International donor organisations

Within our set of RBOs, there is a specific category where international donor organisations have been the crucial actors pushing the concept of RBOs, and the associated principles of organising water management on a basin scale (bioregions), and undertaking such management in a participatory manner (referred to as 'the holy trinity' by Warner and Thomas, Chapter 10). In their chapter on global water governance and RBOs (Chapter 2), Jaspers and Gupta mention the financial assistance and related conditionalities of aid and development projects as one of several factors that explain why certain water management discourses are appealing.

Donor organisations, in turn, are influenced by international epistemic communities, such as the World Water Council, Global Water Partnership, the International Law Association and the engineering community (Jaspers and Gupta, Chapter 2). In our collection of case studies, Ukraine, Afghanistan, Mongolia and Thailand are clear examples of the influence of international donor organisations on RBO formation. The authors of the Afghan case are probably most critical of such organisations (Warner and Thomas, Chapter 10). They point to the lack of a proper problem and institutional analysis, and the inaccurate assumption that water governance in Afghanistan is a 'blank slate'. They illustrate the lack of a proper problem analysis with the example that 'Despite no inter-sectoral conflicts being reported, the Panj-Amu River Basin programme unhelpfully insisted in multi-user mapping'. According to Warner and Thomas, the main reason for the Afghan government officials accepting the founding of RBOs as proposed by the European Union (EU) was their interest in EU funding of water control infrastructure projects so that they could carry on with their longstanding 'hydraulic mission' (which is in many ways the opposite of what the EU strives for in its own legislation such as the Water Framework Directive). International donors often do not work alone. For instance, in the Mongolian case study (Dombrowsky et al., Chapter 11) international actors promoted the RBO concept, but the founding of RBOs was also supported by Mongolian water experts who were trained in Europe and Russia – thereby suggesting the importance of education in forming an epistemic community. The same applies to the Thai case study (Ganjanapan and Lebel, Chapter 12), where the Asian Development Bank, World Bank and Global Water Partnership pushed the RBO concept, but the National Water Resources Committee equally endorsed RBO development.

The empirical findings about actors pushing or hindering RBO foundation are in accordance with our hypothesis that RBO formation is not a straightforward technical exercise but a highly political enterprise.

14.4 THE PERFORMANCE OF RIVER BASIN ORGANISATIONS

14.4.1 Introduction

In the preceding sections, we have compared the institutional design of the RBOs and explored the political dimension of their founding and reform; we now shift our focus to the RBOs' performance. In the introductory chapter we discussed the criteria which may be used to assess the RBOs' performance, and we questioned whether these organisations enhance: (1) the coordination across levels of government, policy sectors, and between public and private parties; (2) the accountability of water policies; (3) the legitimacy of water policies; and (4) the environmental effectiveness of policies. Table 14.4 summarises the findings about RBO performance for each case study. Based on the rich case study data provided by the chapter authors, we have made a qualitative assessment, which distinguishes between poor, average and good performance. If no data are available this is indicated with 'unknown'. The main findings for each criterion will be discussed below.

Table 14.4 Findings on the RBOs' performance

Name of the RBO	Coordination	Accountability	Legitimacy	Environmental effectiveness
Mackenzie River Basin Board	Poor	Average	Average	Poor
Oregon Watershed Enhancement Board	Good	Good	Good	Average
Westcountry Rivers Trust	Good	Good	Good	Good
Erftverband	Good	Good	Good	Good
Portuguese River Basin District Authorities and Councils	Average	Average	Average	Unknown
Breede–Overberg Catchment Management Agency (BOCMA)	Good	Average	Average	Unknown
Western Bug River Basin Administration (RBA) and Council (RBC)	Poor	Average	Average	Unknown
Lower Kunduz and Taloquan River Basin Agencies and Councils	Poor	Poor	Poor	Poor
Mongolian River Basin Administrations (RBAs) and Councils (RBCs)	Poor	Poor	Average	Unknown
Ping River Basin Committee Mae Kuang Sub-basin Working Group	Poor	Average	Poor	Poor
Murray–Darling Basin Authority (MDBA); Murray–Darling Basin Ministerial Council (MDBMC); Basin Community Committee	Average	Average	Average	Unknown

14.4.2 Coordination

A main argument for founding RBOs is that these organisations may enhance the coordination between different levels, sectors and interest, which is key to realising integrated water resources management (IWRM). In this respect, the RBOs which we included in our study yield mixed results. The Mackenzie River Basin Board, as an example of a coordinating RBO, performs relatively poorly when it comes to establishing coordination across the basin states. De Loë and Morris (Chapter 3) explain how the states may easily bypass the commission, such as in case of a new joint monitoring programme for the oil sands. Also in the Thai case study, the coordinating RBOs perform poorly, and decisions which are crucial for the watershed are taken elsewhere (Ganjanapan and Lebel, Chapter 12). The Mackenzie River Basin Board remains dependent on the willingness of contracting parties to conclude bilateral agreements. According to de Loë and Morris (Chapter 3) this is due to the institutional design of the RBO, which in turn reflects the political characteristics of the federal state of Canada. This is quite similar to the Australian case study of the Murray–Darling River Basin Commission (Ross and Connell, Chapter 13) in which the states used to have veto power. However, since the establishment of the Murray–Darling Basin Authority, power has shifted to the federal state government, and coordination may be achieved by imposing policies on the basin states. Ross and Connell doubt whether this more coercive approach will yield better results.

The partnership-type RBOs, the Westcountry Rivers Trust and the Oregon Watershed Enhancement Board and related watershed initiatives, perform rather well. The same goes for the two RBOs, which share characteristics with the autonomous RBO-type, the German Erftverband, and the Portuguese River Basin District Authorities. In the latter case, the RBOs have enhanced coordination within hydrographic basins, but this came at the cost of less coordination across basins on the national level (Thiel and de Brito, Chapter 7).

The agency-type RBOs, which were founded in Ukraine, Afghanistan and Mongolia with the help of international donor organisations, have been less successful in achieving coordination so far. Reasons for this are the still-powerful hierarchical governmental institutions in place, which hinder cross-sector exchange of information, and the failure to take account of traditional water management practices, such as the *mirab* and *abandâz* systems in Afghanistan. The Breede–Overberg Catchment Management Agency, however, performs relatively well (Meissner and Funke, Chapter 8).

14.4.3 Accountability

Whereas in democracies, general-purpose governments such as provinces, municipalities or state governments mostly have a directly elected council which is accountable to the public (the electorate), accountability relations of RBOs are more complicated. River basin boards or councils are not directly elected and often consist of representatives of lower levels of government, sometimes supplemented with representatives of NGOs or industries. In these cases the RBO is accountable to the representatives of the member organisations, which in turn are accountable to their constituencies. Furthermore, the case studies reveal differences between formal accountability and accountability in practice. As de Loë and Morris (Chapter 3) explain, the Mackenzie River Basin Board is accountable to the ministers, but the ministers are not held accountable for progress made. The case studies of the Breede–Overberg Catchment Management Agency (Meissner and Funke, Chapter 8), Murray–Darling Basin Authority (Ross and Connell, Chapter 13), and the transformation of the Portuguese water district administrations into agencies (Thiel and de Brito, Chapter 7), all illustrate that centralisation enhances accountability of the RBO to the national ministers and government, but may come at the cost of accountability to sub-national governments (or states, within a federal state).

In the Ukraine (Hagemann and Leidel, Chapter 9), Afghanistan (Warner and Thomas, Chapter 10), and Mongolia (Dombrowsky et al., Chapter 11) the RBOs have a poor performance in terms of accountability. In Ukraine, members of the River Basin Council are not held accountable; in Afghanistan informal institutions and practices are far more important than the formal ones, and in both the Ukrainian and the Mongolian case studies, the authors report that corruption in these countries is widespread (see also Jaspers and Gupta, Chapter 2). Again, the Westcountry Rivers Trust (Cook et al., Chapter 5), and the Oregon Watershed Enhancement Board and related watershed initiatives (Lach and Calvert, Chapter 4), perform rather well.

14.4.4 Legitimacy

A next question raised in the introductory chapter was to what extent the RBOs enhance the legitimacy of water management policies. The answer to this question is positive for the Oregon Watershed Enhancement Board (Lach and Calvert, Chapter 4), the Westcountry Rivers Trust (Cook et al., Chapter 5), and the German Erftverband (Hüesker and Bernhardt, Chapter 6). The respective chapter authors all point to high input and output legitimacy of the RBOs, and perhaps it is no accident that these are all RBOs that operate on a relatively small scale and emphasise collective action. The results for the cases studies on the Mackenzie River Basin Board (de Loë and Morris, Chapter 3) and the Breede–Overberg Catchment Management Agency (Meissner and Funke, Chapter 8) show a mixed picture. In the former case, input legitimacy falls short because the contracting parties have historically failed to engage Aboriginal people and to make effective use of traditional ecological knowledge, and output legitimacy is questioned by many. In the latter, on the one hand input legitimacy is considered high because of the involvement of a wide range of stakeholders, on the other hand low because of persistent major inequalities between stakeholders. The case studies where international donor organisations played a key role in initiating the founding of RBOs – Ukraine, Afghanistan, Mongolia and Thailand – show poor to average performance. Hagemann and Leidel (in Chapter 9) refer to low NGO participation in the Ukrainian Western Bug basin, and Warner and Thomas (Chapter 10) refer to a lack of support for newly established RBOs and greater support for the much less democratic yet still effective institutions in place. Dombrowsky et al. (Chapter 11) are slightly more positive where they conclude that Mongolian River Basin Councils to some extent have enhanced the legitimacy of river basin policies. Ganjanapan and Lebel (Chapter 12) point to the lack of public participation in the Mae Kuang watershed, and Ross and Connell (Chapter 13), in their chapter on the Murray–Darling Basin Authority, come to the conclusion that the recently adopted centralised, top-down approach to river basin management has to some extent reduced input legitimacy at regional and local levels.

14.4.5 Environmental Effectiveness

The criterion of environmental effectiveness turned out to provide the biggest challenge for the authors participating in this book project. In their assessments, the chapter authors have primarily relied on expert judgements, but were faced with several complications, not least of which is the attribution problem. As RBOs operate within complex institutional environments in which many other organisations impact upon the state of water systems through their policies or activities, it is not possible to isolate the impact of contributions made by the RBOs. The authors who have studied recent examples of the founding or reform of RBOs rightly argued that they have not been able to assess the environmental effectiveness yet, as there is a time lag between the founding of an RBO and the first policies and action plans being drafted, let alone the actual implementation of these plans. But others did make some suggestions. Although the new Murray–Darling Basin plan includes comprehensive measures to improve environmental performance, Ross and Connell (Chapter 13) point to the serious risk of implementation failure in the RBO due to distrust between the main actors and reduced financial contributions by the states. De Loë and Morris (Chapter 3) argue that the environmental effectiveness of the Mackenzie River Basin Board is quite low, and that this should not be attributed to a poorly functioning staff, but to the institutional design of the Board, which leaves all decision-making power with the basin jurisdictions. The environmental performance of the Oregon Watershed Enhancement Board is less clear. Although the Board is generally perceived as being successful in stimulating numerous habitat restoration projects, there is no proof of recovery of the salmon population yet (Lach and Calvert, Chapter 4). As for the other criteria, the German Erfverband performs well again in the view of more or less all decision makers and stakeholders. According to Hüesker and Bernhardt (Chapter 6), the problem-solving capacity of the Erftverband and similar water boards in the state of North Rhine-Westphalia is higher compared to the environmental performance of state's regions where such water boards are absent. The Westcountry Rivers Trust (Cook et al., Chapter 5) also performs well: best management practices are introduced, advice is produced for farmers, landscape restoration is achieved, and farm efficiency savings made.

Ganjanapan and Lebel (Chapter 12), however, are rather negative about the effectiveness of the Thai RBOs they studied since in their eyes they avoid difficult allocation issues, and decisions that really matter are taken elsewhere, including in the previously existing institutions. Finally, in the Afghan case study, environmental issues simply have not been recognised at all (Warner and Thomas, Chapter 10). However, as in the Thai case study, the Afghan case study does demonstrate how older institutions sometimes continue to work in spite of the new layer, so perhaps there is hope still.

14.5 RELATIONSHIP BETWEEN RBO DESIGN AND PERFORMANCE: LESSONS LEARNED

14.5.1 Introduction

Now that we have described both the institutional design and the performance of RBOs, we explore the interrelationship between design and performance. In other words: which design features inhibit or enhance an RBO's performance? This question is as difficult to answer as it is relevant, since there are many factors which potentially impact upon an

RBO's performance, and it is difficult to disentangle these factors let alone to assess their relative importance. Having said that, we have observed several salient patterns that suggest connections between RBO design and performance, which deserve attention. In the rich collection of case studies presented in this volume, four factors stand out as being particularly relevant to (re)designing RBOs: (1) institutional interplay, or the relationship between the RBO and its broader (previously existing) institutional environment; (2) the ability to generate sufficient resources not only for staffing the RBO, but also for realising its policies; (3) the dilemma of centralisation and decentralisation of basin policies; and (4) the time required for RBO reform and impact. Each of these aspects will be treated in the next parts of this section.

14.5.2 Institutional Interplay

RBOs are examples of special-purpose organisations, which are organised on a basin scale. With the exception of the somewhat deviating case study of the Oregon Watershed Enhancement Board, the geographical scope of each of the RBOs analysed in this book was at least partially based on hydrological boundaries. Although we saw that administrative, financial, economic or other pragmatic considerations also play a role in carving out river basin boundaries (see also section 14.2.2 on boundary rules) for most cases discussed in this book a claim could be made that the geographical scope of the RBOs does correspond at least to some degree with the scale on which water management issues manifest themselves. In other words, there is a certain level of fit between the institutions and the bioregions they aim to manage – an outcome in line with the international discourse on RBOs. But at the same time, and quite revealingly, the case studies presented in this volume give every indication that the realisation of spatial fit between problems and institutions is no guarantee of good performance. In fact, a more influential factor for performance is the relation of the new RBOs to the existing institutional environment; which may be called 'institutional interplay'. Our case studies show that, as suggested in Chapter 1, discussions about the foundation of RBOs do not take place in a vacuum: many tasks potentially taken up by a new RBO are normally already performed by existing organisations. And whereas many proponents of RBOs hope for 'strong' RBOs that take over many of these tasks and integrate them –'institutional displacement' in Thelen's (2004) terms – the reality is that 'weak' RBOs with relatively few powers are much more politically feasible in practice. The consequence is that new RBOs are layered on top of existing organisations, and previously existing organisations take varying degrees of control over the new RBO. Therefore, the RBOs discussed in this book are almost always highly dependent on other organisations for realising their objectives, and their performance is highly dependent on the effective management of institutional interplay. Problematic institutional interplay stands out as a key factor that causes poor RBO performance.

This is particularly evident in the cases in which international donor organisations instead of national or local actors have been the main instigators of institutional change. It would seem that the 'institutional engineering' by such actors is characterised by a lack of knowledge about the existing structures, or a lack of awareness that the relations between these and the new RBO should be defined. Indeed Hagemann and Leidel (Chapter 9) conclude that in Ukraine 'the RBOs have been installed within an existing system without having their interrelationship to the other authorities clearly defined'; and on these other authorities state that, 'They keep working on their assigned tasks and follow the old institutional structures from Soviet times. This means they are strictly subordinated to their superiors, particularly the ministries.' Similar observations have been made regarding the Afghan case study (Warner and Thomas, Chapter 10), where a new formal system has been introduced that reflects many traits of the modern IWRM discourse (for example, the devolution of power and stakeholder participation), but which ignores the traditional *mirab* and *abandâz* systems that continue to influence decision making on the ground. As Warner and Thomas argue, this ' pure', imported version of an RBO immediately failed when water allocation problems surfaced, and was replaced by 'a sporadically emerging ad hoc variety mixing formal and informal institutions', which performed reasonably well. These findings accord with Jaspers and Gupta's (Chapter 2) observation that 'many of these [water management and governance, SM and DH] discourses are imported into the developing countries and may, on occasion, even be seen as alien to local cultures'. Dombrowsky et al. (Chapter 11) recognise the relevance of institutional interplay as well. They conclude that the Mongolian democratisation and decentralisation process, which started in the 1990s, has not been completed yet – a factor that substantially influences RBO performance. Therefore they suggest that the evolution of the overall governance context is highly relevant to understanding the functioning of the RBOs. Ganjanapan and Lebel (Chapter 12) also emphasise the relevance of institutional interplay for understanding RBO performance, where they argue that: 'As local administrative organisations have become increasingly important in terms of budget allocation and natural resources management responsibilities within their jurisdictions, it is difficult to plan or manage based on watershed units alone.'

Interplay is an issue for Western RBOs as well. De Loë and Morris (Chapter 3) basically conclude that the success of the Mackenzie River

Basin Board, as an example of a coordinating type RBO, is in the hands of the governments of the participating provinces and territories. One may argue that this is due to the limited competencies granted to the commission by the basin states, but even in the case of the Murray–Darling Basin organisation, which has more competencies with which to see through its basin policies, Ross and Connell (Chapter 13) point to the relevance of institutional interplay. They argue that it remains to be seen whether the institutional redesign of basin institutions will be successful in the end as the many different parties involved in the implementation of river basin policies have ample opportunities to frustrate the implementation process. These findings relate to the fundamental question posed by Jaspers and Gupta (Chapter 2) 'whether water is a "sector" in itself or a cross-cutting issue and whether it should be dealt with in relation to its nexus with other sectors and priorities in society'. As the drivers of change in the water sector arise often in other sectors with other priorities 'This may lead to questioning the very basis of the RBO whose fundamental logic is the hydrological entity encapsulated in a river' (ibid.). The experts engaged in RBO design seem to embrace modern water discourses (which were described in Chapter 2 of this volume) so enthusiastically that they tend to forget a sound problem analysis (see also Warner and Thomas, Chapter 10), do not take account of the specificities of the institutional environment in which they are working, and are prone to underestimate the inertia and path-dependent nature of institutional change processes.

14.5.3 Resources

A second factor which deserves our attention is also related to political battles that predate the foundation of RBOs. In practice, the ability of the RBOs to generate resources for both staffing the RBO and realising its river basin policies turns out to be very limited. This relates to the pay-off rules, which we discussed in section 14.2.2. The ability to generate resources is a crucial success factor for RBO performance. The Mackenzie River Basin Board has a secretariat with a staff of one person with a rather limited budget, which reflects the (un)willingness of the states to share powers with an RBO (de Loë and Morris, Chapter 3). The limited resources and the unwillingness to share power may explain the overall poor performance of this coordinating type RBO. This straightforward argument presents a fundamental structure–agency, or 'chicken and egg' problem. Of course, one might dream about powerful integrative and inclusive RBOs, which are also able to generate the necessary resources, but such organisations may only be realised once other actors are willing to share their powers. The case studies show that they generally are not, hence institutional interplay remains crucial (see section 14.5.2).

We have seen some interesting examples of resource acquisitions in the cases studied. The Oregon Watershed Enhancement Board has 27 staff members, and receives 7.5 per cent of lottery revenues in Oregon, which makes it a relatively powerful and successful organisation (Lach and Calvert, Chapter 4). The relatively strong performance of the Westcountry River Trust could partly be attributed to the partnership's ability to successfully mobilise (mainly EU) financial resources (Cook et al., Chapter 5). In the German Erftverband, good performance should partly be attributed to the RBO's ability to generate its own resources through the collection of user and pollution fees (Hüesker and Bernhardt, Chapter 6). In sharp contrast with these successful cases is the case of Ukraine in which collected fees from water users are all sent to the national government in Kiev, which makes only 30 per cent of the fees available for investments in water resources management (Hagemann and Leidel, Chapter 9). In the Mongolian case study the staffing of the RBO is secured through the national budget, but no financial resources have been allocated for drafting and implementing plans. Although new legal arrangements for water pricing and collecting fees were made recently, their implementation is uncertain as water monitoring institutions are still underequipped and understaffed (Dombrowsky et al., Chapter 11). Ganjanapan and Lebel (Chapter 12) also point to the relevance of sufficient resources, where they state that: 'The first decade has been largely spent filling and refilling committees and working groups which rarely meet and have only very modest budgets, little dedicated office support and limited authority'. The Australian case is a very specific one because of the increasingly influential role of the water market. The federal government continues to buy water rights as a means to return water resources to environmentally sustainable levels of extraction, and the Commonwealth Environmental Water Holder is expected to become an even more powerful actor in the future. In short, the case studies reveal different mechanisms for generating resources for staffing RBOs and for implementing basin policies, and because such mechanisms are highly relevant for RBO performance, they deserve attention to be paid to in institutional design.

14.5.4 Dilemma of (De)centralisation

A third common theme running through the case studies relates to the dilemma of (de)centralisation. The dilemma, which affects both the effectiveness and the legitimacy of the RBOs, is that centralised and closed decision-making processes may enable the RBO to make good progress, and to develop 'rational' policies taking into account the hydrological relations within and across large basins. However, such a centralised approach also typically comes at the cost of (input) legitimacy. On the other hand, decentralised and open decision-making processes may enhance legitimacy, but slow down decision making and lead to suboptimal outcomes at higher levels of scale.

In the case studies presenting longitudinal analyses of RBO development, we have seen parties struggling with this dilemma, and quite a bit of oscillation between the two ends of the spectrum. In the South African case study, after major efforts made to decentralise basin policies and to organise more inclusive processes, the poor progress made with this approach recently triggered a process of increased centralisation again (Meissner and Funke, Chapter 8): 'there now seems to be a move towards reducing the numbers of stakeholders when establishing CMAs [catchment management agencies] and instead only drawing in stakeholders where and when necessary. While this development may speed up the CMA development process countrywide, it may have negative consequences for more marginalised stakeholders to participate in the CMA process'. In Australia, the lack of results from the interstate Murray–Darling Basin Commission was the main reason for the Australian federal state to centralise water management competencies and to establish new Murray–Darling Basin organisational arrangements (Ross and Connell, Chapter 13). Likewise, their poor performance in establishing coordination across basins (besides the predominant hope for financial savings) was an important reason for the Portuguese national government to start reforming the River Basin District Authorities, which are an example of autonomous RBOs, into agency-type organisations again, so as to increase the possibilities for steering and controlling these organisations (Thiel and de Brito, Chapter 7). The researchers who investigated the Portuguese and Australian case studies express serious doubts about the effectiveness of these changes, mainly because support for the reform by parties who need to implement the new structures is often lacking. Based on these observations we would suggest that parties involved in reform processes should develop a better awareness of the inherent dilemmas of either decentralising or centralising basin policies. They would do better to try to carefully compensate for perceived drawbacks of either one of the approaches rather than to radically reorganise the system once again. This brings us to our final observation, which is on the relevance of time in institutional design processes and assessing RBOs' performance.

14.5.5 Time

If the design of RBOs already shows many signs of political bickering, several case studies discussed in this book show that this is far from the end of such politicking. In fact, new organisations that are layered on top of existing ones can eventually transform an institutional regime. To prevent that risk from actually materialising, opponents of RBOs, and those with a different view of their appropriate design, try to redesign the institutions if they get the chance. But this means that a certain design has very limited time in which to settle and to prove itself. Indeed, several authors questioned the speed with which RBOs are being reformed. Basically, their argument is that the system in place was not granted sufficient time to prove effective. As one of the informants on the Mongolian case study states: 'The work with the four RBCs had just started . . . people were aware of river basin management and integrated water resources management, and then this work was suddenly stopped' (Dombrowksy et al., Chapter 11). Likewise, Thiel and de Brito (Chapter 7) question why the seemingly promising River Basin District Authorities were not granted more time to prove themselves successful, and Ross and Connell (Chapter 13) question whether the Murray–Darling Basin Commission was given enough time to become successful: 'the effectiveness of new initiatives has been reduced because initiatives are overtaken or abandoned before they have had time to prove themselves or before they have been properly evaluated' (Ross and Connell, Chapter 13, referencing Ross and Dovers). These observations underscore our claim that those involved in reform process should be careful with introducing frequent radical changes.

Returning to the fourth question from Chapter 1, the implications of these comparative analyses taken as a whole for those involved in water reform are fourfold. Perhaps most important is to realise that the design of RBOs is not like starting from a blank page. The design of RBOs, from their mandate through resources to levels and areas of authority needs to take into account pre-existing institutions. Secondly, the introduction of new RBOs (or their reform) is invariably a political process and thus the RBO's roles and tasks will need to be negotiated and interplay between the RBO and other institutions managed. Thirdly, the performance of RBOs is influenced by their design, in particular their ability to generate sufficient resources, and newly established RBOs should be granted sufficient time to prove successful. Finally, there is a need to manage expectations: whilst RBOs can play a role in enhancing coordination, legitimacy, accountability and environmental effectiveness, they are not a panacea for improving water management.

REFERENCES

Comodomos, G. and J. Ferrer (2011), 'Incremental institutional change and Australian universities', *International Review of Business Research Papers*, **7** (1), 330–339.

Huitema, D., E. Mostert, W. Egas, S. Moellenkamp, C. Pahl-Wostl and R. Yalcin (2009), 'Adaptive water governance: assessing adaptive management from a governance perspective', *Ecology and Society*, **4** (1), 26.

Huitema, H. and S. Meijerink (2009), *Water Policy Entrepreneurs, A Research Companion to Water Transitions around the Globe*, Cheltenham, UK and Northampton, MA, USA: Edward Elgar.

Kingdon, J.W. (1995), *Agendas, Alternatives and Public Policies*, 2nd edn, New York: HarperCollins.

Molle, F. and P. Wester (2009), *River Basin Trajectories: Societies, Environments and Development*, Wallingford, UK: International Water Management Institute.

Schlager, E. and W. Blomquist (2008), *Embracing Watershed Politics*, Boulder, CO: University Press of Colorado.

Sebenius, J.K. (1992), 'Challenging conventional explanations of international cooperation: negotiation analysis and the case of epistemic communities', *International Organization*, **46** (1), 323–365.

Skelcher, C. (2006), 'Does democracy matter? A transatlantic research design on democratic performance and special purpose governments', *JPART*, **17**, 61–76.

Thelen, K. (2004), *How Institutions Evolve: The Political Economy of Skills in Germany, Britain, the United States, and Japan*, Cambridge: Cambridge University Press.

Waterman, R. and K. Meier (1998). 'Principal–agent models: an expansion?', *Journal of Public Administration Research and Theory*, **8**, 173–202.

Index